Intelligent Sensing Technologies for Nondestructive Evaluation

Special Issue Editors

Seunghee Park
Aimé Lay-Ekuakille
Octavian Postolache
Pedro Manuel Brito da Silva Girão

MDPI • Basel • Beijing • Wuhan • Barcelona • Belgrade

MDPI

Special Issue Editors
Seunghee Park
Sungkyunkwan University
Korea

Aimé Lay-Ekuakille
University of Salento
Italy

Octavian Postolache
Instituto de Telecomunicacoes
Portugal

Pedro Manuel Brito da Silva Girão
Instituto Superior Técnicoo
Portugal

Editorial Office
MDPI AG
St. Alban-Anlage 66
Basel, Switzerland

This edition is a reprint of the Special Issue published online in the open access journal *Sensors* (ISSN 1424-8220) from 2017–2018 (available at: http://www.mdpi.com/journal/sensors/special_issues/ISTNE).

For citation purposes, cite each article independently as indicated on the article page online and as indicated below:

Lastname, F.M.; Lastname, F.M. Article title. *Journal Name* **Year**, *Article number*, page range.

First Edition 2018

ISBN 978-3-03842-877-0 (Pbk)
ISBN 978-3-03842-878-7 (PDF)

Table of Contents

About the Special Issue Editors

Seunghee Park received his B.S. in 2002, his M.S. in 2004 and his Ph.D. in 2008 from KAIST, majoring in Structural Engineering/Civil Engineering. Currently, he is an Associate Professor of Civil Engineering in Sungkyunkwan University. Previously he was a Postdoctoral Researcher (2008) working on Smart Structures and Systems at KAIST. He was a Postdoctoral Research Fellow at the Center for Intelligent Material Systems and Structures (CIMSS), Dept. of Mechanical Engineering, Virginia Polytechnic Institute and State University. He was also a visiting professor at the same institution. His main research interests include Smart Materials and Sensors, Smart Structures, Smart Space, Structural Health Monitoring, Nondestructive Testing/Evaluation, Energy Harvesting, Building Information Modeling, Energy Performance Simulation/Optimization, RFID, QR code, Augmented Reality(AR)/Virtual Reality(VR) Application, Information/Data Mining, Signal Processing, Probabilistic/Statistical Pattern Recognition, System Identification, Experimental Modal Analysis and Vibration Testing. He has already received over 31 honors and awards for his accomplishments from different countries.

Aime Lay-Ekuakille received his M.D. in Electronic Engineering from the University of Bari, Italy, a further M.D. in Clinical Engineering from the University of L'Aquila, Italy, and a Ph.D. in Electronic Engineering from the Polytechnic of Bari, Italy. He is currently working at the University of Salento, where he is the Director of the Instrumentation and Measurement Laboratory I. He has authored and co-authored over 230 papers in international journals and proceedings. He co-edited three international books, and also authored one. His main research regards environmental, industrial, and biomedical instrumentation and measurements, Renewable energy, including use of nanotechnology devices. He is an Associate Editor of the IEEE Sensors Journal, and Measurement (Elsevier) and a board member of Sensors (MDPI). He Chairs the IEEE I&M Society TC34 Nanotechnology in Instrumentation and Measurement, and is part of the IEEE Nanotechnology Council AdCom. He also serves as Chairman of Imeko TC19 Environmental Measurements.

Octavian Postolache, graduated in Electrical Engineering at the Gh. Asachi Technical University of Iasi, Romania, in 1992 and received his Ph.D. in 1999 from the same university. He received his university habilitation in 2016 from Instituto Superior Tecnico, Universidade de Lisboa, Portugal. He is a senior researcher at the Instituto de Telecomunicacoes and A. Professor of Instituto Universitario de Lisboa/ISCTE-IUL, Lisbon. His fields of interests include smart sensors, pervasive sensing and computing, wireless sensor networks, signal processing with application in non-destructive testing and diagnosis. Dr. Postolache has authored and co-authored 10 patents, 10 books, 18 book chapters, and over 330 papers in international journals, proceedings of international conference. He is an IEEE Senior Member I&M Society, Distinguished Lecturer of IEEE IMS, chair of IEEE I&MS TC-13 and chair of IEEE IMS Portugal Chapter. He is an Associate Editor of the IEEE Sensors Journal and he was awarded the IEEE best reviewer and the best associate editor in 2011, 2013 and 2017.

Pedro M. B. Silva Girão is a Full Professor of the Department of Electrical Engineering, Instituto Superior Técnico (IST), University of Lisbon (UL), Senior Researcher, Head of the Instrumentation and Measurements Group, and the Coordinator of the Basic Sciences and Enabling Technologies of the Instituto de Telecomunicações. His main research interests include instrumentation, transducers, measurement techniques, and digital data processing, particularly for biomedical and environmental applications. Dr. Girão is a Senior Member of the IEEE, Distinguished Lecturer of IEEE IMS (2015–2018), Honorary Chairman of the International Measurement Confederation (IMEKO) TC19—Environmental Measurements, and Vice-President of the Portuguese Metrology Society.

Preface to "Intelligent Sensing Technologies for Nondestructive Evaluation"

"Intelligent Sensing Technologies for Nondestructive Evaluation" is a wide area of research and their applications have a great impact on our daily life. However, there are new areas to explore where intelligent sensing technologies related to NDE (nondestructive evaluation) can be applied. The main aim of this book is to stimulate discussions regarding technological "start-up" options to discover new frontiers in NDE. The concept of intelligent sensing is encompassed in new methods used especially in industry and for environmental protection. Industry 4.0 is a new approach connecting intelligent sensing and intelligent production using artificial intelligence. It was used for the first time in about 2008 in Germany, and is now widespread thanks to new technologies using intelligent sensing with NDE process. Further benefits include saving money, increasing employment and reducing errors in production/fabrication processes. A further area to explore, as one example, relates to self-repairing and self-rehabilitating systems capable of sensing and repairing defects, detecting failures, etc. We hope the book will stimulate a wide audience, given the fact that it does not/cannot cover all the technological issues connected with NDE.

Seunghee Park, Aimé Lay-Ekuakille , Octavian Postolache , Pedro Manuel Brito da Silva Girão
Special Issue Editors

sensors

MDPI

Review

A Review of Microwave Thermography Nondestructive Testing and Evaluation

Hong Zhang [1,*], Ruizhen Yang [2], Yunze He [3,*], Ali Foudazi [4], Liang Cheng [5] and Guiyun Tian [5]

[1] School of Electronic and Information Engineering, Fuqing Branch of Fujian Normal University, Fuzhou 350300, China

[2] Department of Civil and Architecture Engineering, Changsha University, Changsha 410022, China; xbaiyang@163.com

[3] College of Electrical and Information Engineering, Hunan University, Changsha 410082, China

[4] Electrical and Computer Engineering Department, Missouri University of Science and Technology, Rolla, MO 65409, USA; ali.foudazi@mst.edu

[5] School of Electrical and Electronic Engineering, Newcastle University, Newcastle upon Tyne NE1 7RU, UK; liangcheng85@gmail.com (L.C.); g.y.tian@newcastle.ac.uk (G.T.)

* Correspondence: zhhgw@hotmail.com (H.Z.); yhe@vip.163.com (Y.H.); Tel.: +86-157-1591-8377 (H.Z.); +86-134-6769-8133 (Y.H.)

Received: 21 March 2017; Accepted: 10 May 2017; Published: 15 May 2017

Abstract: Microwave thermography (MWT) has many advantages including strong penetrability, selective heating, volumetric heating, significant energy savings, uniform heating, and good thermal efficiency. MWT has received growing interest due to its potential to overcome some of the limitations of microwave nondestructive testing (NDT) and thermal NDT. Moreover, during the last few decades MWT has attracted growing interest in materials assessment. In this paper, a comprehensive review of MWT techniques for materials evaluation is conducted based on a detailed literature survey. First, the basic principles of MWT are described. Different types of MWT, including microwave pulsed thermography, microwave step thermography, microwave pulsed phase thermography, and microwave lock-in thermography are defined and introduced. Then, MWT case studies are discussed. Next, comparisons with other thermography and NDT methods are conducted. Finally, the trends in MWT research are outlined, including new theoretical studies, simulations and modelling, signal processing algorithms, internal properties characterization, automatic separation and inspection systems. This work provides a summary of MWT, which can be utilized for material failures prevention and quality control.

Keywords: infrared thermography; NDT; microwave thermography; volumetric heating; material

1. Introduction

Infrared (IR) thermography plays an important role in structural health monitoring (SHM) [1] and non-destructive testing (NDT) [2]. IR thermography has great potential and advantages, including fast inspection time, high sensitivity and spatial resolution owing to commercial IR cameras' ability to detect inner defects as a result of heat conduction. It can be split into two categories: passive and active. For the passive approach, the IR camera is used to measure the temperature of materials under test without any external excitation source. The passive thermography configuration is illustrated in Figure 1a. In many industrial processes, passive thermography has been used in production and predictive maintenance [3]. While passive thermography allows qualitative analyses to be performed, active thermography is both qualitative and quantitative [4].

Contrary to the passive approach, an external thermal excitation is required for active thermography. The known characteristics of this external excitation enable depth quantification in composites' debonding detection [5]. As shown in Figure 1b, the configuration of the active infrared approach is similar to that of the passive approach, except for the utilization of an excitation source to generate a distinctive thermal contrast. As illustrated in Figure 1b, the IR camera is situated on the same side of the excitation source in reflection configuration. For transmission configuration, the IR camera is situated on the opposite side of the excitation source. The IR camera is synchronized with the excitation source by a control unit. A computer is required to process and display the obtained thermal images. To improve contrast and quantify defects, active thermography is often performed with advanced signal processing methods. Normally, the reflection mode is suitable for detecting defects situated near the surface, while deeper defects can be detected in the transmission mode. However, the transmission approach cannot be used in some cases where the target is inaccessible [6–8].

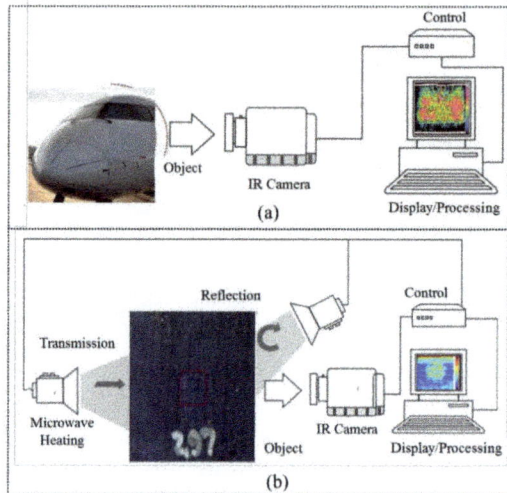

Figure 1. MWT setup for (**a**) the passive approach and (**b**) the active approach.

Depending on the external thermal excitation, different active thermography methods have been developed, such as pulsed thermography (PT) [9], step thermography (ST) [10] and modulated thermography (MT) or so-called lock-in thermography (LT) [11]. Finally, there is pulsed phase thermography (PPT) [12], developed by Maldague and Marinetti in 1996, which combines the advantages of PT and MT [13,14].

Various physical heating sources have been adopted as thermal stimulation sources, such as thermal lamps, lasers, ultrasound devices, and electromagnetic waves. Accordingly, laser thermography [15], ultrasonic thermography and eddy current thermography [16,17] were developed. Taking eddy current thermography as an example, it combines the advantages of IR thermography and eddy current testing, such as being fast and non-contact [18–21]. Eddy current thermography can heat many materials such as metals and carbon fiber reinforced polymer (CFRP) with eddy current heating. However, it only works for conductive materials [22–24], therefore, the excitation source needs to be chosen according to the specific problem. In the last decade, researchers have shown an increased interest in microwave heating techniques. Microwave heating has been exhibited advantages of rapid heat transfer (due to volumetric heating), efficiency, heating uniformity, compact equipment, and being easy to control, etc. Meanwhile, microwave heating has emerged as a powerful platform due to dielectric loss and eddy current heating with different materials under test. So far, built microwave thermography devices have shown some unique advantages such as: (1) microwaves will produce reflection, scattering, transmission at a discontinuous interface. With microwave signal reflection and scattering in the defect area, less microwave energy can be used for heating, and the temperature raise in a defect area is slower than in a non-defect area during microwave heating. IR cameras will capture this abnormal thermal image which will strengthen the effectiveness of defect detection; (2) the heating pattern of microwave heating is relatively uniform, volumetric and selective, and can be achieved in a short time; (3) microwave heating is easy to control, and it is easy to implement different heating function modulations. However, it is necessary to restrict microwave leakage as they are dangerous to human health, therefore, the leakage of the microwaves needs to be kept below a certain recommended level. Generally, in industry microwave heating is operated from 890 MHz to 2.45 GHz to minimize any possible interference with communication services [25].

Thermography has been associated with microwaves in numerous applications. MWT has been employed by some scientists to detect wet rotten wood [26], mines and surrogate signatures [27,28]. MWT has also been used to inspect and characterize various kinds of materials and phenomena, such as debonding and delamination in composite materials [29]. So far, although there are several review works [30–44], they have been limited to a specific field such as composite or renewable energy, so a review of MWT in the material detection field which includes the principles, advantages and disadvantages, developments and research trends is still needed. In this paper, a comprehensive review of MWT techniques for material evaluation has been provided, based on a detailed literature survey.

The overall structure of this paper includes the following: the principle of MWT is presented in Section 2. Then, typical types of MWT applications are summarized in Section 3. Section 4 reviews the development of MWT with case studies. Then, a comparison and discussion are provided in Section 5. Trends are shown in Section 6. Finally, the conclusions are outlined in Section 7.

2. Principle of Microwave Thermography

The principles of MWT mainly include microwave heating and 3D heat conduction. These are analyzed theoretically in the following subsections.

2.1. Microwave Based Heating

The heating style of microwave thermography can be divided into volume heating and surface heating [45], therefore the heating process can be divided into volumetric heating (i.e., dielectric loss heating) and surface heating (i.e., eddy current heating).

2.1.1. Dielectric Loss Heating

For dielectric materials, such as glass fiber composite materials, microwave heating is volumetric heating (i.e., dielectric loss heating). Considering glass fiber composites for instance, material dielectric loss in the microwave radiation field will generate heat. The dissipated power P per unit volume can be expressed as follows [46]:

$$P = 2\pi f \varepsilon_0 \varepsilon'' E^2 \tag{1}$$

where f is the frequency of an electric field, E is the RMS value of the electric field, ε_0 is the permittivity of air, and ε'' is the relative loss factor. Without considering the heat diffusion, the temperature change per unit at heating time t with a continuous microwave source is [46]:

$$T(t) = \frac{Pt}{\rho Cp} = \frac{\omega \varepsilon_0 \varepsilon'' E^2}{\rho Cp} t \tag{2}$$

where ρ is the density of the material and Cp is heat capacity. Obviously, with constant microwave parameters and constant properties of the material under test, the temperature increases linearly with time (during a short period of time).

The basic principles of MWT volumetric heating are shown in Figure 2a: firstly, a microwave excitation module is used to generate a microwave radiation field; secondly, microwaves penetrate the material under test, and the medium molecules will move at the frequency of the electric field which generates heat and eventually is converted into Joule heat, which then will transfer around the material based on the diffusion equation; finally, an IR camera is utilized to obtain the temperature variation in the material under test. Due to the differences in density and heat capacity between the materials under test and defects, information about surface and internal defects can be obtained. Thus, the processes of MWT for dielectric materials evaluation are based on microwave radiation, dielectric loss to generate Joule heat, heat transfer, and IR radiation.

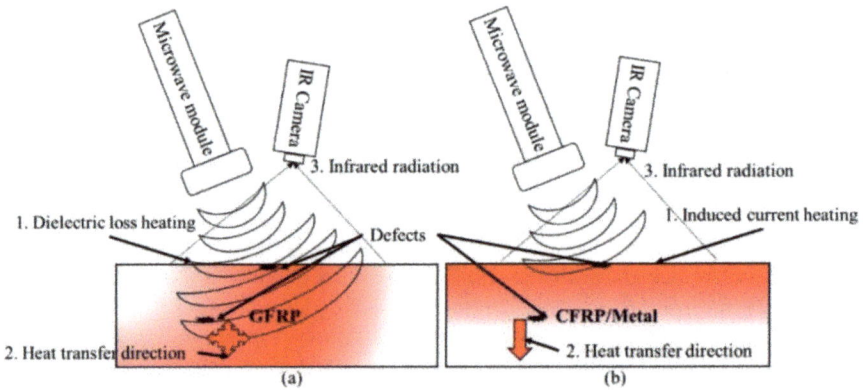

Figure 2. Basic schematic of MWT: dielectric material (**a**) and conductive material (**b**).

2.1.2. Eddy Current Heating

For conductive materials, like metals and carbon fiber composites, microwave heating is eddy current heating. Since the material under test is electrically conductive, microwaves cannot penetrate the conductor material. Thus, the principle of microwave heating is that the energy is radiated to the conductive material surface by microwaves, an alternating electric field is generated, then, induced surface currents are excited from the alternating electric field, resulting in an alternating magnetic field. Next, a vortex electric field is generated by this alternating magnetic field, the vortex electric field promotes the movement of electrons which will generate Joule heat. Finally, the conduct material is heated by Joule heat as shown in Figure 2b. Power P and heat Q generated by eddy current heating can be expressed as [47]:

$$P \sim I_{inductor}^2 \sqrt{\frac{\mu f}{\sigma}} \tag{3}$$

$$Q = Pt \sim I_{inductor}^2 \sqrt{\frac{\mu f}{\sigma}} t \tag{4}$$

where, $I_{inductor}$ is the current flowing through the inductor, σ is the conductivity (S/m), μ is the magnetic permeability of the material under test and f is the frequency of the induced current. It is observed that with a stable induced current and induced current frequency, the generated heat is directly proportional to the microwave excitation time and it is inversely proportional to the square root of the conductivity. However, due to the presence of heat transfer and dissipation problems during actual applications, MWT must be corrected in order to minimize the measurement error.

Due to the skin effect, induced current depth (within the skin depth) in the conductive material is an extremely important factor. The skin depth can be obtained by the following equation [48]:

$$\delta = \frac{1}{\sqrt{\pi \mu \sigma f}} \tag{5}$$

where f is the microwave's frequency, σ is the conductivity (S/m), and μ is the magnetic permeability (H/m) of the material under test. A typical conductivity of CFRP is probably 1000 S/m, and the permeability is around 1. With 2.4 GHz microwave excitation, the skin depth is about 0.002 mm. Therefore, MWT belongs to the surface heating category as only the surface of the CFRP is heated.

2.2. 3D Heat Transfer and Temperature Field

Heat Q generated by dielectric loss or the Joule heat will be conducted from inside to the surrounding material. The heat conduction equation is a time-dependent heat diffusion equation [49]:

$$\frac{\partial T}{\partial t} = \underbrace{\frac{k}{\rho C_p}\left(\frac{\partial^2 T}{\partial x^2} + \frac{\partial^2 T}{\partial y^2} + \frac{\partial^2 T}{\partial z^2}\right)}_{\text{Thermal diffusion}} + \underbrace{\frac{1}{\rho C_p}Q(x,y,z,t)}_{\text{Microwave heating}} \tag{6}$$

where, $T = T(x,y,z,t)$ is the temperature distribution of the surface, k is the material thermal conductivity (W/m × K), C_p is specific heat capacity (J/kg × K), ρ is the density (kg/m^3), and $Q(x,y,z,t)$ is the heat generation function with microwave heating (the dielectric loss heating or eddy current heating). A surface temperature distribution will eventually reflect disturbances of the electromagnetic and thermal fields. Therefore, MWT has the potential to characterize and track the property variations of the material, such as magnetic permeability, electrical conductivity, permittivity, thermal conductivity, thermal diffusivity, etc. In addition, the depth of defects can be quantified. The heat generated by Joule heat will propagate a certain distance within the material in the form of heat waves. The penetration depth δ_{th} of these heat waves is [50]:

$$\delta_{th} \approx 2\sqrt{\alpha t} \tag{7}$$

$$\alpha = (k/\rho Cp) \tag{8}$$

where, α is the thermal diffusivity, and t is the observation time. α can be expressed as a function of the density of the material ρ, heat capacity Cp and thermal conductivity k, as shown in Equation (8). It shows that the penetration depth of heat δ_{th} is proportional to the square root of t and α [11]. In the case of a modulated thermal wave, the length of thermal diffusion decides the penetration depth, which can be found from the following equation [11]:

$$\mu_t = \sqrt{\frac{2k}{\omega\rho Cp}} = \sqrt{\frac{\alpha}{\pi f}} \tag{9}$$

where, ρ is density, k is thermal conductivity, α is thermal diffusivity, Cp is heat capacity, and f is the frequency of the thermal wave. The penetration depth is proportional to the reciprocal of the square root of f and α. In other words, the detection depth varies according to the modulation frequency. In summary, the detection ability of microwave thermography is closely related to the electrical, dielectric and thermal properties of the material under test.

3. Types of Microwave Thermography

3.1. Classification of Excitation Configuration

According to the excitation configurations of microwave heating, MWT can be divided into microwave pulsed thermography (MPT), microwave pulsed phase thermography (MPPT), microwave lock-in thermography (MLT) [51], or microwave step thermography (MST), also known as microwave time-resolved thermography [52]. With MPT, the material under test is heated by a small period of microwave excitation as shown in Figure 3a. The variation of temperature is observed in the heating phase and the cooling phase. For MST, the sample is step heated by a long pulse as shown in Figure 3b, and the variation of temperature is observed in the heating phase. As shown in Figure 3c, the material under test is heated by a periodic amplitude modulated microwave with MLT and the periodic temperature change is captured. A square pulsed modulated excitation is used to derive phase information from multiple thermal waves with one inspection. The influence of non-uniform heating is been reduced and deeper defects can be displayed with a higher contrast. A pulse excitation signal is used by MPPT. Phase analysis is carried out in the frequency domain [13]. Predictably, microwave frequency modulated thermography will employ a frequency modulated microwave excitation in order to derive phase information. From the above analysis, the conclusion can be reached that: MPT and MST analyze the temperature of thermal imaging in the time domain, which is affected by surface emissivity variations and non-uniform heating; MLT obtains information in the frequency domain such as phase, which can suppress the influence of the surface emissivity variations and non-uniform heating. However, the MLT inspection system requires a long measurement time and it is relatively complex.

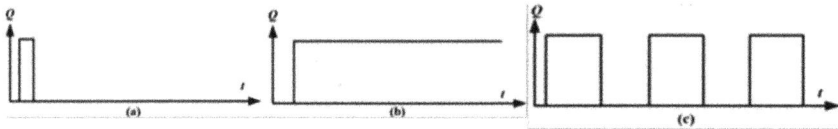

Figure 3. Excitation functions of MWT: (**a**) MPT; (**b**) MST; (**c**) MLT.

Comparisons among MPT, MST, MLT, and MPPT are listed in Table 1. Due to the use of an IR camera, all of them exhibit high sensitivity, high resolution, full-field detection and good visibility. In addition, quantification information can be achieved based on the heat conduction:

1. MPT can be fast and easily deployed. Surface temperature gradients will be introduced not only from defects, but also local variations in surface emissivity and non-uniform heating. A long inspection time is required for a thick material. In addition, the material could be damaged due to the high heating energy.
2. MST is a time-resolved method and it can be used to quantify defect depth. However, the radiation from the heat source in the continuous heating process could deteriorate the temperature measurements. Also, the non-uniform heating and surface emissivity variation have adverse effects on defect evaluation.
3. MLT generally required less excitation energy than MPT. MLT exhibits a higher sensitivity than MPT. The phase data can be extracted which is independent of surface emissivity and heating variations. Tests are repeated with various frequencies and it becomes a time-consuming process to detect defects with various depths. However, MLT offer a compromise with a better depth resolution.
4. MPPT combines the advantages of MPT and MLT. MPPT is less sensitive to non-uniform heating and surface emissivity than MLT as only phase information can be obtained. Moreover, MPPT employed a short excitation pulse which is faster than MLT and wide frequency spectra can be obtained. However, with increasing frequency, the transferred energy is decreased with MPPT. With post processing algorithm, MPPT exhibits better detect ability and resolution than MPT for deeper defects.

Table 1. Comparisons among MPT, MST, MLT and MPPT.

Techniques	Strength	Limitation
Microwave pulsed thermography [13]	Full-field, high resolution, high sensitivity, good visibility, quantification, fast, easy deployment	Small depth, long time for thick material, emissivity and non-uniform heating dependence, high power
Microwave step thermography [52]	Full-field, high resolution, high sensitivity, good visibility, quantification, fast, easy deployment, time-resolved	The radiation from heat source in the continuous heating process could have a negative influence on temperature measurements of MUT, emissivity, and non-uniform heating dependence
Microwave lock-in thermography [51]	Full-field, high resolution, higher sensitivity, good visibility, quantification, low power, emissivity independence, elimination of non-uniform heating	Compromise between depth and depth resolution, time-consuming,
Microwave pulsed phase thermography	Full-field, high resolution, high sensitivity, good visibility, quantification, fast, emissivity independence, elimination of non-uniform heating, greater depth and resolution, better detectability	Post signal processing, energy attenuation with frequency

3.2. Classification by Heating Style

MWT can be classified into surface heating thermography, volume heating thermography, and abnormal heating thermography [14,45]. In Figure 4a, for eddy current heating, MWT can be also called surface heating thermography (SHT). Due to the great permeability, the skin depth is very small [53–55]. Thus, it can be classified as part of the SHT family. With reflection mode, the heat conduction from surface to inside is used to quantify the depth of defect.

MWT exhibits volumetric heating for dielectric material inspection. MWT can be called volume heating thermography (VHT), as illustrated in Figure 4b [45,55]. For the transmission and reflection modes with VHT, the characterizations of defects are similar [54]. Only interesting areas are heated without heating the host material in some cases. Abnormal heat will caused by defects. Furthermore, this abnormal heat is used to quantify depth information of the related defect. In Figure 4c, these methods are called abnormal heating thermography (AHT). For detecting water in concrete structures, MWT can be considered as a kind of AHT.

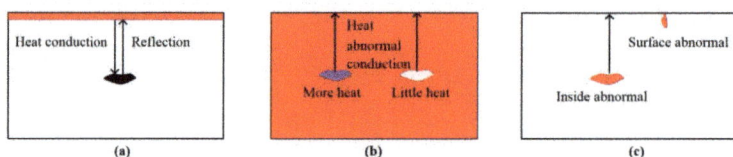

Figure 4. Styles of heating: (**a**) SHT; (**b**) VHT; (**c**) AHT.

4. Developments and Case Studies

4.1. A History of MWT Development

Developed countries have taken the lead on the use of MWT to carry out related researches and achieved some interesting results [56–58]. Levesque and Ambrosio did a preliminary study on MWT in the 1990s. Levesque et al. employed X and Ku bands (range from 8 GHz to 18 GHz) horns and parabolic antennas to excite the sample under test [59]. The excitation of the antenna is provided by a 100 W amplifier. Thermal images are produced by an AGEMA (model 782 LWB) IR camera with 8 μm–12 μm range. An area of 300 $\times$ 300 mm^2 is been measured due to the limited excitation area. Several 10 mm thick glass-epoxy composites have been tested to identify the inserted carbon particles area. The proposed method were used to characterize artificial defects' size and location under different depths in composite materials. Ambrosio et al. used a cavity with a minimum 600 W to detect artificial defects in non-metallic composites [60]. The cavity was 300 $\times$ 200 $\times$ 280 mm^3 which is excited by a 2.45 GHz magnetron and an Alter VPG1540 power supply unit (maximum power 1200 W). They used a Cober PM45 power meter to monitor the incident and reflected microwave powers. Two cavity applicators (an open cavity applicator and a large cavity applicator) have been proposed to avoid the aperture edge effects and to improve the field uniformity over the sample. To estimate the defect permittivity and depth influence on surface temperature perturbation, theoretical modeling and numerical simulations were carried out.

In 1999, Takahide et al. at Osaka University applied MWT to detect surface breaking cracks in reinforced concrete structures and they pointed out that the tip of crack will produce more heat during the test [61]. Heat was introduced by a time-gated microwave source into the homogeneously reinforced concrete structure. With the microwave penetration features, the subsurface of the reinforced concrete structure could be inspected. Moreover, wet cracks could be selectively heated. Therefore, subsurface cracks will be immediately identified with MWT. Analyses of time and spatial dependence features of measured thermal images are possible to quantify cracks' depth and thermal diffusivity information [61]. In the 21st century, groups in the USA, UK, France, Poland, Italy, Korea, etc. have developed MWT for metal, dielectric material, cement/concrete, GFRP, CFRP, honeycomb and cement-based composites detection problems. These works have been simply summarized in Table 2 and are introduced in detail in the following Sections 4.2–4.9.

Table 2. Summary of researches with MWT.

Hardware Development					Software Development		Experimental Study	
Operation Frequency	Antenna/Sensor	Power	IR Camera	Simulation Study	Sampling Software	Signal Processing	Material under Test	Defects
8 GHz to 18 GHz [59]	Horns and parabolic antennas [59]	100 W [59]	AGEMA 782 LWB [59]	Not available	CEDIP PTR-5010 system [59]	Normalized the infrared images pixel by pixel [59]	Glass-epoxy composites [59]	Delaminations
2.45 GHz [60,61]	Magnetron [60]	600 W–1700 W [60]	AGEMA AGA 880 [60]	Influence of defect permittivity and depth has been estimated [60].	HP 8757 C network analyzer [60]	Open cavity applicator and large cavity applicator have been designed [60]	Kevlar or fiberglass slabs and sandwich samples [60]	Defects
	Microwave oven [61]	1400 W [61]	Nikon LAIRD-3 [61]	Not available	Not available	Thermal image was taken at 20 s [61]	Mortar block [61]	Cracks
2 GHz to 3 GHz [62], 2.45 GHz [63,64]	TEM horn antenna [62,64]	50 W [62,64], 10 W [63]	DRS Tamarisk 320 [62,64], FLIR SC 500 [63]	CST Microwave Studio and MPHYSICS Studio [62].	Not available	The surface thermal profile was taken after 10 s of heating [62]. The surface thermal profile was taken after 5 s and 15 s of heating [63].	Reinforced steel bars [62], Rebar in air [63]; Embedded in cement [63,65], CFRP [64]	Corrosion, delaminations, debonding and crack
2.45 GHz [65]	Magnetron generated horn antenna [65]	600 W [65]	Not available	Not available	ALTAIR software	A contrast algorithm is used to analyze the thermogram series with 5 min of heating [65].		Steel bar corrosion
5 GHz to 10 GHz [66]	Horn antenna [66]	2.3 W [66,67]	Mikron 6T61 and SantaBarbara IR camera [66]	Not available	Macintosh microcompute [66]	A 2.7 s microwave pulse was used [66]	Multilayered plexiglass-water-teflon specimen [66]	Debonding
9 GHz [67]	A single flare horn antenna [67]		Santa Barbara Focalplane [67]	Not available	LabVIEW	The surface thermal profile was taken after 8 s and 10 s of heating [67]	Carbon fibers in different epoxy structures [67]	Embedded fibers
18 GHz [57]	Flann 18 094-SF40 waveguide [57]	1 W [57]	FLIR SC7500 [57]	COMSOL Multiphysics and CST Microwave Studio [57]	Rohde & Schwarz SMF 100 A signal generator [57]	The surface thermal profile was taken after 10 min of heating [57]	GFRP [57]	Defects
	WR430 waveguide [57]	1000 W [57]				The surface thermal profile was obtained after 10 s to 15 s of heating [57]		Debonding and delamination
2.45 GHz [56,68]	Magnetron [68]	500 W [68]	Flir A325 [57,68]	Not available	Not available	A sequence of 180 thermograms was obtained and processed by using normalized, standardized contrast and cosine transform [68]	Composite materials with adhesive bounded joints [68]	Defects
	Pyramidal horn antenna [56]	360 W [56]	Not available	Not available	ALTAIR program [56]	The surface thermal profile was taken after 150 s of heating [56]	CFRP [56]	Defects
1 GHz [69]	Coaxial-type probes [69]	0.1 W [69]	FLIR T620 [69]	CST Microwave Studio [69]	Spectrometer [69]	Normalized to the maximum value of field intensity [69]	Carbon-fiber composite materials [69]	Conductivity measurement

4.2. Metals and Corrosion

Foudazi et al. from the Missouri University of Science and Technology proposed the use of MWT for the detection and characterization of corroded reinforced steel bars [62]. In Figure 5a, the measurement setup for steel bars is illustrated. They employed a 14×24 cm^2 horn antenna to irradiate steel bars with a 50 W microwave signal for 10 s. The distance between the steel bars and the antenna was 1 cm. During the experiment, four pieces of corroded material were mounted on steel bars and spaced 1 cm apart. A DRS Tamarisk 320 thermal camera was utilized to obtain the thermal profile of the steel bars with a 0.05 K sensitivity. In Figure 5b, the surface thermal profiles for steel bars after 10 s as detected with 2 GHz, 2.5 GHz, and 3 GHz energy are provided. The steel bars were a smooth rebar without ribs. The radius of this rebar was 4.8 cm which contained a 1 cm corrosion area. In Figure 5b, the visible hot spots are the corroded areas on these steel bars. Because of the relatively low thermal conductivity of corroded steel, heat quickly dissipates in uncorroded steel. Moreover, these temperature differences between the corroded areas indicated that different amounts of corrosion absorb different amounts of microwave energy. With a preliminary simulation and experimental study of the MWT, it demonstrated that a higher excitation microwave frequency will lead to a higher temperature for corrosion detection in steel bars. Moreover, increased corrosion leads to absorption of microwave energy increasing results in a greater difference in the measured IR images.

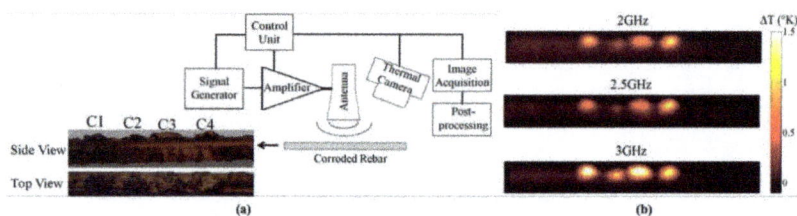

Figure 5. (a) MWT measurement setup for corroded rebar detection; (b) temperature profile for corroded steel bar [62]. Reprinted/reproduced with permission from IEEE. .

Pieper et al. demonstrated an active MWT for inspection of large corrosion areas on reinforcing steel bars for cement-based structures [63]. They used CST Microwave studio and MultiPhysics studio to construct a coupled microwave and thermal simulation model. The effects of steel bars have been investigated in the air and in concrete. To detect the change in temperature with a thermal camera, the incident microwave power should be increased because some of the microwave energy is absorbed by the concrete. The effect of different polarization has been studied with CST. With 10 W incident power, the parallel polarization generates the most uniform heating in that of orthogonal polarization and circular polarization, but the temperature increase was the lowest (only 0.014 °C). The highest temperature increase was generated by orthogonal polarization (around 0.025 °C), therefore, orthogonal polarization is the best choice for thin corrosion detection.

During the experimental study, two steel (AISI 1008) bars (each of length 150 mm and radius 4.8 mm) were measured, which have been embedded in parallel in a concrete block ($170 \times 150 \times 50$ mm^3). The sample was heated for 5 s by a microwave oven operating at 2.45 GHz. A FLIR Thermacam SC 500 was used to provide thermal images with a sensitivity of 0.1 °C. In Figure 6a, the two steel bars have been heated by microwave energy for 15 s. One (top) with localized significant corrosion (on the order of 1 mm–4 mm) on a portion of its length, the one below with light corrosion (on the order 0.2 mm or less) along half of its length. As illustrated in Figure 6b,c, the corroded areas in the two steel bars have been identified in the thermal images as indicated by the black circle. Moreover, in Figure 6c, the uncorroded area in the steel bar is also visible, appearing as a relatively lower temperature line (in the left of the black circle). In Figure 6e, the corroded steel bar is still detectable when embedded in a concrete block as highlighted by the black circle. These results show that steel corrosion in concrete will lead to a higher loss tangent and the thermal properties are changed. With microwave heating, the physical property changes due to corrosion become measurable in the thermograms. Depending on the orientation and thickness of corrosion, the polarization of the incident microwave signal can be optimized through simulation. The effect of corrosion layer thickness on microwave heating has been investigated.

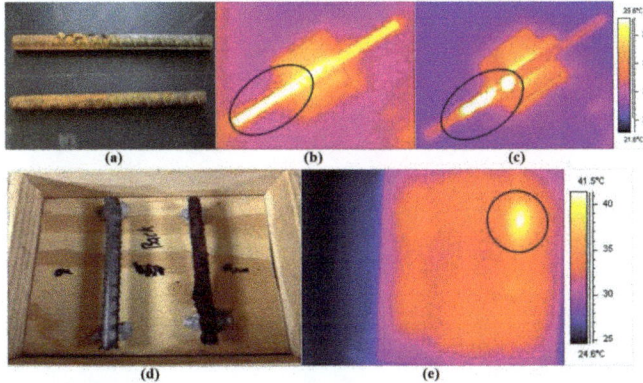

Figure 6. (a) Photograph of localized corrosion (top) and light corrosion (bottom); (b) thermal images of a steel bar with light corrosion and (c) localized corrosion; (d) photograph of clean (left) and corroded (right) steel bars; (e) thermal images of clean (left) and corroded (right) steel bars embedded in the concrete block after microwave heating [63]. Reprinted/Reproduced with permission from AIP Publishing LLC.

Keo et al. presented a pyramidal horn antenna based MWT to detect corroded steel in a reinforced concrete wall [65]. During the experimental study, a commercial magnetron operating at 2.45 GHz was used to irradiate a maximum 800 W of microwave energy. The steel reinforcements (each one with a diameter of 12 mm, located at a regular spacing of 10 cm) with a 38 mm concrete covering were embedded in a concrete block ($1000 \times 1000 \times 65$ mm^3). The specimen was heated by 600 W microwave energy for 5 min. The reflection mode was used as the thermal camera was placed on the same side as the excitation source. A data analysis method based on a contrast algorithm was used to analyze the thermogram series to reduce the effect of non-uniform excitation. As shown in Figure 7a, the infrared camera was composed of a 320×256 matrix detector of indium antimonide (InSb) with a sensitivity range of 3–5 μm. The excitation antenna was placed in a 45° direction to heat the largest surface area of the specimen. In addition, the infrared camera was placed at 2.32 m away from the specimen in a 30° direction which can detect the largest possible surface area of the specimen. Thermograms were recorded at 1 image per s by using the ALTAIR software (developed by FLIR for obtaining measurement data). Figure 7b shows the obtained thermograms at the 250 s instant. The abnormal temperature areas correspond to the corrosion area in the steel reinforcements. In the preliminary experimental study of the MWT, it was demonstrated that the detection depth of MWT was about 3.8 cm in a thick concrete block.

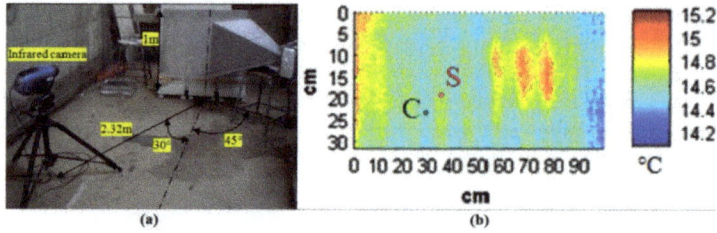

Figure 7. Photograph of microwave thermography measurement setup (**a**) and temperature profile for corroded steel bar (**b**) [65]. Reprinted/reproduced with permission from Elsevier.

4.3. Dielectric Materials

Osiander et al. employed a time-resolved MWT system for surface and sub-surface inspection on Plexiglass-water-Teflon specimens with absorbers at different depths (0.15, 0.30 and 0.45 mm) [66]. Figure 8a shows a diagram of the experimental setup. An HP 6890B oscillator was used to generate microwave pulses with a frequency range from 5 GHz to 10 GHz. A Hughes 1277 X-band traveling wave tube amplifier with 2.3 W maximum power was utilized as a power amplifier. An excitation horn antenna was placed in a 45° direction with respect to the sample surface. Two infrared cameras were used to monitor the surface temperature of the sample under test. One was a Mikron 6T61 infrared scanner with an HgCdTe detector. The second one was a SantaBarbara infrared camera with a 128 × 128 pixels InSb focal plane array. The temperature resolution of the first camera was 0.1 K. For the second IR camera, the temperature resolution is about 0.003 K. The specimen under test is illustrated in Figure 8b. It is structured with a Teflon layer of varied thickness and a water layer with a constant thickness backed with Plexiglass layers. Considering the power, the microwave horn was working in the near field. The measured data was normalized to the peak amplitude to eliminate the effects of non-uniform microwave distribution. Figure 8c shows IR images of the test sample before, during and after a 2.7 s microwave pulse. Three water layers corresponding to the Teflon layer thicknesses appear in the IR images in a temporal order. With MWT, the potential of surface temperature versus time has been shown for subsurface defects quantitative characterization. Compared with laser beam heating, dry epoxy coated steel samples exhibit a very small response with microwave heating. When a debonding region contains water, the whole structure of the debonding region can be illustrated. The wavelength independent resolution has been demonstrated to be 30 μm. An analytical model has been provided to extract the time dependence of the surface temperature where quantitative data such as the depth of the defect can be extracted.

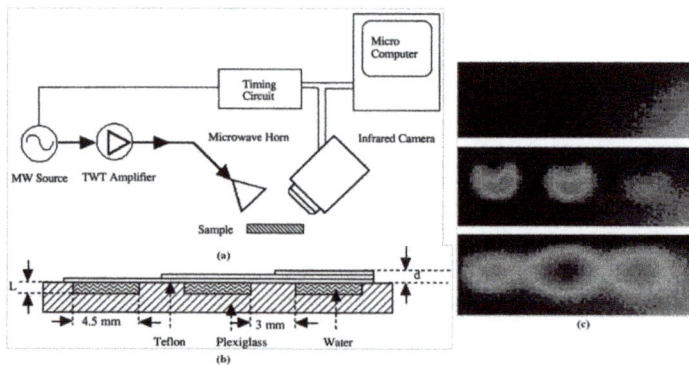

Figure 8. Schematic diagram of the experimental setup (**a**), the specimen under test (**b**) and IR images (**c**) [66]. Reprinted/reproduced with permission from SPIE.

4.4. Cement and Concrete

Takahide et al. proposed the use of MWT for detecting surface-breaking cracks in concrete [61]. Surface-breaking cracks can be penetrated with water. The crack opening distance was 0.2 mm and 0.4 mm. The size of the cracks was 10 mm in depth and 40 mm in width. Since the microwave absorptivity of the concrete is much smaller than that of water, cracks containing water can be selectively heated. By applying microwaves to the concrete structure, the thermal conduction of these heated cracks will generate localized high-temperature regions. The experimental MWT setup is illustrated in Figure 9a. A commercially available 2.45 GHz magnetron was used to irradiate microwaves into the mortar block with a maximum 1400 W output power. The specimen under test was placed in the microwave oven. A Nikon LAIRD-3 with PtSi array was used to measure the temperature distribution of the mortar block. In Figure 9b, four artificial cracks have been introduced to a rapid hardening mortar block specimen. By injecting water into crack C and applying microwaves for 5 s, crack C with water was selectively heated. Compared with the non-crack area, the temperature of the crack rose about 6 °C to 8 °C after microwave heating. As shown in Figure 9c, the position of the crack C can be easily detected from the thermal image taken immediately after the microwave excitation. Meanwhile, the authors found that the temperature region of the crack degraded after 20 s due to the thermal diffusion from the crack into the surrounding area. The relation between the crack size and temperature rise should be determined for quantitative investigations.

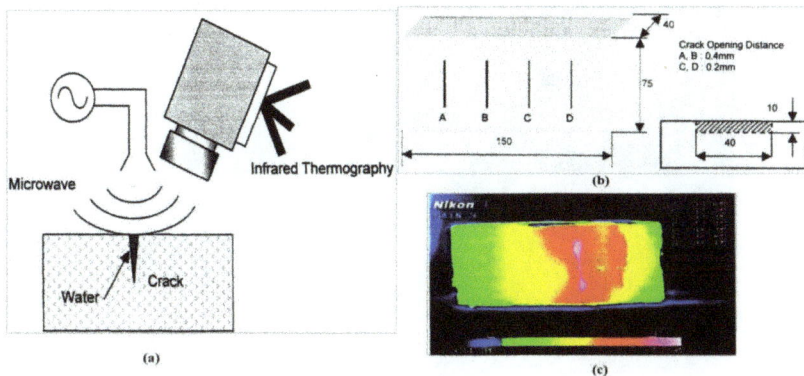

Figure 9. Schematic diagram of MWT (**a**), mortar block specimen with artificial cracks (**b**) and thermal images taken after microwave heating (**c**) [61]. Reprinted/reproduced with permission from SPIE.

4.5. Glass Fiber Composite Structures

Bowen et al. introduced microwave-source time-resolved infrared radiometry for detecting and characterizing microwave absorption in dielectric materials [67]. An HP 6890B oscillator was used to excite a 9 GHz microwave signal. A Hughes 1277 X-band traveling wave tube amplifier was used to amplify this signal to a maximum 2.3 W power. A single flare horn antenna with about 50° beam width was placed 15 cm from the sample under test. A Santa Barbara camera with 128 × 128 InSb focal plane array was used to detect the IR radiation. The temperature resolution is about 0.003 K. With embedded fibers at different locations, a fiberglass-epoxy specimen was measured. The depth of the fibers was 0.25 mm and 0.75 mm. Due to the loss tangent and Joule heating, there is a very high contrast for defects in embedded epoxy materials. The size and orientation of these embedded fibers in the measured thermal images have been studied. The interaction is strongly dependent on the length of the fiber (9 mm to 30 mm) and orientations. The temperature shows a modal distribution along the fiber for fibers longer than 12 mm.

Cheng et al. [57] developed a microwave pulsed thermography system for glass fiber composite measurement. Figure 10a illustrates the setup of the experimental system: (1) an adaptor connected with the waveguide; (2) GFRP wind turbine blade; (3) microwave generator linked with cable; (4) IR camera. GFRP wind turbine blade with 4 holes (4.5 mm in radius) at the root was heated by an open-ended waveguide connected with a Flann 18 094-SF40 adaptor. The adaptor was linked to a signal generator (Rohde & Schwarz SMF 100 A) with the maximum output power of 30 dBm (1 W) at 18 GHz. The waveguide was placed above the sample with roughly a 45 degree illumination direction. A FLIR SC7500 IR camera was used to obtain the temperature distribution. Artificial holes with 4.5 mm radius were investigated. During the experiment, the signal generator provided 1 W microwave power, although only 115.2 mW microwave power was emitted from the waveguide due to the cable loss. In order to maximize the actual illumination power and minimize the power loss during the transmission, 22.7 mm was selected as the standoff distance (distance between the microwave antenna aperture and sample under test) due to the impedance matching (this distance should be around $\lambda/4$ to maximum power transfer, where λ is the wavelength of microwave). In their primary experimental results, a discontinuous temperature was captured in the defect region (mainly at the edges). A high power heating system is required for a better contrast of heat patterns between defect and non-defect regions.

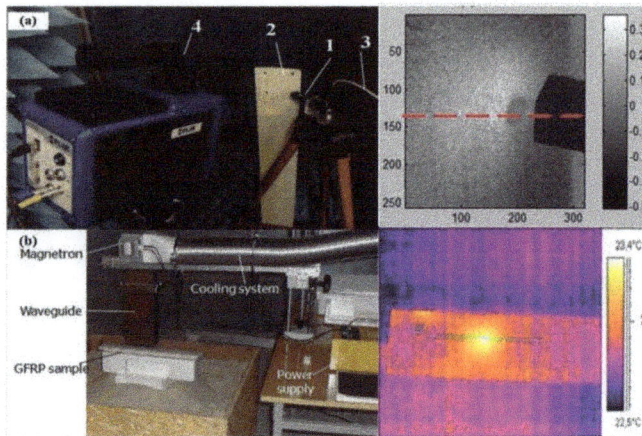

Figure 10. MWT setup in Newcastle University (**a**) and West Pomeranian University of Technology (**b**) [57]. Reprinted/reproduced with permission from IEEE.

A magnetron-based microwave pulsed thermography system has been developed by West Pomeranian University of Technology for glass fiber composite debonding and delamination detection [57]. A GFRP wind turbine blade (BLADES200W from Navitron) with a 4.5 mm hole at the middle was the study object. Figure 10b illustrates system setup for high power microwave pulsed thermography. With a maximum 1 kW power, the microwave excitation was a 2.45 GHz magnetron. An open-ended aperture was matched impedance with a WR-430 waveguide. Dipolar Magdrive 1000 was used to control the output power. A Flir A325 device was used to obtain thermograms. A 0.238 °C maximum temperature raise can be obtained from 2 s heating. In their primary experiment results, the location and the size information of the defect have been detected from the continuities in the line-scan results.

Ryszard et al. also employed MWT for inspecting adhesive joints in composite materials [68]. Due to the dielectric property differences between defects (lack of adhesive or delamination) and the background, the induced thermal energy in the defect region and the background in the material under test is different. The experimental setup is illustrated in Figure 11a: (1) an IR camera in a protective box to prevent damage caused by the high power microwaves; (2) a 500 W magnetron working at 2.45 GHz; (3) microwave absorbers; (4) an open-ended rectangular waveguide; (5) the sample under test; (6) a cooling system for the magnetron. The MWT observation time was set to 180 s to obtain 180 thermograms in one sequence. The measured results were obtained by using dedicated image processing algorithms. As shown in Figure 11b–e, the original thermogram (b), cosine transform (c), standardized contrast (d) and contrast enhancement (e) were used to obtain images of the defect. In the primary experiment results, approximate location and the size information about delamination can be obtained in the thermograms after processing.

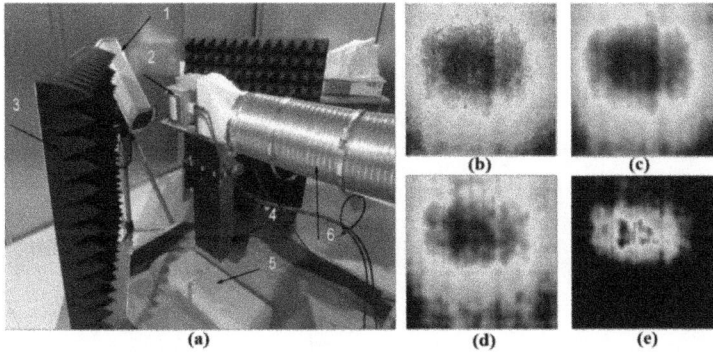

Figure 11. Experimental setup for MWT (**a**) and measurement results (**b–e**) [68]. Reprinted/reproduced with permission from authors.

We also investigated MWT for defect detection in a glass fiber wind turbine blade, as shown in Figure 12. A horn antenna (ETS Lindgren 3115, frequency range: 750 MHz–18 GHz.) was used for microwave excitation. The antenna was connected to Rohde & Schwarz SMF 100 A signal generator with a maximum output power of 1 W. The waveguide was placed above the sample with roughly 90 degree illumination direction. A FLIR SC7500 IR camera was used to obtain the temperature distribution of sample under test. During the experiment, the signal generator provided 1 W at 0.915 GHz and 2.45 GHz. In order to detect defects during the experiment, the microwave heating duration was selected as 1 min (as shown in Figure 13a,c) and 2 min (as shown in Figure 13b,d). In our initial experiment results, a discontinuous temperature was captured in the defect region. Furthermore, the heating effect of 2.4 GHz is not as good as 0.915 GHz due to the penetration ability difference. With the captured IR images, defects around 1 mm radius with 0.2 mm in depth could be obtained in the GFRP sample. With a longer heating duration, the number of defects became hard to identify due to the heat diffusion, therefore, the heating time must be optimized for different situations.

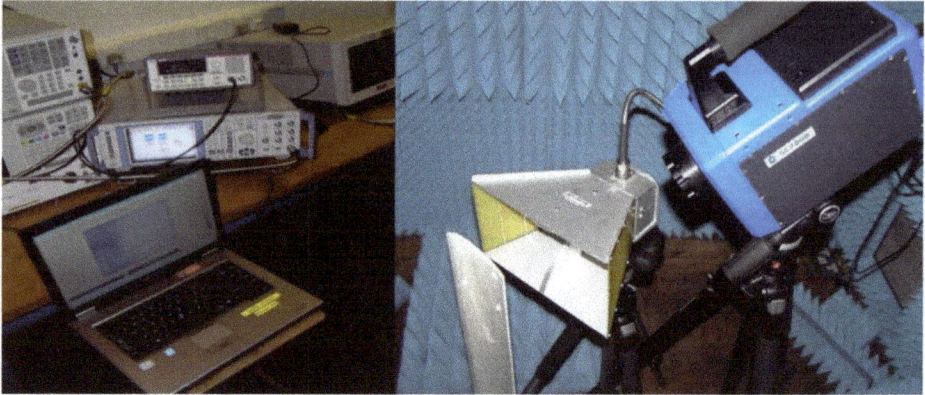

Figure 12. Experimental setup for defect detection with MWT.

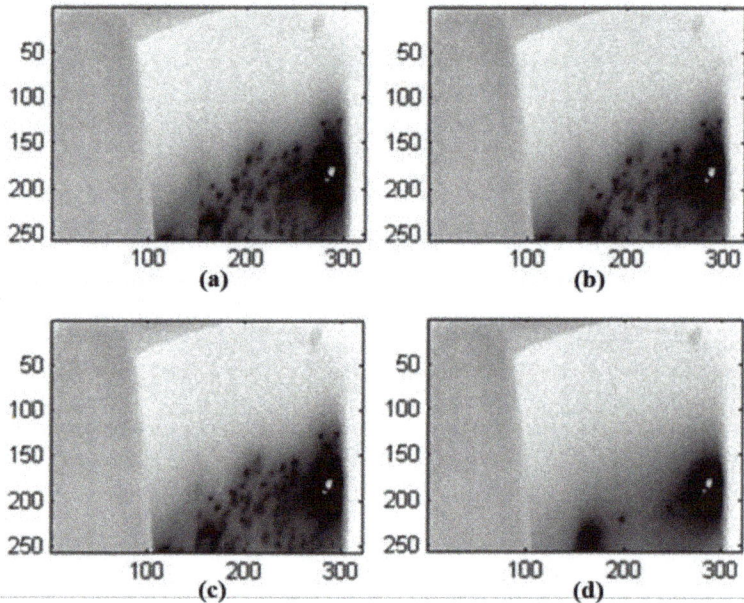

Figure 13. IR results with MWT for defect detection in glass fiber wind blade after 1 min microwave excitation with 0.915 GHz (**a**), after 2 min microwave excitation with 0.915 GHz (**b**), after 1 min microwave excitation with 2.45 GHz (**c**) and after 2 min microwave excitation with 2.45 GHz (**d**).

4.6. Carbon Fiber Composite Materials

In France, Keo et al. developed a microwave pulsed thermography for CFRP detection [56]. They used a commercial 2.45 GHz magnetron to generate microwave signals. With a detector with a 320 × 256 InSb matrix, the sensitivity of the IR camera is in a range of 3 μm–5 μm. To capture the whole inspection area, the sample was placed 1.5 m away from IR camera at 55°. The CFRP sample was $40 \times 40 \times 4.5$ cm^3. A 10×10 cm^2 defect was created at the middle of the CFRP sample by the absence of adhesive. The antenna was placed in the 45° direction as shown in Figure 14a. In Figure 14b, the microwave beam was guided by a 59×56 cm^2 pyramidal horn antenna onto the sample under test. With the ALTAIR program provided by FLIR, a computer was used to record thermograms at 1 Hz. A 360 W microwave was used to heat the specimen for 150 s. The same testing procedure was carried out on samples with/without defect. As shown in Figure 15, the thermograms at the 100 s were obtained. In Figure 15a, a non-defect specimen has been shown in the thermogram at the 100 s instant. In Figure 15b, a specimen with a defect has been shown in the thermogram. In Figure 15c, the thermogram of the sample with a defect (Figure 15a) has been subtracted with the thermogram of the sample without defect (Figure 15b) at the same instant. In Figure 15d, the thermogram in Figure 15c is subtracted from its initial thermogram in order to highlight the defect. In the primary experiment results, the CFRP defect area can be clearly found. It is hotter than the non-defect area due to the absence of adhesive. The defect area can be estimated directly from the thermograms (about 10×10 cm^2) [56].

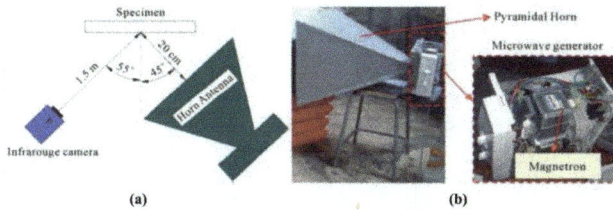

Figure 14. MWT setup (**a**) and Horn (**b**) in University Institute of Technology of Bethune [56]. Reprinted/reproduced with permission from Scientific Research Publishing Inc.

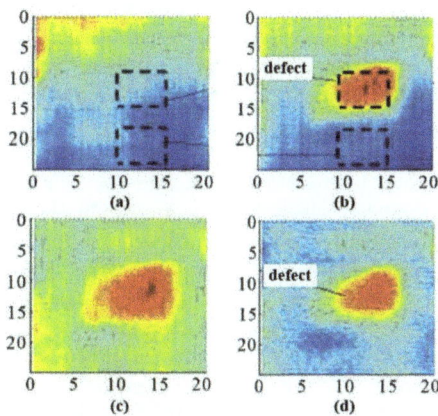

Figure 15. MWT results for the absence of adhesive in CFRP: the thermogram of the sample without defect (**a**), the thermogram of the sample with the defect (**b**), the subtraction between the thermograms of the sample with and without the defect (**c**), thermogram subtracted from its initial thermogram (**d**) [56]. Reprinted/reproduced with permission from Scientific Research Publishing Inc.

In 2014, Foudazi et al. investigated MWT for the inspection of rehabilitated cement-based structures [64]. The experimental setup is shown in Figure 16a,b, 2.4 GHz microwaves have been transmitted with 50 W power. A sweep oscillator (HP8690B) and a power amplifier (OphirRF 5303084) are used to generate high power microwave signals. The sample under test was illuminated by a horn antenna. With a sensitivity of 0.05 K, the thermal profile of the sample's surface was captured by an IR camera (DRS Tamarisk 320). In Figure 16c, the sample is a $20 \times 20 \times 4$ cm^3 mortar sample covered with a 13×13 cm^2 CFRP. There is a 2×2 cm^2 and 2 mm depth delamination in the center of the sample [58]. With 5 s heating time, measurements were performed at a 6 cm standoff distance. Meanwhile, measurements were also performed at a 45 cm standoff distance with 15 s heating time.

Thermal profiles were captured before and after heating. In Figure 16d, the thermal image before heating is provided. Figure 16e shows the thermal image after 5 s heating. In their primary experiment results, the delamination in CFRP can be easily identified. Both the level of power and heating time should be optimized to detect small size disbands. In addition, they found that the measured result is evidently affected by edge effects with 45 cm standoff distance, as shown in Figure 16f.

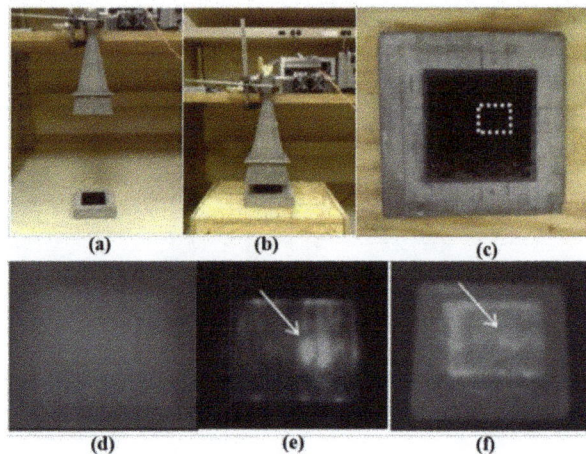

Figure 16. MWT measurement system setup used for 45 cm (**a**) and 6 cm (**b**) standoff, photograph of mortar sample with delamination (**c**). Measurements on mortar sample before heating (**d**), 6 cm standoff after 5 s heat (**e**) and 45 cm standoff after 15 s heat (**f**) [58]. Reprinted/reproduced with permission from IEEE.

Recently, Foudazi has shown that the MWT method can be used for detection of delaminations, debonding and cracks in rehabilitated cement-based materials [70]. The experimental setup is shown in Figure 17a. Three CFRP-strengthened cement-based specimens have been measured (a unbonded defect was made by placing a 5 mm thin sheet of foam with dimensions of 6 cm × 8 cm in a sample with $52 \times 38 \times 9$ cm^3, several delaminations (with thicknesses ranging from 1 mm to 3 mm and areas ranging from 10 cm^2 to 100 cm^2) have been formed in a sample with dimensions of $52 \times 38 \times 7.8$ cm^3, the crack sample had dimensions of $52 \times 38 \times 9$ cm^3), as shown in Figure 17b. Thermal profiles for these specimens are illustrated in Figure 17c. The approximate location and the size information about defects in these specimens have been obtained. Meanwhile, the temperature of the defect will be affected by the orientation of the carbon fibers (due to the electric field difference). In their initial experimental results, for the case of unidirectional carbon fiber, if the electric field is along the fiber direction, the temperature change for the defected area will be smaller compared to the case of perpendicular polarization. This is due to the different electromagnetic response of the carbon fiber at different polarizations. In other words, if the wave is parallel to the fiber orientation, it will react as an electric conductor, while for the perpendicular case it is similar to high loss dielectric materials. Additionally, the thermal contrast (TC) between healthy and defective areas is much greater. Moreover, increasing defect dimensions also led to a greater TC.

Furthermore, the authors used MWT to monitor debonding in CFRP by using different excitation antennas [64]. A horn antenna linked with a double-ridged waveguide operating at frequency ranged from 0.75 GHz to 18 GHz was employed. The aperture size of this antenna is 14×24 cm^2. In Figure 18a, the CFRP sample is $30 \times 30 \times 0.2$ cm^3 backed with $40 \times 40 \times 5$ cm^3 Al sheet which containing six debonds with dimensions of 6×6 cm^2, 5×5 cm^2, 4×4 cm^2, 3×3 cm^2, 2×2 cm^2, and 1×1 cm^2, respectively. Figure 18b shows top and side views of the CFRP backed by a layer of Al sheet. During experiment, the sample under test was illuminated at 2.4 GHz microwave signal of different power (50, 100, 150 and 200 W). In addition, the effect of the heating time (from 10 s to 30 s) has been investigated by using CST Microwave Studio and MultiPhysics Studio. As shown in Figure 18c,d, normalized thermal images of CFRP with different excitation power levels at 10 s and 30 s have been demonstrated. In the preliminary experimental results, they found that a higher excitation microwave power is needed for a smaller debond detection. In addition, the debonding becomes clearly visible with increasing excitation power due to the temperature difference increase. Therefore, increasing the incident power improves the detection of small disbonds. Meanwhile, an increase of the heating time leads to an increased temperature throughout the sample under test, thereby the possibility of detecting the disbonds is reduced, therefore, the heating time must be optimized according to the actual situation.

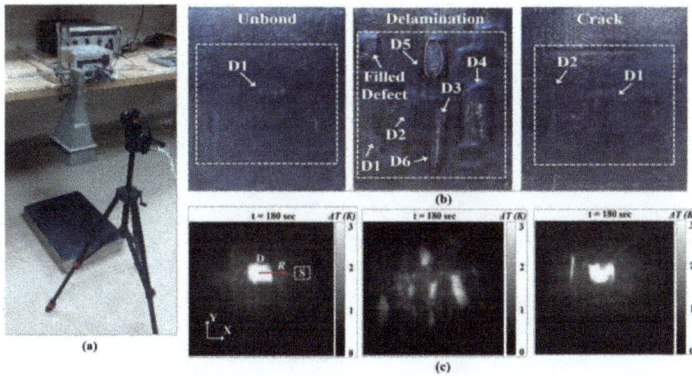

Figure 17. MWT measurement setup (**a**), samples under test (**b**) and thermal profiles for samples under test (**c**) [70]. Reprinted/reproduced with permission from IEEE.

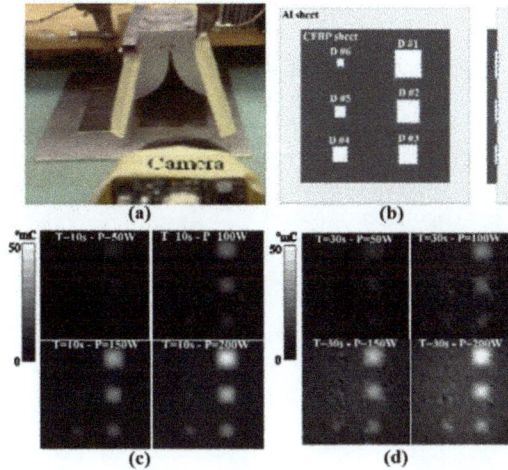

Figure 18. MWT experimental setup (**a**), sample (**b**), measurement results for 10 s heating (**c**) and 30 s heating (**d**) [64]. Reprinted/reproduced with permission from IEEE.

A microwave time-resolved infrared radiometry system was proposed by scientists in The John Hopkins University [71]. The experimental setup is shown in Figure 19a. 9 GHz microwave signals were produced by an HP 6890B Oscillator. A Hughes 1277 X-band traveling wave tube amplifier was used to amplify these signals and then to feed them into a rectangular waveguide linked with a single flare horn antenna. The beam width of this horn antenna is about 50° and the sample was placed about 15 cm away. 2.3 W input power was transmitted to the antenna and a 20 mW/cm^2 power density was created for the experimental study. Both the polarization of the microwave and the angle of incidence were controlled. By operating in the 3 μm–5 μm band, a 128 × 128 InSb focal plane array infrared camera was used to monitor the surface temperature of the sample under test. Before, during and after the application of the microwave step heating pulse, a series of frames were recorded for time-dependent measurements. In the initial experimental results, the authors measured carbon fibers with two different depths (0.25 mm and 0.75 mm) in fiberglass-epoxy. The temperature at the center of the fiber is shown in Figure 19b. Moreover, they found that the experimental data can be fitted with a solid line. The determination of surface layer thickness and thermal diffusivity can be almost independent from the surface properties of the layer.

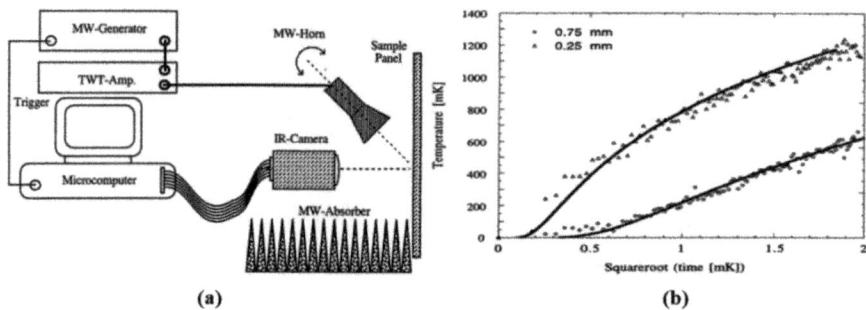

Figure 19. Schematic diagram of the experimental setup in The John Hopkins University (**a**) and Surface temperature as a function of square root time for a single point on fibers in 0.25 mm and 0.75 mm depth (**b**) [71]. Reprinted/reproduced with permission from SPIE.

Lee et al. proposed a microwave probe pumping technique to characterize the anisotropic electrical conductivity in carbon-fiber composite materials [69]. They used CST Microwave Studio to investigate the electromagnetic field distribution with different anisotropic conductivity. Two 10 mm coaxial probes were used to pump and scan the microwave field. The pumping probe was fixed on the backside of the sample under test. The near-field distribution was scanned by the scanning probe. A spectrometer was used to measure the microwave power. A network analyzer was used as the microwave source with a continuous mode. A FLIR T620 was used for measurement. A 1 GHz microwave source with 20 dbm power was used during the measurements. A 50 × 50 × 0.1 mm^3 carbon-fiber/PEEK composite sheet with a defect (characteristic length is 5 mm) has been measured. In the initial experimental results, they obtained an intense area around the scanning probe. Moreover, they found that the conductivity of the carbon-fiber/PEEK has an elliptical distribution.

4.7. Honeycomb Structures

Microwave pulsed thermography for water measurement in honeycomb materials was developed by scientists in Poland [72]. They introduced 2 GHz antennas with a 30° beam width. The surface of a honeycomb sample was illuminated with a 30 mW/cm^2 power density. The antenna and IR camera were arranged in a reflection configuration. The proposed MWT detection system is shown in Figure 20a. The antenna was located at 1 m away from the sample under test and the IR camera was placed at 0.7 m. The sample was a 290 × 215 mm^2 sandwich panel with two 0.7 mm thickness Fibredux face skins. Different quantities of water (5.0, 2.5, 1.2 and <1 mL) are introduced in the sample to form four defects. Microwaves were used to irradiate the sample for 5 s, and then the IR camera was used to capture thermal images for 20 s at a rate of 1 Hz. In their experiment results, as shown in Figure 20b, the location of water can be well visible even in small quantities with the reflection and transmission arrangement, but it is difficult to qualify precisely the water content due to the phenomenon of longitudinal heat diffusion.

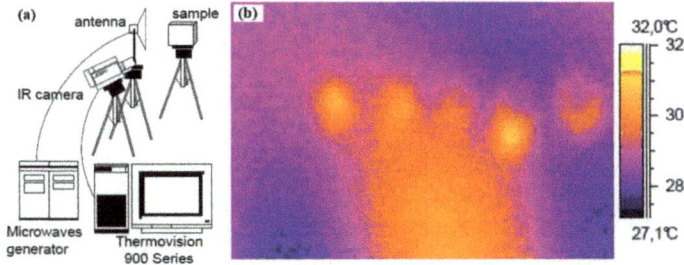

Figure 20. MWT setup at the Military Institute of Armament Technology (**a**) and results for water in the honeycomb material (**b**) [72].

4.8. Cement Based Composite

Foudazi et al. proposed active MWT to evaluate steel fiber distribution in cement-based mortars [73]. $200 \times 200 \times 200$ mm^3 fiber-reinforced cement-based mortar (FRCM) samples were measured. The steel fibers have diameters of 0.55 mm and lengths of 30 mm. The effects of clumping, dielectric properties and fiber depth have been evaluated with a full-wave coupled electromagnetic-thermal simulation which was conducted by using CST MultiPhysics Studio^{TM}. As illustrated in Figure 21a, microwave signals were generated by a signal generator at the desired 2.4 GHz operational frequency. A 50 W power amplifier was used to amplify the excitation power level. A horn antenna was used to radiate a relatively uniform microwave excitation toward the sample's surface. A DRS Tamarisk 320 thermal camera was utilized to capture the surface thermal profile of the sample.

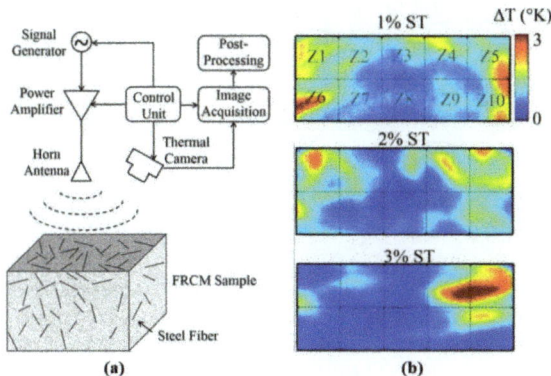

Figure 21. MWT measurement setup (**a**) and specimen surface temperature variation with different steel fiber contents (**b**) [73]. Reprinted/reproduced with permission from Springer.

During measurements, microwave energy was applied for 30 s and an additional 90 s of thermal profile measurement followed to capture the cooling period. As shown in Figure 21b, surface temperature differences after 30 s microwave heating were provided for 1%, 2% and 3% of steel fiber contents. In their primary experiment results, they found that a larger temperature difference was contributed by the induced surface current on areas containing steel fibers. Therefore, with increasing volume content of steel fibers, the temperature will increase. Due to variation in heating associated with induced surface current and dielectric heating, fiber depth and dielectric properties of mortar have a significant influence on the temperature difference at the surface of samples. They found that MWT is capable of determining the presence of the fiber clumping in the cement-based composite structures: 1% and 2% steel fibers are shown to have higher surface temperature difference compared to the sample made with 3% fiber content.

4.9. Advanced Signal and Image Processing Methods

Microwave lock-in thermography has been developed by scientists at Politecnico di Bari [74]. A function generator was used to control the power by switching the oven off/on. The frequency of excitation was 0.1 Hz. As shown in Figure 22a, the IR camera was positioned at 15 cm from the oven to allow it to focus on and frame the whole specimen. The dimensions of the specimen were 76 mm in length, 76 mm in width and 8 mm in thickness. Due to the low heat diffusion velocity and thermal conductivity, a strong drift in the temperature evolution over time is noted. Moreover, this setup can avoid damage due to possible microwave leakage. To obtain the amplitude and the phase information, Fast Fourier Transform-based algorithms were used to process the thermal image data which acquires thermal images frame by frame over time. Figure 22b shows the phase image and Figure 22c shows amplitude image. In the experimental results, the problem of the specimen's edges was observed. Due to the interaction between the specimen geometry and the electromagnetic field, a high temperature was exhibited near the edge of the specimen. Nevertheless, the authors could clearly identify the damage area in the captured thermograms. In addition, quantitative analysis of the damaged areas showed a good agreement between the defect area results obtained with microwave thermography (1693 mm^2) and X-ray scanning (1385 mm^2).

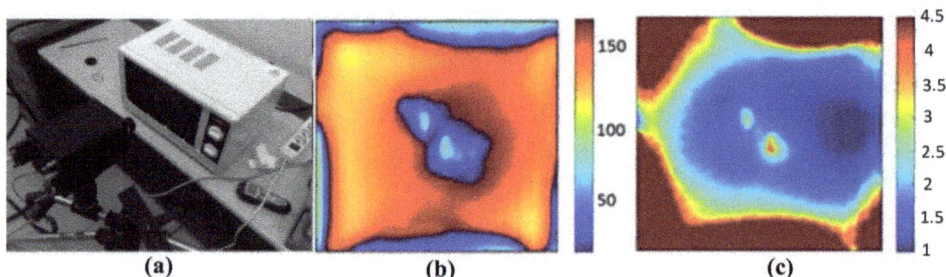

Figure 22. The experimental set-up used for relay (**a**), lock-in phase image (**b**) and amplitude image (**c**) [74].

In 2014, Palumbo et al. investigated MWT for quantitative evaluation of damaged areas in CFRP through experiments and numerical simulations [75]. Damaged CFRP samples were inspected by microwave pulsed thermography. The dimensions of the four sandwich tested specimens were 76 × 76 mm^2 with 8 mm thickness. Non-linear heating behavior was characterized and the undamaged area exhibited a higher slope during the heating and cooling phases. A new approach was proposed to process the obtained thermographic data. Numerical simulations were carried out to assess the sensitivity. An IR camera was located at 150 mm from the samples under test. To acquire the heating phase and subsequent cooling phase, the IR camera recorded for 7 s at 100 Hz. This proposed technique offers advantages in term of testing time (only 2 s of heating and a very fast data processing). In Figure 23a, X-ray images of the specimens under test are provided for comparison with the MWT results. As shown in Figure 23b, T_{max} analysis can be used to quantitatively indicate damaged areas located at a greater depth. The authors proposed a new algorithm based on the non-linear heating and cooling thermal behavior of damaged and undamaged areas for the quantification of damaged areas. The image after the data binarization is shown in Figure 23c. The location and size information of the damaged area can be obtained from these measured results. Quantitative analysis of the damaged areas shows a good agreement between results obtained with microwave thermography (784.3 mm^2 with heating slope), X-ray (769.4 mm^2) and lock-in thermography (779.3 mm^2). The size of the minimum resolvable defect was $2 \times 2 \times 0.5$ mm^3, with a depth of 1.3 mm.

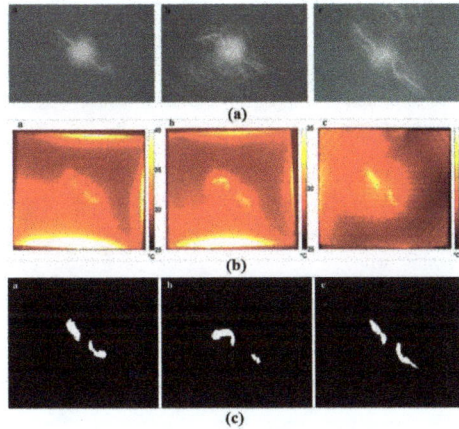

Figure 23. X-ray images of three samples (**a**), T_{max} maps during all acquisition sequence (**b**), and binary images obtained by T_{max} maps (**c**) [75]. Reprinted/reproduced with permission from Springer.

Usamentiaga et al. proposed an artificial neural network for automatic energy estimation of impact damages in carbon fiber composite materials, but the data acquisition time required for each inspection area was at least 30 s [76]. Three specimens were measured: specimen 1 was a carbon fiber composite made up of 12 plies with 2.5 mm in thickness. Specimens 2 and 3 were composite panels with two stiff facings made of carbon fiber and a low density material bonded between them. These two specimens have two valid sides, side A made up of six plies (1.5 mm in thickness) and side B made up of 12 plies (2.5 mm in thickness). A FLIR SC5000 IR camera was used to acquire infrared images for damages caused by impacts of different energy, ranging from 6 J to 50 J. With 12 bits per pixel, 320×256 pixel thermal images are produced by the FLIR SC5000 and its' thermal sensitivity is 0.02 K. To cut down the number of obtained images, the experimental acquisition rate is 50 Hz while the maximum frame rate is 383 Hz. To improve the signal-to-noise ratio in the thermal images, a post-processing method was applied. With a Discrete Fourier Transform (DFT), the temperature-time history of each pixel during the heating period is transformed into the frequency domain and the phase information can be calculated based on it. As shown in Figure 24, the value below each defect is the estimated impact energy and the real impact energy is the number in brackets. In their primary experiment results, the impact energy of nearly half of the defects was estimated with an error of less than 2.5 J. This percentage increases to 80% when the considered error is 5 J, and nearly 100% when the estimated error is 10 J. The results indicate that an artificial neural network (ANN) is able to quantitatively characterize impact damages.

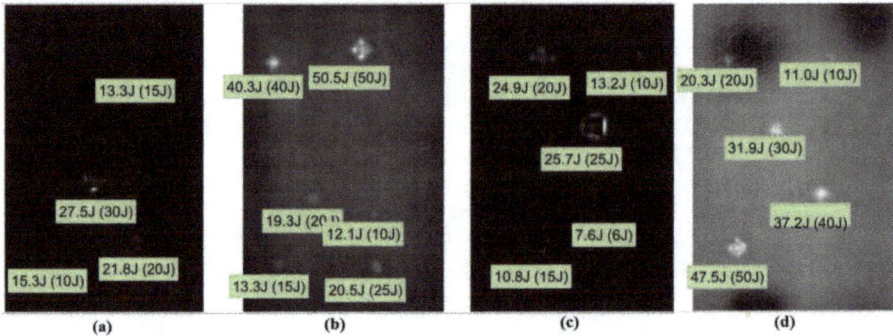

Figure 24. Estimated impact energy using artificial neural networks: specimen 2 front side (**a**), specimen 2 backside (**b**), specimen 3 front side (**c**), specimen 3 backside (**d**) [76]. Reprinted/reproduced with permission from Elsevier.

In summary, the purpose of adopting advanced signal processing methods in MWT is to eliminate noise, increase the signal to noise ratio for extracting the abnormality information of defects, enhance the contrast of detection images and improve the assessment accuracy. Meanwhile, the inspection time can be further reduced.

5. Discussion

5.1. Comparison with Other Excitation Sources

As mentioned before, a thermal contrast between surface/subsurface defects and the surrounding material can be produced by many energy sources. Some works are focused on the comparison between heating methods. For example, Keo et al. [56] compared microwave infrared thermography with CO_2 laser thermography. MWT was found to be more suitable for the detection of deeper defects while CO_2 laser thermography was more suitable for the detection of surface/near surface defects in CFRP.

According to a literature review [14,30,56,77], a comparison between MWT and other thermography excitation sources has been provided in Table 3. Due to the use of an IR camera, most of the thermography methods can offer fast inspection, non-contact, full-field, great sensitivity, high resolution, and quantitative analyses. As listed in Table 3, energy sources can be divided into:

1. Flash/lamp: halogen or IR lamps are commonly employed for a long period in large inspection areas with an x-y scanner/robot. In all cases, the measurement surface is been illumined by light to transfer heat and to propagate inside the specimen (containing a wavelength range from the ultraviolet, visible and infrared spectrum).
2. Laser: the heat is introduced into the material under test by the laser. Compared with lamps, a scan of the inspection area is needed.
3. Mechanical: sound or ultrasound waves are injected by transducers. With waves propagating through the specimen, heat is produced by slapping and rubbing of the surfaces (mostly in the defect areas). Compared with optical excitation, non-uniform heating is considerably reduced and the visibility of sub-surface defects is improved.
4. Induction: eddy currents are generated by an excitation coil. The penetration depth varies inversely with the operation frequency. The induction heating is limited to conductive materials. Compared with mechanical heating, heating non-uniformities have less influence in induction heating since heat is produced locally.

5. Microwave: the heat is introduced into the specimen by a time-gated microwave excitation source. The sub-surface microwave absorbing features can be used for measurements (such as metal bars/fibers, water-filled areas). By analyzing the time of the thermal images, quantitative defect information can be extracted (such as depth and size).

Table 3. Comparison of thermography inspection methods with different excitation sources.

Heating Sources	Strengths	Limitations
Flash, lamp	Non-contact, full-field, low cost, methods are mature	Surface heating, impact of surface condition on heating, heating reflection
Laser	Non-contact, remote heating from a far distance, high sensitivity, great resolution, quantification, fast	Heating area relies on excitation source, scanning is required, more suitable for surface defect detection
Mechanical	Full-field, high resolution, high sensitivity, quantification, fast, selective heating	Contact, know-how, specimen needs to be fixed, lack of quantitative information
Induction	Non-contact, relatively low-cost of excitation system, full-field, high resolution, great sensitivity, quantification, fast, inner heating	Limited to conductive material, non-uniform heating, complex heating system, near-field heating, heating area is limited to the excitation coil
Microwave	Non-contact, remote excitation, full-field, high resolution, great sensitivity, quantification, fast, uniform heating, selective heating	Complex and expensive microwave excitation system, electromagnetic radiation

5.2. Comparison with Other NDT Methods

Eight major categories of NDT techniques are listed in Table 4. A comparison of these technologies is provided and an overview of each method given to identify the advantages and limitations of current NDT techniques.

Table 4. Summary and comparison for MWT with major NDT methods.

NDT Techniques	Strength	Limitation
Ultrasound-echo/Phased array/Linear array	Great depth, high resolution, many deployment options	Sound attenuation, coupling for contact testing, non-sensitive to surface defects
Guide wave	Large areas	Sound attenuation, coupling for contact testing
Acoustic emission	In-service, passive, large areas,	Noise, bad quantitation, non-sensitive to static defects
Shearography	Non-contact, full-field, fast, high sensitivity	Sensitive to part movement, small thickness/stiffness, require unique test set-ups, expensive, hard to quantitatively analyze
Eddy current	Non-contact, low-cost, no surface treatment	Conductive material, scanner required, sensitive to lift-off, low resolution
Microwave	Non-contact, high resolution, suitable for dielectric material	Scanner required, near-field, lift-off influence
Microwave thermography	Non-contact, full-field, high resolution, high sensitivity, quantification, fast, uniform heating, selective heating	Heating system complex, electromagnetic radiation
X-ray/Gamma-ray	High resolution, non-contact	X-ray radiation hazards, operation complex, scanner required

For materials inspection, there is no universally applicable method. Selection of a particular NDT technique requires more consideration than the detection capabilities. Meanwhile, the application, portability of equipment, inspection schedule, inspection area, types of materials, accessibility, costs and expected defects types are also important.

5.3. Shortcomings of MWT

From the MWT literature, it can be found that MWT is not always perfect for quantitative material detection. There are still many shortcomings in existing studies, which can be mainly summarized as follows:

1. Inadequate theoretical studies

Multi-physics coupling mechanisms of metal/composite materials detection with microwave thermography are not deeply studied. For example, composite materials are typically composed of a variety of materials, and the microwave heating principles of composite materials are different from those of conductive and dielectric materials. Meanwhile, the physical processes of MWT for material evaluation are very complex, which includes microwave heating, heat conduction, and heat diffusion. For example, microwave heating represents a dielectric loss in glass fiber composite materials, where it is a volumetric heating method; while microwave heating is Joule heat for conductive materials which is affected by skin effects, and it is a surface heating method.

2. Lack of study on excitation signal modulation and corresponding data processing methods

As mentioned above, IR thermography techniques can be subdivided into pulse thermography, lock-in thermography, pulse phase thermography and step heating thermography [14]. Many scholars have studied microwave pulse thermography, microwave step heating thermography and microwave lock-in thermography. However, no researcher has studied pulse phase microwave heating in the frequency domain. Microwave pulse phase thermography combines the advantages of MPT and MLT which can inhibit change in the surface emissivity and other negative factors; the pulse width can bring a stronger contrast and a deeper defect can be detected.

3. Lack of systematic research on microwave excitation module optimization

Microwave excitation module is an important part of MWT, and the heating effect is directly reliant on it. Furthermore, the subsequent thermal imaging is directly affected by rapid and uniform heating. British and Polish researchers have investigated waveguides as excitation modules [57,72]; French scholars have studied the pyramidal horn antenna as an excitation module [56]; South Korean scholars have studied the coaxial setup as excitation module [69]. However, the advantages and disadvantages of these microwave excitation modules are not thoroughly studied, which results in is it being difficult to achieve the optimal detection ability with MWT systems.

4. Lack of internal properties characterization and defect quantification methods

Temperature variation from the infrared camera is a result of joint action by the surface properties of materials (emissivity), internal thermal properties (thermal conductivity, diffusivity, interlayer reflection coefficient), electrical properties (conductivity and permittivity) and other factors. How to extract these features from the surface temperature response and to further quantitatively characterize the material properties is important and difficult in the current studies. Existing studies did not provide an effective method for property characterization and defect quantification.

5. Lack of automatic separation and damage area quantification methods

Some scholars have studied several prefabricated macroscopic defects in composite materials (lamination defects, cracks and debonding, etc.), and experimental empirical formulas have been established; but there is a lack of related research on automatic separation of different defects and damage area quantification. The data acquired with MWT is an image sequence or a three-dimensional matrix. Matrix analysis method is theoretically possible to achieve fast imaging, automatic separation and damage area quantification. However, existing studies employed advanced matrix decomposition methods for MWT data processing.

6. Trends

1. Multiple physics and new physics

The physical properties of the specimens are different which results in the physics of MWT being different. For example, composites are multi-layered and their parameters are anisotropic as a result of fiber reinforcement. In addition, several composite materials are often included in a composite structure (such as a sandwich structure). Hence, the physics of metals differ from those of composites. Furthermore, the effects of the electromagnetic field, microwave propagation and other multiple-physical field also need to be investigated, such as thermal pattern interpretation [78] in thermal optical flow [79], and the spatial-, time-, frequency-, and sparse-pattern domains. Thus, multiple physics and new physics-based MWT methods are required for materials evaluation.

2. Computer simulation and modeling

Over the last several years, computer modeling and simulation (such as method of moments or MoM and finite element method or FEM) have been employed for understanding the physics during MWT measurements (such as microwave radiation and propagation, heat generation and diffusion). In the past, researchers have investigated three different approaches to resolving the electromagnetic phenomena of microwave propagation and heating processes. Firstly, time-domain solvers have been applied to microwave heating problems [80]. These use a time-marching algorithm to predict the electric and magnetic fields at the next time step. Secondly, frequency-domain methods have been investigated [81], where the numerical solution strategy uses a particular frequency to predict the electric and magnetic fields. Lastly, a method that combines an efficient time-domain solver with the power of a frequency-domain solver, has been used to predict the power distribution generated in a lossy medium during microwave heating [82]. Moreover, the operational frequency and radiation pattern of microwave excitation system can be optimized with simulation and modeling for better detection performance. In addition, the influence of materials' properties (such as conductivity, dielectric, size and shape) can be investigated with simulation and the total cost of experiments will be reduced. What's more, the parameters of defects (such as location, size, orientation and shape) can be examined. For composite materials, the influence of different fiber orientations in the microwave EM field can be investigated too. Therefore, simulation and modeling are needed to improve the reliability and accuracy of MWT systems.

3. Microwave excitation system optimization

MWT is based on microwave heating. The thermal profile of a material under test is created by an IR camera after microwave excitation. With MWT, a large amount of microwave excitation systems can be used to introduce the heat, however, the heating efficiency of microwave excitation systems is not only dependent on the properties of the microwave system (such as operational frequency, radiation pattern and power, etc.) but also relies on the physical properties of the material under test (such as size, shape, conductivity, dielectrics and microwave energy absorbing ability, etc.). Thus, the optimization of microwave excitation systems is required to improve the ability and sensitivity of MWT systems.

4. Signal processing algorithms

To extract useful features from the captured thermal images, advanced signal processing algorithms have been used. These algorithms includes wavelet transform [83], independent components analysis (ICA) [84], principal components analysis (PCA) [85,86], pattern recognition [87], support vector machine [88,89] and Tucker decomposition [90]. With suitable signal processing algorithms, the inspection results for size and depth identification, subsurface defect detection, emissivity variation reduction and defect dimension quantification can be significantly improved.

Therefore, more advanced signal processing algorithms are needed to further improve the sensitivity and quantification ability of MWT systems.

5.   Intelligent inspection systems

The efficiency of a MWT system can be improved by implementing an intelligent inspection system with artificial intelligence. As various types of defects can be acquired during material measurement, the treatment for different types of defects is different. Taking a composite material for example, the most common embedded defects are delamination, adhesive debonding and out-of-plane waviness. These defects are the most typical defects observed during manufacturing which need to be identified to improve the manufacturing quality of the composite. Therefore, it is important to classify the defect type with an intelligent inspection system. As computers become increasingly capable, artificial intelligence methods can be used in MWT to reduce the inspection time and improve the reliability of MWT systems. For example, Moomen et al. employed machine learning for feature selection in microwave NDT [91].

6.   Mobile inspection systems

For a large material under test, the MWT needs to be placed in a mobile robot or a vehicle. In Figure 25, a MWT inspection system has been combined with a vehicle [61]. The inspection time of MWT for a large material can be significantly reduced. The whole inspection can be performed autonomously. The safety and efficiency of the MWT systems are being improved too. However, lightweight equipment and advanced detection algorithms including compressed sensing are also required in order to provide automatic inspection capability.

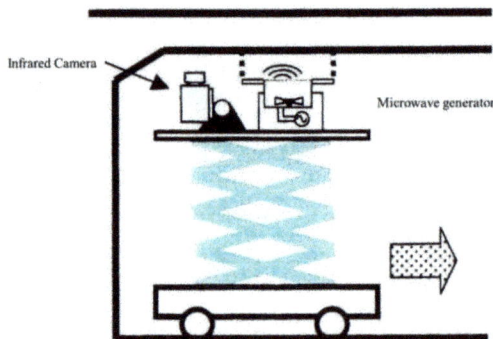

Figure 25. MWT combined with a vehicle [61]. Reprinted/reproduced with permission from SPIE.

7. Conclusions

The basic principles and types of MWT have been reviewed in this paper. MWT exhibits great potential, including fast heating, high resolution, fast inspection and high sensitivity, no contact requirement and better detectability for inner defects. Moreover, the manufacturing quality and reliability of materials can be improved to prevent failures. In this work, a comprehensive review of MWT techniques for material inspection has been reported based on a detailed literature survey. Firstly, the theory of MWT has been presented and MWT has been classified into four categories. Then, the development of MWT has been outlined through case studies. Next, limitations in current MWT research have been outlined based on detailed comparisons. Finally, some research trends in MWT are predicted. It is concluded that:

1.   MWT combines the advantages of microwave technology and infrared thermography. A higher heating efficiency and uniform heating pattern can be expected. A full-field, non-contact, fast detection can be performed.

2. MWT can be divided into MPT, MPPT, MST and MLT. In the near future, microwave frequency modulated thermography and microwave pulsed phase thermography will be achieved.

3. MWT is a fast and effective non-destructive method for material inspection, especially for water/defects identification in concrete/composite structures.

Acknowledgments: This work was supported by National Natural Science Foundation of China (61601125 and 61501483), National Key Research and Development Program of China (2016YFF0203400), Natural Science Foundation of Fujian Province (2016J05152) and Fujian Province Young and Middle Age Education Research Fund (JA15577).

Conflicts of Interest: The authors declare no conflict of interest.

Abbreviations

AHT	Abnormal heating thermography
ANN	Artificial neural networks
CFRP	Carbon fiber reinforced polymer
DFT	Discrete Fourier transform
ECLT	Eddy current lock-in thermography
ECPT	Eddy current pulsed thermography
ECPPT	Eddy current pulsed phase thermography
ECST	Eddy current step thermography
FEM	Finite element method
FRCM	Fiber-reinforced cement-based mortars
GFRP	Glass fiber reinforced polymer
LT	Lock-in thermography
MLT	Microwave lock-in thermography
MoM	Method of moments
MPT	Microwave pulsed thermography
MPPT	Microwave pulsed phase thermography
MST	Microwave step thermography
MT	Modulated thermography
MUT	Material under test
MWT	Microwave thermography
NDT	Nondestructive testing
PMC	Polymer matrix composites
PPT	Pulsed phase thermography
PT	Pulsed thermography
SHM	Structural health monitoring
SHT	Surface heating thermography
ST	Step thermography
VHT	Volume heating thermography

References

1. Bakht, B.; Mufti, A. *Bridges: Analysis, Design, Structural Health Monitoring, and Rehabilitation*; Springer: Cham, Switzerland, 2015.
2. Hellier, C. *Handbook of Nondestructive Evaluation*; Mcgraw-hill: New York, NY, USA, 2001.
3. Maldague, X.P. Introduction to NDT by active infrared thermography. *Mater. Eval.* **2002**, *60*, 1060–1073.
4. Vergani, L.; Colombo, C.; Libonati, F. A review of thermographic techniques for damage investigation in composites. *Fract. Struct. Integr. Ann.* **2014**, *8*. [CrossRef]
5. Sun, J.G. Analysis of Pulsed Thermography Methods for Defect Depth Prediction. *J. Heat Transf.* **2005**, *128*, 329–338. [CrossRef]
6. Czichos, H. *Handbook of Technical Diagnostics*; Springer: Berlin, Germany, 2013; Volume 40, pp. 43–68.
7. Shalin, R.E.E. *Polymer Matrix Composites*; Springer Science & Business Media: Berlin, Germany, 2012.

8. Zalameda, J.N.; Burke, E.R.; Parker, F.R.; Seebo, J.P.; Wright, C.W.; Bly, J.B. Thermography inspection for early detection of composite damage in structures during fatigue loading. *Proc. SPIE* **2012**, *8354*. [CrossRef]
9. Waugh, R.C.; Dulieu-Barton, J.M.; Quinn, S. Modelling and evaluation of pulsed and pulse phase thermography through application of composite and metallic case studies. *NDT E Int.* **2014**, *66*, 52–66. [CrossRef]
10. Roche, J.M. Common tools for quantitative time-resolved pulse and step-heating thermography—Part I: Theoretical basis. *Quant. Infrared Thermogr. J.* **2014**, *11*, 43–56.
11. Maldague, X.P. *Theory and Practice of Infrared Technology for Nondestructive Testing*; John Wiley Interscience: New York, NY, USA, 2001.
12. Castanedo, C.I. *Quantitative Subsurface Defect Evaluation by Pulsed Phase Thermography: Depth Retrieval with the Phase*; Université Laval: Québec City, QC, Canada, 2005.
13. Maldague, X.P.; Marinetti, S. Pulsed phase infrared thermography. *J. Appl. Phys.* **1996**, *79*, 2694–2698. [CrossRef]
14. Yang, R.; He, Y. Optically and Non-optically Excited Thermography for Composites: A review. *Infrared Phys. Technol.* **2016**, *75*, 26–50. [CrossRef]
15. Li, T.; Almond, D.P.; Rees, D.A.S. Crack imaging by scanning pulsed laser spot thermography. *NDT E Int.* **2011**, *44*, 216–225. [CrossRef]
16. Wilson, J.; Tian, G.Y.; Abidin, I.Z.; Yang, S.; Almond, D. Pulsed eddy current thermography: System development and evaluation. *Insight Non-Destr. Test. Cond. Monit.* **2010**, *52*, 87–90. [CrossRef]
17. Riegert, G.; Zweschper, T.; Busse, G. Eddy-current lockin-thermography: Method and its potential. *J. Phys. IV Fr.* **2005**, *125*, 587–591. [CrossRef]
18. He, Y.; Tian, G.; Pan, M.; Chen, D. Eddy current pulsed phase thermography and feature extraction. *Appl. Phys. Lett.* **2013**, *103*, 084104. [CrossRef]
19. Yang, R.; He, Y.; Gao, B.; Tian, G.Y.; Peng, J. Lateral heat conduction based eddy current thermography for detection of parallel cracks and rail tread oblique cracks. *Measurement* **2015**, *66*, 54–61. [CrossRef]
20. He, Y.; Pan, M.; Tian, G.Y.; Chen, D.; Tang, Y.; Zhang, H. Eddy current pulsed phase thermography for subsurface defect quantitatively evaluation. *Appl. Phys. Lett.* **2013**, *103*, 144108. [CrossRef]
21. He, Y.; Pan, M.; Chen, D.; Tian, G.Y.; Zhang, H. Eddy current step heating thermography for quantitatively evaluation. *Appl. Phys. Lett.* **2013**, *103*, 194101. [CrossRef]
22. He, Y.; Tian, G.; Pan, M.; Chen, D. Impact evaluation in carbon fiber reinforced plastic (CFRP) laminates using eddy current pulsed thermography. *Compos. Struct.* **2014**, *109*, 1–7. [CrossRef]
23. Zhang, H.; Gao, B.; Tian, G.Y.; Woo, W.L.; Bai, L. Metal defects sizing and detection under thick coating using microwave NDT. *NDT E Int.* **2013**, *60*, 52–61. [CrossRef]
24. Qaddoumi, N.N.; Saleh, W.M.; Abou-Khousa, M. Innovative Near-Field Microwave Nondestructive Testing of Corroded Metallic Structures Utilizing Open-Ended Rectangular Waveguide Probes. *IEEE Trans. Instrum. Meas.* **2007**, *56*, 1961–1966. [CrossRef]
25. Meredith, R. *Engineers' Handbook of Industrial Microwave Heating*; The Institution of Electrical Engineers: London, UK, 1998.
26. Wyckhuyse, A.; Maldague, X. A Study of Wood Inspection by Infrared Thermography, Part II: Thermography for Wood Defects Detection. *Res. Nondestruct. Eval.* **2001**, *13*, 13–22. [CrossRef]
27. Balageas, D.; Lemistre, M.; Levesque, P. Mine detection using the EMIR® method-Improved configuration using a mobile detection system. In Proceedings of the 7th International Conference on Quantitative Infrared Thermography (QIRT), Bordeaux, France, 7–11 July 2004.
28. Swiderski, W.; Hłosta, P.; Szugajew, L.; Usowicz, J. Microwave enhancement on thermal detection of buried objects. In Proceedings of the 11th International Conference on Quantitative InfraRed Thermography, Naples, Italy, 11–14 June 2012.
29. Vinson, J.R. Adhesive bonding of polymer composites. *Polym. Eng. Sci.* **1989**, *29*, 1325–1331. [CrossRef]
30. Yang, R.; He, Y.; Zhang, H. Progress and Trends in Nondestructive Testing and Evaluation for Wind Turbine Composite Blade. *Renew. Sustain. Energy Rev.* **2016**, *60*, 1225–1250. [CrossRef]
31. Ibarra-Castanedo, C.; Maldague, X. Review of pulse phase thermography. *Proc. SPIE* **2015**, *9485*. [CrossRef]
32. Kylili, A.; Fokaides, P.A.; Christou, P.; Kalogirou, S.A. Infrared thermography (IRT) applications for building diagnostics: A review. *Appl. Energy* **2014**, *134*, 531–549. [CrossRef]

33. Bagavathiappan, S.; Lahiri, B.B.; Saravanan, T.; Philip, J.; Jayakumar, T. Infrared thermography for condition monitoring—A review. *Infrared Phys. Technol.* **2013**, *60*, 35–55. [CrossRef]

34. Yang, B.; Zhang, L.; Zhang, W.; Ai, Y. Non-destructive testing of wind turbine blades using an infrared thermography: A review. In Proceedings of the International Conference on Materials for Renewable Energy and Environment, Chengdu, China, 19–21 August 2013; pp. 407–410.

35. Ibarra-Castanedo, C.; Maldague, X. Pulsed phase thermography reviewed. *Quant. Infrared Thermogr. J.* **2004**, *1*, 47–70. [CrossRef]

36. Hung, Y.; Chen, Y.S.; Ng, S.; Liu, L.; Huang, Y.; Luk, B.; Ip, R.; Wu, C.; Chung, P. Review and comparison of shearography and active thermography for nondestructive evaluation. *Mater. Sci. Eng. R Rep.* **2009**, *64*, 73–112. [CrossRef]

37. Omar, M.A.; Zhou, Y. A quantitative review of three flash thermography processing routines. *Infrared Phys. Technol.* **2008**, *51*, 300–306. [CrossRef]

38. Ghosh, K.K.; Karbhari, V.M. A critical review of infrared thermography as a method for non-destructive evaluation of FRP rehabilitated structures. *Int. J. Mater. Prod. Technol.* **2006**, *25*, 241–266. [CrossRef]

39. Vavilov, V.P.; Burleigh, D.D. Review of pulsed thermal NDT: Physical principles, theory and data processing. *NDT E Int.* **2015**, *73*, 28–52. [CrossRef]

40. Banerjee, D.; Chattopadhyay, S.; Tuli, S. Infrared thermography in material research—A review of textile applications. *Indian J. Fiber Text. Res.* **2013**, *38*, 427–437.

41. Usamentiaga, R.; Venegas, P.; Guerediaga, J.; Vega, L.; Molleda, J.; Bulnes, F.G. Infrared thermography for temperature measurement and non-destructive testing. *Sensors* **2014**, *14*, 12305–12348. [CrossRef] [PubMed]

42. Meola, C.; Carlomagno, G.M. Application of infrared thermography to adhesion science. *J. Adhes. Sci. Technol.* **2006**, *20*, 589–632. [CrossRef]

43. Meola, C.; Carlomagno, G.M. Recent advances in the use of infrared thermography. *Meas. Sci. Technol.* **2004**, *15*, R27. [CrossRef]

44. Yang, R.; Zhang, H.; Li, T.; He, Y. An investigation and review into microwave thermography for NDT and SHM. In Proceedings of the IEEE Far East NDT New Technology & Application Forum, Zhuhai, China, 28–31 May 2015; pp. 133–137.

45. He, Y.; Yang, R.; Zhang, H.; Zhou, D.; Wang, G. Volume or inside heating thermography using electromagnetic excitation for advanced composite materials. *Int. J. Thermal Sci.* **2017**, *111*, 41–49. [CrossRef]

46. Meredith, R. Engineers' Handbook of Industrial Microwave Heating [Book Review]. *Power Eng.* **1999**, *13*, 3.

47. Vrana, J.; Goldammer, M.; Bailey, K.; Rothenfusser, M.; Arnold, W. Induction and Conduction Thermography: Optimizing the Electromagnetic Excitation towards Application. *AIP Conf. Proc.* **2009**, *1096*, 518–525.

48. Shao, K.; Lavers, J.D. A skin depth-independent finite element method for Eddy current problems. *IEEE Trans. Magn.* **1986**, *22*, 1248–1250. [CrossRef]

49. Niliot, C.L.; Gallet, P. Infrared thermography applied to the resolution of inverse heat conduction problems: Recovery of heat line sources and boundary conditions. *Revue Générale Thermique* **1998**, *37*, 629–643. [CrossRef]

50. Salazar, A. Energy propagation of thermal waves. *Eur. J. Phys.* **2006**, *27*, 1349. [CrossRef]

51. Liu, J.; Yang, W.; Dai, J. Research on thermal wave processing of lock-in thermography based on analyzing image sequences for NDT. *Infrared Phys. Technol.* **2010**, *53*, 348–357. [CrossRef]

52. Osiander, R.; Spicer, J.W. Time-resolved infrared radiometry with step heating. A review. *Revue Générale Thermique* **1998**, *37*, 680–692. [CrossRef]

53. Yang, R.; He, Y.; Gao, B.; Tian, G.Y. Inductive pulsed phase thermography for reducing or enlarging the effect of surface emissivity variation. *Appl. Phys. Lett.* **2014**, *105*, 184103. [CrossRef]

54. Yang, R.; He, Y. Eddy current pulsed phase thermography considering volumetric induction heating for delamination evaluation in carbon fiber reinforced polymers. *Appl. Phys. Lett.* **2015**, *106*, 234103. [CrossRef]

55. He, Y.; Yang, R. Eddy Current Volume Heating Thermography and Phase Analysis for Imaging Characterization of Interface Delamination in CFRP. *IEEE Trans. Ind. Inf.* **2015**, *11*, 1287–1297. [CrossRef]

56. Keo, S.-A.; Defer, D.; Breaban, F.; Brachelet, F. Comparison between Microwave Infrared Thermography and CO_2 Laser Infrared Thermography in Defect Detection in Applications with CFRP. *Mater. Sci. Appl.* **2013**, *4*, 600–605.

57. Cheng, L.; Tian, G.Y.; Szymanik, B. Feasibility studies on microwave heating for nondestructive evaluation of glass fibre reinforced plastic composites. In Proceedings of the IEEE International Instrumentation and Measurement Technology Conference, Hangzhou, China, 10–12 May 2011; pp. 1–6.

58. Foudazi, A.; Donnell, K.M.; Ghasr, M.T. Application of Active Microwave Thermography to delamination detection. In Proceedings of the IEEE International Instrumentation and Measurement Technology Conference, Montevideo, Uruguay, 12–15 May 2014; pp. 1567–1571.

59. Levesque, P.; Deom, A.; Balageas, D. Non destructive evaluation of absorbing materials using microwave stimulated infrared thermography. In *Review of Progress in Quantitative Nondestructive Evaluation*; Springer: Berlin, Germany, 1993; pp. 649–654.

60. D'Ambrosio, G.; Massa, R.; Migliore, M.D.; Cavaccini, G.; Ciliberto, A.; Sabatino, C. Microwave excitation for thermographic NDE: An experimental study and some theoretical evaluations. *Mater. Eval.* **1995**, *53*, 502–508.

61. Sakagami, T.; Kubo, S.; Komiyama, T.; Suzuki, H. Proposal for a new thermographic nondestructive testing technique using microwave heating. *Proc. SPIE* **1999**, *3700*. [CrossRef]

62. Foudazi, A.; Ghasr, M.T.; Donnell, K.M. Characterization of Corroded Reinforced Steel Bars by Active Microwave Thermography. *IEEE Trans. Instrum. Meas.* **2015**, *64*, 2583–2585. [CrossRef]

63. Pieper, D.; Donnell, K.M.; Ghasr, M.T.; Kinzel, E.C. Integration of microwave and thermographic NDT methods for corrosion detection. *AIP Conf. Proc.* **2014**, *1581*, 1560–1567.

64. Foudazi, A.; Ghasr, M.T.; Donnell, K.M. Application of active microwave thermography to inspection of carbon fiber reinforced composites. In *Autotestcon*; IEEE: Washington, DC, USA, 2014; pp. 318–322.

65. Keo, S.A.; Brachelet, F.; Breaban, F.; Defer, D. Steel detection in reinforced concrete wall by microwave infrared thermography. *NDT E Int.* **2014**, *62*, 172–177. [CrossRef]

66. Osiander, R.; Spicer, J.W.M.; Murphy, J.C. Thermal imaging of subsurface microwave absorbers in dielectric materials. *Proc. SPIE* **1994**, *2245*. [CrossRef]

67. Bowen, M.W.; Osiander, R.; Spicer, J.W.M.; Murphy, J.C. Thermographic Detection of Conducting Contaminants in Composite Materials Using Microwave Excitation. In *Review of Progress in Quantitative Nondestructive Evaluation*; Thompson, D.O., Chimenti, D.E., Eds.; Springer US: Boston, MA, 1995; Volume 14, pp. 453–460.

68. Sikora, R.; Chady, T.; Szymanik, B. Infrared thermographic testing of composite materials with adhesive joints. In Proceedings of the 18th World Conference on Nondestructive Testing, Durban, South Africa, 16–20 April 2012; pp. 1–8.

69. Lee, H.; Galstyan, O.; Babajanyan, A.; Friedman, B.; Berthiau, G.; Kim, J.; Lee, K. Characterization of anisotropic electrical conductivity of carbon fiber composite materials by a microwave probe pumping technique. *J. Compos. Mater.* **2015**, *50*. [CrossRef]

70. Foudazi, A.; Edwards, C.A.; Ghasr, M.T.; Donnell, K.M. Active Microwave Thermography for Defect Detection of CFRP-Strengthened Cement-Based Materials. *IEEE Trans. Instrum. Meas.* **2016**, *65*, 1–9. [CrossRef]

71. Osiander, R.; Spicer, J.W.M.; Murphy, J.C. Microwave-source time-resolved infrared radiometry for monitoring of curing and deposition processes. *Proc. SPIE* **1995**, *2473*. [CrossRef]

72. Swiderski, W.; Szabra, D.; Wojcik, J. Nondestructive evaluation of aircraft components by thermography using different heat sources. In Proceeding of the QIRT Conference, Dubrovnik, Croatia, 24–27 September 2002; pp. 78–84.

73. Foudazi, A.; Mehdipour, I.; Donnell, K.M.; Khayat, K.H. Evaluation of steel fiber distribution in cement-based mortars using active microwave thermography. *Mater. Struct.* **2016**, *49*, 5051–5065. [CrossRef]

74. Galietti, U.; Palumbo, D.; Calia, G.; Pellegrini, M. Non destructive evaluation of composite materials with new thermal methods. In Proceedings of the 15th European Conference on Composite Materials, Venice, Italy, 24–28 June 2012.

75. Palumbo, D.; Ancona, F.; Galietti, U. Quantitative damage evaluation of composite materials with microwave thermographic technique: Feasibility and new data analysis. *Meccanica* **2015**, *50*, 443–459. [CrossRef]

76. Usamentiaga, R.; Venegas, P.; Guerediaga, J.; Vega, L.; López, I. Feature extraction and analysis for automatic characterization of impact damage in carbon fiber composites using active thermography. *NDT E Int.* **2013**, *54*, 123–132. [CrossRef]

77. Pickering, S.; Almond, D. Matched excitation energy comparison of the pulse and lock-in thermography NDE techniques. *NDT E Int.* **2008**, *41*, 501–509. [CrossRef]

78. Gao, B.; Woo, W.L.; Tian, G.Y. Electromagnetic Thermography Nondestructive Evaluation: Physics-based Modeling and Pattern Mining. *Sci. Rep.* **2016**, *6*, 25480. [CrossRef] [PubMed]

79. Gao, B.; He, Y.; Woo, W.L.; Tian, G.Y.; Liu, J.; Hu, Y. Multidimensional Tensor-Based Inductive Thermography With Multiple Physical Fields for Offshore Wind Turbine Gear Inspection. *IEEE Trans. Ind. Electron.* **2016**, *63*, 6305–6315. [CrossRef]

80. Shankar, V.; Mohammadian, A.H. A Time-Domain, Finite-Volume Treatment for the Maxwell Equations. *J. Microw. Power Electromagn. Energy* **1990**, 128–145. [CrossRef]

81. Harms, P.H.; Chen, Y.; Mittra, R.; Shimony, Y. Numerical Modeling of Microwave Heating Systems. *J. Microw. Power Electromagn. Energy* **1996**, *31*, 114–121. [CrossRef]

82. Vegh, V.; Turner, I.W. A hybrid technique for computing the power distribution generated in a lossy medium during microwave heating. *J. Comput. Appl. Math.* **2006**, *197*, 122–140. [CrossRef]

83. Liu, T.; Zhang, W.; Yan, S. A novel image enhancement algorithm based on stationary wavelet transform for infrared thermography to the de-bonding defect in solid rocket motors. *Mech. Syst. Signal Process.* **2015**, *62–63*, 366–380. [CrossRef]

84. Cheng, L.; Gao, B.; Tian, G.Y.; Woo, W.; Berthiau, G. Impact Damage Detection and Identification using Eddy Current Pulsed Thermography through Integration of PCA and ICA. *IEEE Sens. J.* **2014**, *14*, 1655–1663. [CrossRef]

85. Liang, T.; Ren, W.; Tian, G.Y.; Elradi, M.; Gao, Y. Low energy impact damage detection in CFRP using eddy current pulsed thermography. *Compos. Struct.* **2016**, *143*, 352–361. [CrossRef]

86. Edis, E.; Flores-Colen, I.; de Brito, J. Quasi-quantitative infrared thermographic detection of moisture variation in facades with adhered ceramic cladding using principal component analysis. *Build. Environ.* **2015**, *94*, 97–108. [CrossRef]

87. Gao, B.; Woo, W.L.; He, Y.; Tian, G.Y. Unsupervised sparse pattern diagnostic of defects with inductive thermography imaging system. *IEEE Trans. Ind. Inform.* **2016**, *12*, 371–383. [CrossRef]

88. Wang, H.; Hsieh, S.J.; Peng, B.; Zhou, X. Non-metallic coating thickness prediction using artificial neural network and support vector machine with time resolved thermography. *Infrared Phys. Technol.* **2016**, *77*, 316–324. [CrossRef]

89. Zou, H.; Huang, F. A novel intelligent fault diagnosis method for electrical equipment using infrared thermography. *Infrared Phys. Technol.* **2015**, *73*, 29–35. [CrossRef]

90. Gao, B.; Yin, A.; Tian, G.; Woo, W.L. Thermography spatial-transient-stage mathematical tensor construction and material property variation track. *Int. J. Therm. Sci.* **2014**, *85*, 112–122. [CrossRef]

91. Abdelniser, M.; Abdulbaset, A.; Ramahi, O.M. Reducing Sweeping Frequencies in Microwave NDT Employing Machine Learning Feature Selection. *Sensors* **2016**, *16*, 559.

![sensors](sensors logo) *sensors*

MDPI

Article

A New Electromagnetic Acoustic Transducer Design for Generating and Receiving S0 Lamb Waves in Ferromagnetic Steel Plate

Jianpeng He [1], Steve Dixon [2], Samuel Hill [2] and Ke Xu [3,*]

[1] National Engineering Research Center of Advanced Rolling Technology, University of Science and Technology Beijing, Beijing 100083, China; o832an@hotmail.com

[2] Department of Physics, University of Warwick, Coventry CV4 7AL, UK; S.M.Dixon@warwick.ac.uk (S.D.); samuel.hill@warwick.ac.uk (S.H.)

[3] Collaborative Innovation Center of Steel Technology, University of Science and Technology Beijing, Beijing 100083, China

* Correspondence: xuke@ustb.edu.cn; Tel.: +86-10-6233-2159

Academic Editor: Vittorio M. N. Passaro
Received: 24 March 2017; Accepted: 26 April 2017; Published: 4 May 2017

Abstract: Electromagnetic acoustic transducers (EMATs) are non-contact, ultrasonic transducers that are usually kept within 5 mm from the sample surface to obtain a sufficient signal-to-noise ratio (SNR). One important issue associated with operation on a ferromagnetic plate is that the strong attraction force from the magnet can affect measurements and make scanning difficult. This paper investigates a method to generate fundamental, symmetric Lamb waves on a ferromagnetic plate. A coil-only, low-weight, generation EMAT is designed and investigated, operating at lift-offs of over 5 mm. Another design of an EMAT is investigated using a rectangular magnet with a much higher lift-off than the coil, of up to 19 mm. This results in a much lower force between the EMAT and sample, making scanning the EMAT much easier.

Keywords: electromagnetic acoustic transducer; EMAT; NDT; Lamb wave; magnetostriction

---

1. Introduction

An electromagnetic acoustic transducer (EMAT) can generate and detect ultrasonic waves on electrically-conductive samples without making physical contact, making it possible to take measurements on moving or hot objects. Another advantage of EMATs is the capability of generating various ultrasonic wave modes by careful design of the coils and magnets [1–9]. However, EMATs have some disadvantages: the signal-to-noise ratio is relatively low compared with piezoelectric transducers and the gap between the bottom of the EMAT and the conductive material is typically limited to less than 5 mm, even after signal averaging and optimized design [10,11].

Traditional EMATs consist of a coil and a magnet, and are used to generate ultrasonic waves in metallic samples that are under test. EMATs generate and detect these ultrasonic waves mainly through two mechanisms: the Lorentz force and magnetoelastic mechanisms [12–15], depending on the experimental sample properties. The Lorentz force can be produced in any electrically-conducting material. When an alternating current is driven through the coil, it generates a dynamic, time varying magnetic flux density, B_d that, in turn, generates an eddy current, J_E, in the surface region of the metallic plate. The Lorentz force describes the forces generated by the interaction between the eddy current and the magnetic field [16] (both the dynamic magnetic field, B_d, from the coil and static magnetic field, B_s, provided by the permanent or electromagnet). This gives rise to a temporally and spatially varying Lorentz force under the EMAT, described by:

$$f = J_E \times (B_s + B_d) \tag{1}$$

When the dynamic magnetic field is relatively small, there is a linear relationship between the applied static magnetic field and the Lorentz force component of the amplitude of the generated ultrasonic waves. If the applied static magnetic field is parallel to the surface of the sample, then the other main transduction mechanism is typically magnetostriction [12,17]. This occurs only in ferromagnetic media and is highly sensitive to material properties and operational conditions, such as the magnetostriction coefficient λ, relative magnetic permeability μ, and bias magnetic field B_s. Ferromagnetic materials have a structure that is divided into domains, each of which is a region of uniform magnetic polarization. When a magnetic field is applied, the domains tend to align parallel to the total magnetic field, inducing mechanical strains in the material. Magnetostriction is highly non-linear, and depends on the surface conditions, the previous history of magneto-mechanical loads, and the residual stress [18]. The receiving processes are also different for the Lorentz mechanism and the magnetostriction mechanism. For detection of ultrasonic waves via the Lorentz mechanism, when a passing acoustic wave oscillates the atomic lattice beneath the receiving EMAT, the oscillatory movement in the static field causes an oscillation flow of electrons within the metal plate. The alternating electromagnetic field from the electron flow in the surface of the metal induces an alternating voltage in the coil of the receiving EMAT. The inverse magnetostrictive effect (also known as the Villari effect) is used for the detection of ultrasonic waves via the magnetostrictive mechanism. The time-varying stress induced by the ultrasonic wave, beneath the receiving EMAT, induces a change in the magnetisation of ferromagnetic sample. This results in a time varying magnetic flux density change under the EMAT coil, which induces electric potential difference in the coil. The permanent magnetic field provides a magnetic bias that significantly increases the efficiency of the detection mechanism. A third transduction mechanism on ferromagnetic materials is the magnetization force. When providing a tangential bias magnetic field, a major part of the Lorentz force will be cancelled by magnetization force. The contribution of magnetization force is relatively small compared with the Lorentz force and the magnetostriction force in our EMAT design, so it will be neglected in the subsequent analysis [19].

When designing the EMAT, the important point to bear in mind is that the generation efficiency of the ultrasonic wave is dependent on both the sample under test and the configuration of EMATs. Compared with aluminium, it is generally considered more challenging to perform EMAT measurements on un-oxidised steel because of the lower conductivity and higher density of the steel, both of which serve to reduce the EMAT efficiency in the Lorentz mechanism of operation, whilst the magnetostriction force in ferromagnetic material can also produce forces that contribute to the generation of ultrasonic waves [20,21], especially when the bias magnetic is parallel to the sample surface.

In this paper, we designed a coil-only EMAT, which generates Lamb waves in mild steel plate via the interaction between the steel plate and the dynamic magnetic field generated by the coil. Our experimental results demonstrated that the transduction efficiency in the steel plate is higher than that in aluminium plate due to the ferromagnetic nature of the steel and the S0 mode is mainly produced by magnetostrictive mechanism. A rectangular magnet is then applied to provide a bias magnetic field for the coil-only EMAT to enable it to detect ultrasonic signals and also to increase the efficiency of the ultrasonic generation. The results show that the EMATs are able to work over a wide range of lift-offs and the attraction force between the EMAT and the steel plate are reduced dramatically.

2. Lamb Waves

Lamb waves are of practical importance in guided-wave inspection and have been widely utilised in modern engineering for crack detection [22], texture measurement [23], and corrosion monitoring [24]. Lamb waves are capable of propagating over relatively long distances with low attenuation and large areas can be inspected efficiently. However, the multimode and the dispersive nature of Lamb waves can make the interpretation of the received signal difficult. In order to simplify the analysis of the signal in our experiments, the current drive to the EMAT coil is at a sufficiently low frequency to excite only the fundamental modes with any significant amplitude. Whilst the A0 mode may have better detection resolution to defect characterisations because of the shorter wavelength compared with S0 mode at the same frequency, the A0 mode is also highly dispersive at low frequencies. This dispersion has been exploited to measure sample thickness, but in some cases the dispersive nature of the A0 mode can make defect detection more difficult. In contrast, the fundamental symmetric or S0 Lamb wave is less dispersive at low frequencies, making the analysis of the signal simpler, to some extent.

3. Design of a Coil-Only EMAT

3.1. Theoretical Analysis

Linear coils have been used widely to generate both Lamb and Rayleigh waves for a range of applications [25,26]. As shown in Figure 1a, the conventional linear coil can be constructed by wrapping a single layer of insulated (lacquered) copper wire around a cylindrical magnet. EMATs with this configuration are difficult to operate at a lift-off distance over 5 mm because both the magnetic flux density and the sensitivity of the coil to generate or detect eddy currents in the sample surface decrease rapidly with an increase in lift-off distance. The coil, wound around the sides and top of the permanent magnet, can also dissipate energy as eddy currents can also be generated in the magnet, leading to the reduction of EMAT efficiency. Additionally, the magnet attracts ferromagnetic particles, which may cause mechanical damage to the EMAT and the test object. Previous studies have shown that it is possible to achieve ultrasonic generation with the 'self-field' generated by the coil [16]. The contribution of the dynamic field should not be ignored because the displacement due to the dynamic magnetic field may exceed the displacement caused by the static magnetic field when the excitation current is large, such as in the order of hundreds of amps [27]. Compared with the meander coil in [27], a linear coil is more feasible for generating high-intensity dynamic magnetic fields, because the dynamic magnetic field produced by each turn of the coil adds constructively. Compared with aluminium, it is much more practicable to generate a strong magnetic field in steel because of the ferromagnetic nature of the steel.

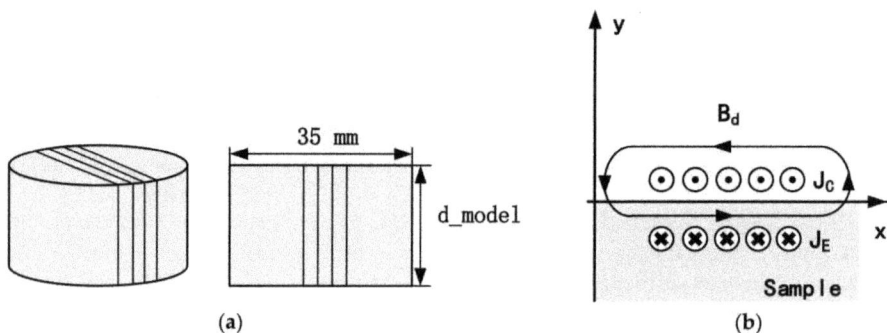

Figure 1. (**a**) Schematic diagram of the EMAT construction. The gap between the coils is depicted much larger than the actual EMAT for clarity. (**b**) The cross-sectional view of the eddy current and dynamic magnetic field generated by linear coils.

When designing a coil-only EMAT, the magnet in Figure 1a is substituted by a 3D-printed, cylindrical plastic model. The diameter is 35 mm, whilst the d_model parameter represents the height of the model. Figure 1b shows the eddy current and dynamic magnetic field produced in the test piece. The coordinate reference frame for the model is placed on the surface of the plate. The x-axis coincides with the surface of the plate. The positive part of y-axis refers to the air domain and the negative part of the y-axis represents the metallic plate. The coil provides a dynamic magnetic field parallel to the surface of the plate. Figure 2a shows the forces that arise due to the interaction of the dynamic magnetic field with the eddy current, which is based on the Lorentz mechanism. Figure 2b depicts the forces generated by magnetostriction. For the Lorentz force, the out-of-plane component is large, while the in-plane component is relatively small. On the contrary, magnetostriction mainly produces in-plane force and the out-of-plane component is relatively small. In the low frequency-thickness regime, the vibration associated with S0 mode is predominantly in-plane [15], meaning that the S0 Lamb wave mode can be efficiently generated by the in-plane, magnetostrictive forces. The A0 mode is mainly generated by the out-of-plane force, which is through the Lorentz mechanism in our design.

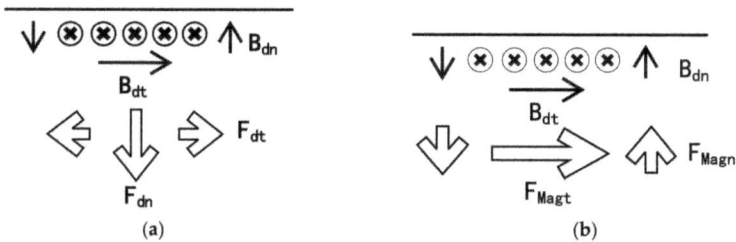

(a)   (b)

Figure 2. Forces that produced by the coil-only EMAT. B_{dt} and B_{dn} are transverse and normal component of dynamic magnetic field. (a) Lorentz forces that arise due to the interaction of the dynamic magnetic with the eddy current. (b) Magnetostriction forces that arise due to magnetostrictive mechanism.

3.2. Simulation of Magnetic Flux Density with Lift-Off

When considering how lift-off will change the coil-only EMAT performance, the dynamic magnetic field generated by the coil is expected to be the most important factor. A finite element (FE) numerical model was firstly implemented in Comsol Multiphysics (5.1 version, COMSOL Inc. Burlington, MA, USA) to simulate the dynamic magnetic field density in the skin depth of the metallic plate. The coils used in the FE model employed a single layer of 10 turns of 0.68 mm diameter wire. The current pulse used in our simulation is broadband pulse with frequency component mainly from 0 to 300 kHz. Low frequency ensures that the wavelength of the Lamb wave is much larger than the sheet thickness, so higher-order wave modes can be suppressed. The pulse through the coil is shown in Figure 3 and the data was collected by measuring the voltage across a resistor in series with the coil-only EMAT. The current peak value is approximately 270 A with several microseconds duration.

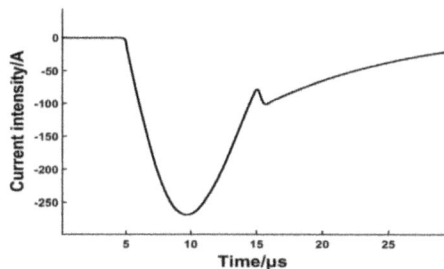

Figure 3. Experimental excitation pulse.

The lift-off distance started from 0 mm and increased to 14 mm in 2 mm increments. The simulation was done on both steel and aluminium plate in order to compare the difference of magnetic flux density due to the ferromagnetic properties. The x component of the magnetic flux density (parallel to the plate surface), at the same position in the skin depth, was collected.

The results are depicted in Figure 4a,b. The magnetic flux density generated by the dynamic field in steel and aluminium plate both decrease exponentially. If the x component of magnetic flux density in steel at a lift-off of 0 mm is defined as 100%, the value at the lift-off of 0 mm on aluminium is only 2.89%. The magnetic flux density in the steel is obviously stronger than that in aluminium at all lift-off distances because of the higher relative permeability. When the coils are placed on a ferromagnetic plate with high magnetic permeability, most of the magnetic flux is confined in the plate and parallel to the surface inside the plate. Although the magnetic flux density in the steel plate drops dramatically at low lift-offs, it is still possible to generate a strong horizontal dynamic magnetic field in the surface region of the steel plate. The y component of the magnetic flux density was also collected and the field intensity is insignificant compared with the x component.

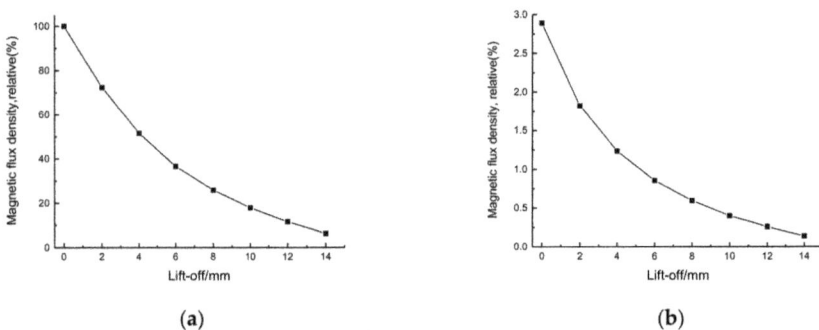

(a)   (b)

Figure 4. The variation of x component of magnetic flux density as the lift-off of coil-only EMAT is varied on steel (**a**) and aluminium (**b**) samples. The results are normalized to aid comparison, with the magnetic flux density at a 0 mm lift-off defined as 100%.

3.3. Experimental Procedure

EMATs were set in a 'pitch-catch' arrangement, with one EMAT generating a signal that is received by a second one a short distance away. Both a non-oxidized mild steel plate and an aluminium plate were used in the experiment, with a thickness of 1 mm in both cases. The generation EMAT was constructed by wrapping a single layer of insulated copper wire around the 35 mm diameter 3D-printed plastic model, which has a height equal to 20 mm (d_model = 20 mm in Figure 1a). Both the diameter of the coil and number of turns were varied in order to find the optimum conditions for the maximum peak-to-peak amplitude in the steel plate, and it was found that the optimized coil consisted of 10 turns of 0.68 mm diameter insulated wire. The receiving EMAT employed a single layer of 40 turns of 0.1 diameter wire wound onto a cylindrical permanent magnet (35 mm in diameter and 20 mm in height), which provided a static magnetic field normal to the plane of the steel plate. The receiving EMAT is predominantly sensitive to in-plane vibration, but will have some out-of-plane motion sensitivity because of the fringing field of the permanent magnet. The lift-off of receiver was kept constantly at 0 mm during the whole experimental process.

A low-pass filter, with a cut-off frequency of 500 kHz, was used in order to reduce the level of noise. The received signal was then transmitted to a broadband pre-amplifier with a gain of 50 dB. The pre-amplifier was connected to a digital oscilloscope with 8-bit resolution to record signals in the time-domain, which were averaged 32 times before storage. The results are shown in Figure 5. The group velocity of S0 mode was calculated and the result corresponds with theoretical calculation.

The S0 mode shows less dispersion and is temporally sharp due to its broadband nature. V_{pp} shown in Figure 5 is the peak-to-peak value of the S0 Lamb wave. A larger V_{pp} value is often considered a higher transduction efficiency in EMAT design and this value will be used to evaluate the EMAT performance in our experiments.

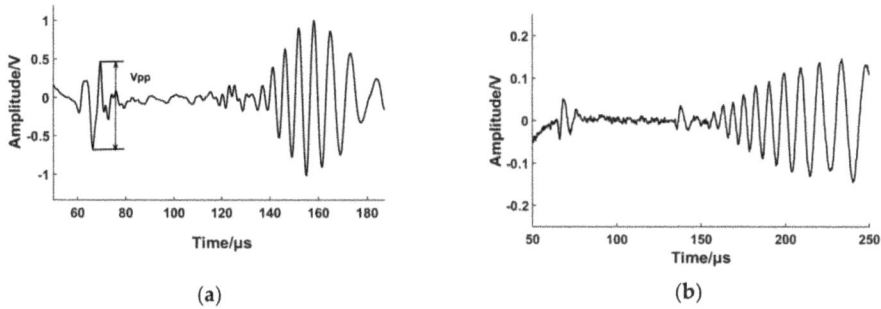

(a)   (b)

Figure 5. Pulse-echo signal from the coil-only EMAT on steel (**a**) and aluminium (**b**) samples. The amplitude of S0 and A0 mode is much larger in the steel sample.

The experimental results illustrate that coil-only EMAT driven by large current could generate fundamental Lamb waves in ferromagnetic plate more efficiently than in aluminium plate. The peak-to-peak amplitudes of S0 Lamb waves are 1.1 V and 0.1 V, respectively. Based on the theoretical analysis in Section 3.1 and simulation results in Section 3.2, the increased amplitude of the S0 mode is caused by the magnetostrictive force on the ferromagnetic plate, whilst the increase in the A0 mode amplitude is due to the enhancement of the x-component of the magnetic flux density. In order to prove this hypothesis, an experimental test of transducer lift-off was carried out. The lift-off distance was increased using plastic spacer in a 1 mm step normal to the steel plate and measurements were taken after each step, up to a maximal lift-off of 9 mm. The thickness of the plastic spacer was measured by a micrometer and the maximal relative error was 1.7%. Figure 6 illustrates how the peak-to-peak amplitude of S0 and A0 mode varies with the increment of the lift-off distance.

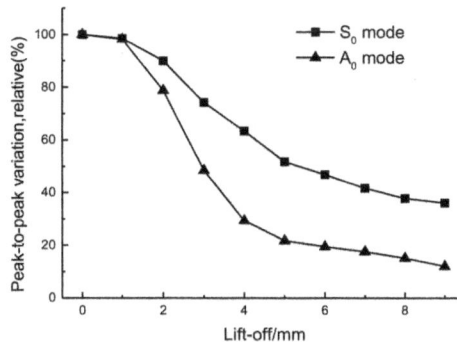

Figure 6. The variation in the peak-to-peak amplitude of S0 and A0 mode. The results are normalized with the peak-to-peak amplitude at a 0 mm lift-off defined as 100%.

We can observe that the amplitude of the A0 mode attenuates faster than that of the S0 mode. The main reason is that, with the growth of lift-off distance, the Lorentz force falls off more rapidly as both the image current and dynamic magnetic field decrease exponentially. The force caused by magnetostriction simply varies with the decrease of dynamic magnetic field, so the Lorentz force decreases more rapidly than the magnetostriction force with the rise of the gap between EMAT and the ferromagnetic plate. In the low lift-off range (lower than 1 mm), the amplitude varies very little because both the Lorentz force and the magnetostriction force contribute to the generation of ultrasonic waves. A portion of the force produced due to magnetostriction may be counteracted by the force generated by the Lorentz force. Additionally, the magnetostrictive effect is non-linear, making the analysis of the signal complicated. When the lift-off distance is higher than 2 mm, the main transduction mechanism for S0 and A0 become the magnetostriction force and Lorentz force, respectively, and the peak-to-peak values decrease almost exponentially. The amplitude is relatively stable after 7 mm, but the peak-to-peak amplitude is smaller. Figure 7 depicts the A-scan of the S0 Lamb wave generated by the coil-only EMAT in our experiment when the lift-off distance is 10 mm. The result demonstrates that it is possible to generate an S0 Lamb wave with a high signal-to-noise ratio at higher lift-off distance via magnetostriction because eddy current generation is not a requisite factor for the magnetostrictive effect.

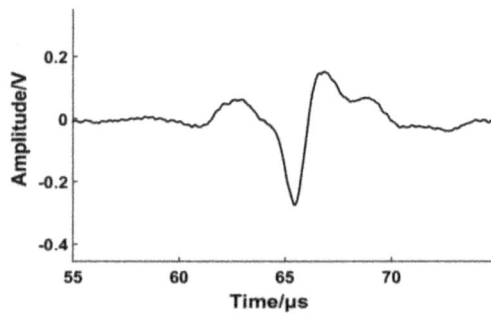

Figure 7. An averaged A-scan of S0 mode Lamb wave when the lift-off distance is 10 mm.

It is important to note that the top layer of the linear coil shown in Figure 1 can also produce a dynamic field in the steel plate in the opposite direction to that from the bottom layer, so we need to ensure the dynamic magnet field from the top layer does not reduce the magnetic field intensity in the surface region of the steel plate. If the maximum operational lift-off distance is D_{max}, the following equation should be satisfied:

$$d\_model + D \geq D_{max} \tag{2}$$

In Equation (2), d_model is the height of the plastic model as shown in Figure 1. D represents the operational lift-off distance of the EMAT, which is the distance between the bottom layer of the coil and the steel plate. (d_model + D) is the lift-off distance of the top layer coil. This equation can ensure that the steel plate is not influenced by the dynamic magnetic field produced by the top layer coil.

Previously-published work has shown that the angle and magnitude of the bias magnetic field has important influence on magnetostrictive transducer efficiency [28,29]. A block NdFeB magnet was employed for the purpose of investigating the influence of the bias magnetic field to the excitation efficiency in the following section.

4. Generation and Reception of S0 Lamb Wave Based on the Magnetostrictive Mechanism

4.1. Design of the Magnetostrictive EMAT

The magnetostrictive EMAT (shown in Figure 8) consists of a magnet and a coil-only EMAT described in Section 3. The magnet used in the experiment is a block NdFeB magnet (N40 grade with a

Sensors **2017**, *17*, 1023

residual magnetization of 1.26 T) with a nickel (Ni-Cu-Ni) coating and it is polarised, such that the magnetic axis is parallel to the steel plate.

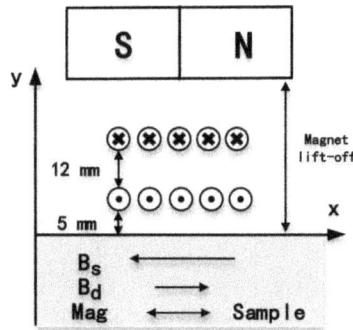

Figure 8. Magnetostrictive EMAT based on the coil-only EMAT described in Section 3. The operational lift-off distance of the coil is 5 mm, so the influence of the Lorentz force can be neglected. The height of the plastic model is set equal to 12 mm (d_model = 12 mm in Figure 1) to avoid interference from the top layer.

This configuration is able to reduce energy dissipation because less eddy currents can be generated in the magnet. Another advantage is that it can reduce the magnetic attraction force between the magnet and the ferromagnetic particles. A finite element method (FEM) model was constructed using COMSOL, consisting of a rectangular magnet and steel plate. After defining the geometry corresponding to the magnet arrangement and ferromagnetic plate, a residual magnetization of 1.26 T was applied to the model along the x-axis. All other magnetization components were set to zero. Figure 9 shows the magnetic force as a function of lift-off distance and the force drops dramatically at low lift-off distance. This reduced magnetic attraction can diminish the interference with signal measurements and decrease the risk of mechanical collision during the testing process.

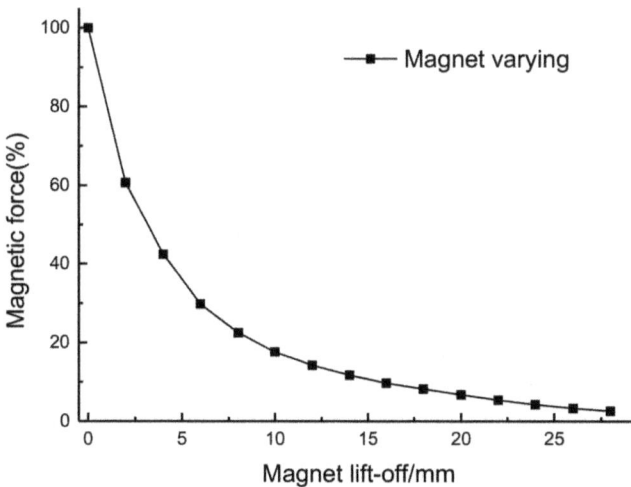

Figure 9. The variation of magnetic force between the magnet and steel plate as the magnet lift-off distance is varied.

The minimum lift-off distance of the magnet is 19 mm due to the present of the plastic model and coils in Figure 8. At this lift-off distance, the magnetic force has dropped to 10% of the maximum value. The magnetic field strength was also calculated using the FE model, with Figure 10 depicting the simulation result at the surface of the steel plate when the magnet lift-off is 19 mm and 25 mm. The static magnetic field provided by the magnet is about 22 kA/m and 17 kA/m respectively. The magnetostrictive strain coefficients are related to the slope of the magnetostriction curve and the slope value in mild steel is large when the bias magnetic field $\overline{H}$ is located in a certain interval [29]:

$$16 \, \text{kA/m} \leq \overline{H} \leq 32 \, \text{kA/m} \tag{3}$$

In this interval, greater dynamic strain can be produced, which leads to a larger ultrasonic signal strength.

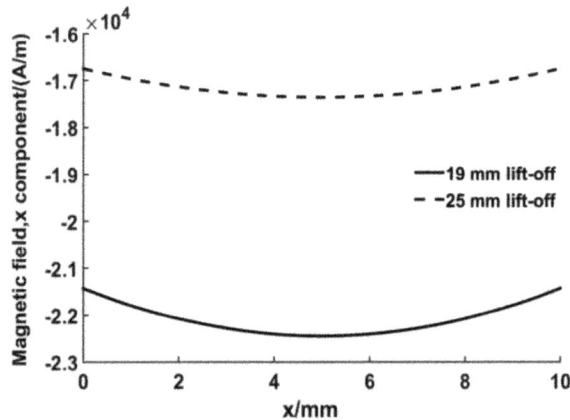

Figure 10. The x component of the magnetic field at the surface when the magnet lift-off is 19 mm and 25 mm.

An experiment was conducted to study the influence of a bias magnet on the excitation efficiency. In order to reduce influence from Lorentz force and design EMATs based on the magnetostriction mechanism, the lift-off of the coil was kept at 5 mm and the lift-off of the magnet varied. The impact from the top layer coil could be neglected at the lift-off values higher than 19 mm.

Figure 11 depicts the comparison between the coil-only EMAT and EMAT with a bias magnet. The lift-off of the Lorentz receiver was kept at 0 mm in order to capture more energy. With a bias magnetic field, the wave packet broadens and the peak-to-peak value triples. For the coil-only EMAT, magnetization can be associated with domain wall movement induced by the dynamic magnetic field. When a bias magnet is utilized, magnetization is expected to change primarily by reversible domain rotations [30]. The experimental result demonstrates that the new configuration is able to generate the S0 Lamb wave more efficiently. This raises the question of whether the same configuration can be used as a receiver. A receiving EMAT was also manufactured by winding a single layer of 20 turns of 0.1 diameter wire, as shown in Figure 8. An experiment was carried out to compare the receiving capability of the EMATs based on Lorentz mechanism and magnetostrictive mechanism. The Lorentz receiver has been introduced in Section 3.2 and the lift-off was set to 0 mm. For the magnetostrictive receiver, the lift-off of the coil and magnet were 5 mm and 19 mm respectively. Figure 12 depicts that the peak-to-peak value from the magnetostrictive receiver was two-fold higher than that from the Lorentz receiver. Furthermore, more peaks can be observed using the magnetostrictive receiver.

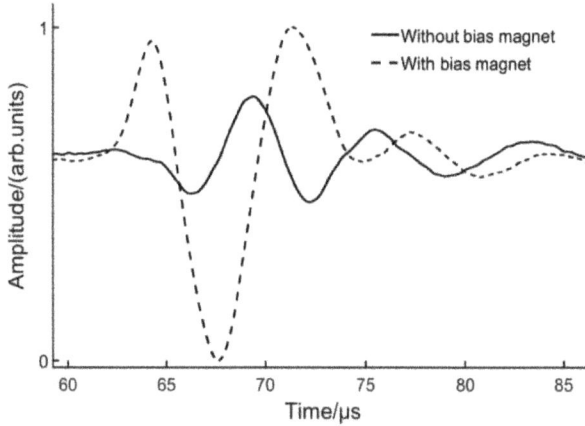

Figure 11. S0 Lamb wave generated by the EMATs with/without a bias magnet.

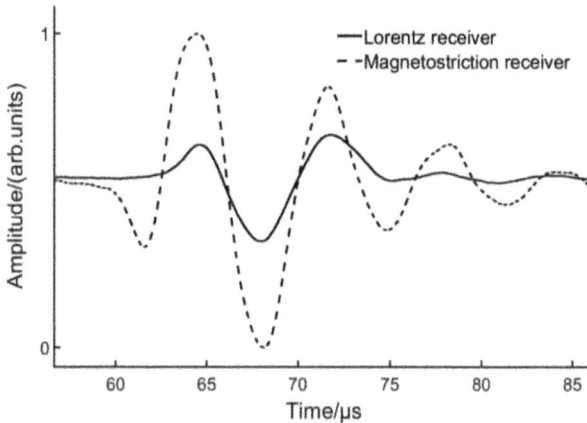

Figure 12. S0 Lamb wave received by the EMATs based on Lorentz/magnetostrictive mechanism.

4.2. Experimental Study of Magnetostrictive EMAT Lift-Off

Obtaining experimental lift-off data on a steel sample is easier in our experiment compared with traditional EMATs because the steel sample could not be lifted by the magnetic attraction. In the experiment, the lift-off distances were controlled by plastic spacers with different thicknesses. The lift-off of the coils was kept at 5 mm consistently, to provide a sufficient dynamic magnetic field and the lift-off of the magnet varied. Two procedures were performed: the magnet lift-off of the generator was maintained at 19 mm and the magnet of the receiver varied, and then the magnet of the receiver was fixed at 19 mm and the magnet of the generator varied. The variable magnet lift-off started at 19 mm and increased to 29 mm in 1 mm increments. The peak-to-peak value of each step was recorded. Figure 13 reveals the relative value of the amplitude as a function of magnet lift-off distance.

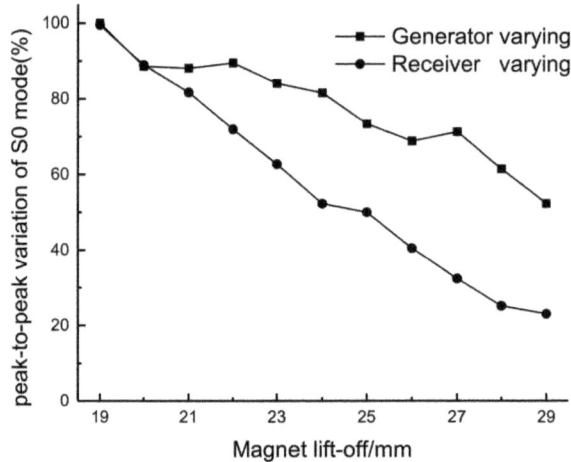

Figure 13. The variation in the peak-to-peak amplitude of the S0 Lamb wave. The lift-off of the coils was kept at 5 mm. The magnet of one EMAT is varied with the other transducer at a fixed magnet lift-off.

The generator output varies slower with changes in the magnet lift-off. For the receiver, the signal attenuates almost linearly. It could be expected that signals could be generated and received at a magnet lift-off of 25 mm. The newly-designed EMATs were tested on different kinds of ferromagnetic materials. The results show that the EMATs can be successfully utilized to inspect various types of ferromagnetic steel plate. However, the magnetostriction curve is dissimilar for different ferromagnetic materials and the results depend significantly on the material properties. The lift-offs of the coil and magnet should be optimized for each kind of ferromagnetic steel sample.

5. Conclusions

Electromagnetic acoustic transducers are one of the most important non-contact methods for defect detection and material characterization. However, traditional linear EMATs, which are usually constructed by wrapping a signal layer of insulated copper wire around a magnet, need to be kept close to the sample surface to obtain a sufficient SNR. When conducting measurements on ferromagnetic samples, things become more complicated because the attraction between the magnet and the sample may disturb the measurements, or even lead to mechanical damage to the transducers. A coil-only EMAT was firstly designed and investigated. The amplitude of both S0 mode and A0 mode are larger in mild steel than in aluminium. The Lorentz force from the interaction of the dynamic magnetic field and eddy current mainly causes out-of-plane force. The enhancement of the magnetic field in mild steel is the primary reason for the A0 mode increment. The magnetostriction force is the main source of the in-plane force, which produce the S0 Lamb wave efficiently. The coil-only EMAT is able to operate over a larger range of lift-offs than conventional EMATs because eddy current generation is not a requisite factor for the magnetostrictive mechanism. Based on the coil-only EMAT, a horizontal bias magnetic field was applied to increase the generating and receiving efficiency. This arrangement is able to generate and receive S0 Lamb waves predominantly via the magnetostrictive mechanism and the generating and receiving efficiency both increased by a factor of about three. The block magnet can provide a bias magnetic field over a 19 mm lift-off distance with the coil fixed at a lift-off of 5 mm, leading to a dramatic decrease of the magnetic force between the magnet and ferromagnetic samples. Compared with traditional EMATs, the higher operational lift-off distance and smaller magnetic force make the new EMAT more suitable for commercial environments.

Sensors **2017**, *17*, 1023

Acknowledgments: This work is sponsored by the China-European Research Cooperation Funding Projects of Small and Medium Sized Enterprises in Energy-saving and Emission-reduction (Project No. SQ2013ZOA000004).

Author Contributions: Jianpeng He built the FE models, conducted the experiments, and wrote this paper. Samuel Hill contributed to the theoretical research and data analysis. Steve Dixon and Ke Xu supervised the research and modified the paper.

Conflicts of Interest: The authors declare no conflict of interest.

References

1. Clough, M.; Fleming, M.; Dixon, S. Circumferential guided wave EMAT system for pipeline screening using shear horizontal ultrasound. *NDT E Int.* **2017**, *86*, 20–27. [CrossRef]
2. Petcher, P.A.; Dixon, S. Weld defect detection using PPM EMAT generated shear horizontal ultrasound. *NDT E Int.* **2015**, *74*, 58–65. [CrossRef]
3. Kang, L.; Zhang, C.; Dixon, S.; Zhao, H.; Hill, S.; Liu, M.H. Enhancement of Ultrasonic Signal Using a New Design of Rayleigh-Wave Electromagnetic Acoustic Transducer. *NDT E Int.* **2017**, *86*, 36–43. [CrossRef]
4. Yi, P.X.; Zhang, K.; Li, Y.H.; Zhang, X.M. Influence of the lift-off effect on the cut-off frequency of the EMAT-generated Rayleigh wave signal. *Sensors* **2014**, *14*, 19687–19699. [CrossRef] [PubMed]
5. Burrows, S.E.; Fan, Y.; Dixon, S. High temperature thickness measurements of stainless steel and low carbon steel using electromagnetic acoustic transducer. *NDT E Int.* **2014**, *68*, 73–77. [CrossRef]
6. Petcher, P.A.; Burrows, S.E.; Dixon, S. Shear horizontal (SH) ultrasound wave propagation around smooth corners. *Ultrasonics* **2014**, *54*, 997–1004. [CrossRef] [PubMed]
7. Huang, S.L.; Wei, Z.; Zhao, W.; Wang, S. A new omni-directional EMAT for ultrasonic Lamb wave tomography imaging of metallic plate defects. *Sensors* **2014**, *14*, 3458–3476. [CrossRef] [PubMed]
8. Hill, S.; Dixon, S. Frequency dependent directivity of periodic permanent magnet electromagnetic acoustic transducers. *NDT E Int.* **2014**, *62*, 137–143. [CrossRef]
9. Wang, Y.G.; Wu, X.J.; Sun, P.F.; Li, J. Enhancement of the Excitation Efficiency of a Torsional Wave PPM EMAT array for Pipe Inspection by Optimizing the Element Number of the array Based on 3-D FEM. *Sensors* **2015**, *15*, 3471–3490. [CrossRef] [PubMed]
10. Edwards, R.S.; Dixon, S.; Jian, X. Enhancement of the Rayleigh wave signal at surface defects. *J. Phys. D Appl. Phys.* **2004**, *37*, 2291–2297. [CrossRef]
11. Huang, S.L.; Zhao, W.; Zhang, Y.; Wang, S. Study on the lift-off effect of EMAT. *Sens. Actuators A Phys.* **2009**, *153*, 218–221. [CrossRef]
12. Ashigwuike, E.C.; Ushie, O.J.; Mackay, R.; Balachandran, W. A study of the transduction mechanisms of electromagnetic acoustic transducers (EMATs) on pipe steel materials. *Sens. Actuators A Phys.* **2015**, *229*, 154–165. [CrossRef]
13. Murayama, R. Study of driving mechanism on electromagnetic acoustic transducer for Lamb wave using magnetostrictive effect and application in drawability evaluation of thin steel sheets. *Ultrasonics* **1999**, *37*, 31–38. [CrossRef]
14. Ribichini, R.; Cegla, F.; Nagy, P.B.; Cawley, P. Study and comparison of different EMAT configurations for SH wave inspection. *IEEE Trans. Ultrason. Ferroelectr. Freq. Control* **2011**, *58*, 2571–2581. [CrossRef] [PubMed]
15. Seher, M.; Huthwaite, P.; Lowe, M.J.S.; Nagy, P.B. Model-based design of low frequency Lamb wave EMATs for mode selectivity. *J. Nondestr. Eval.* **2015**, *34*, 22. [CrossRef]
16. Dixon, S.; Palmer, S.B. Wideband low frequency generation and detection of Lamb and Rayleigh waves using electromagnetic acoustic transducers (EMATs). *Ultrasonics* **2004**, *42*, 1129–1136. [CrossRef] [PubMed]
17. Murayama, R. Driving mechanism on magnetostrictive type electromagnetic acoustic transducer for symmetrical vertical-mode Lamb wave and for shear horizontal-mode plate wave. *Ultrasonics* **1996**, *34*, 729–736. [CrossRef]
18. Ribichini, R.; Cegla, H.; Nagy, P.B.; Cawley, P. Quantitative modeling of the transduction of electromagnetic acoustic transducers operating on ferromagnetic media. *IEEE Trans. Ultrason. Ferroelectr. Freq. Control* **2010**, *57*, 2808–2817. [CrossRef] [PubMed]
19. Hirao, M.; Ogi, H. *EMATs for Science and Industry: Noncontacting Ultrasonic Measurements*; Springer Science & Business Media: Boston, MA, USA, 2003; pp. 13–31.

20. Murayama, R.; Hoshihara, H.; Fukushige, T. Development of an Electromagnetic Acoustic Transducer that can Alternately Drive the Lamb Wave and Shear Horizontal Plate Wave. *Jpn. J. Appl. Phys.* **2003**, *42*, 3180–3183. [CrossRef]

21. Murayama, R.; Mizutani, K. Development of an Electromagnetic Acoustic Transducer with Multi-Wavelength for Lamb Wave. *Jpn. J. Appl. Phys.* **2002**, *41*, 3534–3538. [CrossRef]

22. Dixon, S.; Burrows, S.E.; Dutton, B.; Fan, Y. Detection of cracks in metal sheets using pulsed laser generated ultrasound and EMAT detection. *Ultrasonics* **2011**, *51*, 7–16. [CrossRef] [PubMed]

23. Potter, M.D.G.; Dixon, S. Ultrasonic texture measurement of sheet metals: An integrated system combining Lamb and shear wave techniques. *Nondestr. Test. Eval.* **2005**, *20*, 201–210. [CrossRef]

24. Nagy, P.B.; Simonetti, F.; Instanes, G. Corrosion and erosion monitoring in plates and pipes using constant group velocity Lamb wave inspection. *Ultrasonics* **2014**, *54*, 1832–1841. [CrossRef] [PubMed]

25. Rosli, M.H.; Edwards, R.S.; Fan, Y. In-plane and out-of-plane measurements of Rayleigh waves using EMATs for characterising surface cracks. *NDT E Int.* **2012**, *49*, 1–9. [CrossRef]

26. Potter, M.D.G; Dixon, S.; Davis, C. Development of an automated non-contact ultrasonic texture measurement system for sheet metal. *Meas. Sci. Technol.* **2004**, *15*, 1303–1308. [CrossRef]

27. Wang, S.J.; Kang, L.; Li, Z.C.; Zhai, G.F.; Zhang, L. 3-D modeling and analysis of meander-line-coil surface wave EMATs. *Mechatronics* **2012**, *22*, 653–660. [CrossRef]

28. Ogi, H.; Goda, E.; Hirao, M. Increase of efficiency of magnetostriction SH-wave electromagnetic acoustic transducer by angled bias field: Piezomagnetic theory and measurement. *Jpn. J. Appl. Phys.* **2003**, *42*, 3020–3024. [CrossRef]

29. Thompson, R.B. Mechanisms of electromagnetic generation and detection of ultrasonic Lamb waves in iron-nickel alloy polycrystals. *J. Appl. Phys.* **1977**, *48*, 4942–4950. [CrossRef]

30. Yamasaki, T.; Yamamoto, S.; Hirao, M. Effect of applied stresses on magnetostriction of low carbon steel. *NDT E Int.* **1996**, *29*, 263–268. [CrossRef]

sensors

MDPI

Article

Comparison Study between RMS and Edge Detection Image Processing Algorithms for a Pulsed Laser UWPI (Ultrasonic Wave Propagation Imaging)-Based NDT Technique

Changgil Lee [1], Aoqi Zhang [1], Byoungjoon Yu [2] and Seunghee Park [1,*]

[1] School of Civil, Architectural Engineering and Landscape Architecture, Sungkyunkwan University, Gyeonggi-do, Suwon-si 16419, Korea; tolck81@gmail.com (C.L.); zhangaoqi623@hotmail.com (A.Z.)
[2] Department of Convergence Engineering for Future City, Sungkyunkwan University, Gyeonggi-do, Suwon-si 16419, Korea; mysinmu123@naver.com
* Correspondence: shparkpc@gmail.com or shparkpc@skku.edu; Tel.: +82-31-290-7648

Academic Editor: Aime' Lay-Ekuakille
Received: 13 March 2017; Accepted: 25 May 2017; Published: 26 May 2017

Abstract: In this study, a non-contact laser ultrasonic propagation imaging technique was applied to detect the damage of plate-like structures. Lamb waves were generated by an Nd:YAG pulse laser system, while a galvanometer-based laser scanner was used to scan the preliminarily designated area. The signals of the structural responses were measured using a piezoelectric sensor attached on the front or back side of the plates. The obtained responses were analyzed by calculating the root mean square (RMS) values to achieve the visualization of structural defects such as crack, corrosion, and so on. If the propagating waves encounter the damage, the waves are scattered at the damage and the energy of the scattered waves can be expressed by the RMS values. In this study, notch and corrosion were artificially formed on aluminum plates and were considered as structural defects. The notches were created with different depths and angles on the aluminum plates, and the corrosion damage was formed with different depths and areas. To visualize the damage more clearly, edge detection methodologies were applied to the RMS images and the feasibility of the methods was investigated. The results showed that most of the edge detection methods were good at detecting the shape and/or the size of the damage while they had poor performance of detecting the depth of the damage.

Keywords: pulsed laser scanning; ultrasonic waves; plate-like structures; crack; corrosion; edge detection

1. Introduction

Non-destructive testing (NDT) technology has been used for a number of decades, and NDT techniques have been successfully applied in many practical applications in various fields such as civil, mechanical, and aerospace engineering, etc. [1]. These methods can prolong the lifetimes of structures, and facilitate maintenance of structural health to minimize premature part changes. The method that can easily generate understandable detection results is preferred. With these methods, the cost of training personnel can be reduced, and the risk of human error can be decreased. In this case, the methods for damage detection with imaging capabilities have great potential to fulfill these requirements. Multiple locations on a target structure, which potentially contain various types of damage, can be monitored at the same time by using these methods. The location and degree of damage is very important for making a decision about a maintenance plan [2]. For example, due to the restricted accessibility of some structures such as nuclear power plants and due to the high-precision geometries or other inaccessible parts of a structure, the detection of structural safety is particularly difficult. Therefore, when damage detection is needed in a large-scale structure, it is necessary to develop a non-contact NDT method for damage detection [3].

In NDT, the methods for damage detection should be effective and also have high throughput, because of the increasing size of structures which need to be inspected. Ultrasonic waves are sensitive to most material damage and are not radiation hazards. In addition, they also can provide many features for damage characterization. Therefore, a wide range of inspection methods based on acoustic-ultrasonic waves have imaging capabilities. Not only that, most laser-ultrasonic systems can be integrated into mobile systems, and the laser also provides non-contact remote characteristics. To achieve that goal, some acoustic and ultrasonic wave technologies have been developed, such as full-field laser wave field imaging, laser vibrometry, laser interferometry, and pulsed lasers. As one of the full-field laser wave field imaging techniques, the holography based imaging technique requires a highly diffusive surface of the target structure. However, holography is always regarded as a technique which requires dark rooms. Therefore this method is inappropriate for remote automatic detection, even though it has the capability of noncontact detection [4]. A previous study [5] showed that Lamb waves were generated in an aluminum plate immersed in water. The laser vibrometer was used to scan the target surface, in which the laser beam was perpendicular to the surface. The scan locations can be changed by moving the laser head. It was verified that Lamb waves can be propagated in an aluminum plate, in the previous study [6]. The result was confirmed by using a laser scanning vibrometer, and the propagating waves were visualized in the vicinity of flaw area. Their results showed that the effectiveness of flaw detection depends on the flaw size. Other studies [7–9] have used Lamb waves to detect the damage of aluminum plates by using a laser scanning vibrometer to scan the surface of the target side. Due to their low noise and narrow line widths (on the order of a few millihertz), a single-mode HeNe laser is the preferred light source for the laser Doppler vibrometers (LDVs). The flaws were measured by finding the areas with bigger signal values. However, laser scanning vibrometers still have some disadvantages, such as the limiting factor about capturing a full field, and the signal-to-noise ratio of the photodetector output [10,11]. To overcome these drawbacks, an Nd:YAG pulse laser system has been developed to generate the ultrasonic waves. This pulsed laser could provide many advantages such as fast wave generation with low pulse energy, good detection capability in complex structures, and a high spatial resolution [12,13].

In this study, a non-contact laser ultrasonic wave propagation imaging (UWPI) method using a Nd:YAG pulsed laser system was used to detect the damage on aluminum plates. An Nd:YAG pulse laser was used to generate the ultrasonic waves, and the laser scanner based on a galvanometer was used to scan the target structure. In order to measure the wave responses at this stage, a piezoelectric sensor was installed to the central position on the front or the back side of the scanned surface. The damage can be visualized by obtaining root mean square (RMS) images [14]. Additionally, a series of edge detection methodologies were applied to the RMS images and compared to improve the performance of the damage visualization. To verify the feasibility of the approach, aluminum plates with notch and corrosion damages were tested. In the case of notch damage, different depths and angles were considered while different area and depths were investigated for the corrosion damage.

2. Ultrasonic Wave Propagation Imaging (UWPI) System

As shown in Figure 1, the UWPI system includes an image processor, a high-speed data digitizer, a Q-switch pulsed laser system, a laser mirror scanner based on a galvanometer, and an ultrasound transducer. In this study, a Q-switch diode-pumped high-power solid-state Nd:YAG pulse laser [15] is used, with the wavelength of 532 nm and the maximum pulse repetition rate of 20 Hz. The laser mirror scanner can adjust the scanning location of the target structure, which is designed so that the laser beam can be reflected at the tilting mirrors in the scanner with the wavelength of 532 nm. For ensuring that the laser beam can efficiently scan the two dimensional area of the target structure, the tilting mirrors are designed so that the operating angles are orthogonal to each other. An f-theta lens is installed at the end of the laser scanner system so that the laser beam can be focused on the target area. In this study, the laser beam vertically scans the target structure in the horizontal direction along the scanning coordinate as shown in Figure 1, which is preliminarily designed at the image processor. Also, the measured wave signals from the ultrasonic sensor were saved and treated to obtain ultrasonic wave propagation images (UWPI). The details of the process for the UWPI are explained in the following section.

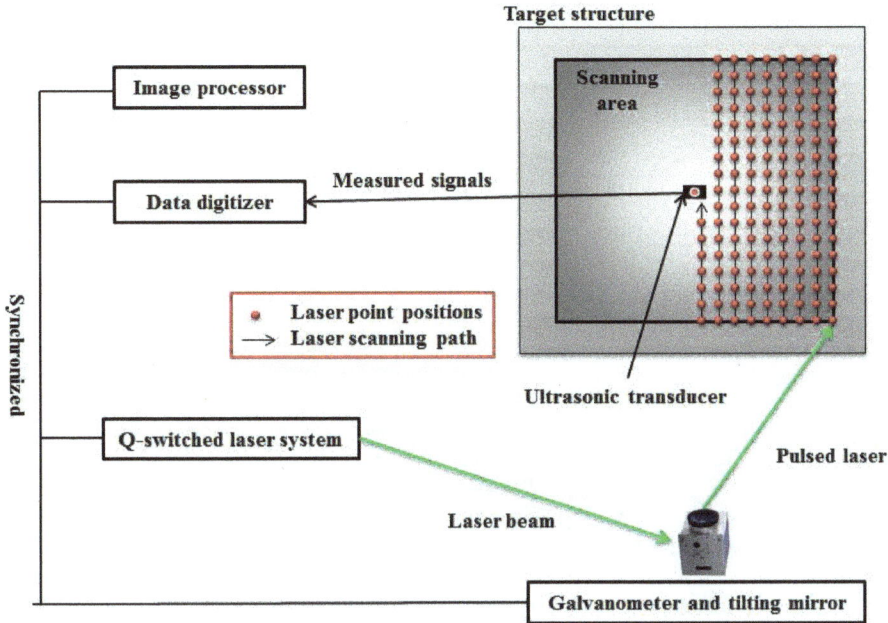

Figure 1. A schematic diagram of the laser-induced ultrasonic wave propagation imaging (UWPI) system.

During the scanning process, due to the thermoelastic mechanism, the ultrasonic waves are generated at the point where the laser beam is impinged on the surface of the target structure and propagated to the ultrasonic sensor. An ultrasonic transducer which is installed at the front or the back side of the scanned surface measures the wave responses. In this study, an acoustic emission sensor (AE sensor) which is made of lead zirconate titanate piezoelectric ceramics is selected as an ultrasonic transducer. The imaging process of ultrasonic wave propagation is shown in Figure 2. Firstly, the time-domain signal is also obtained at each laser scanning point. In addition, a band-pass digital filter is used to filter the noise signals and improve the signal-to-noise ratio. After that, the filtered signal groups in a vertical structure on a spreadsheet can be stacked, for each laser scanning point on the vertical axis. Then, the stacked vertical data need to be stacked repeatedly along the horizontal axis. Finally, 3-D UWPI data can be obtained with the three axes of the vertical scan, horizontal scan, and time frame [16]. A snapshot of the ultrasonic wave propagation image can be captured by slicing the 3-D data at a certain time point. Using these snapshots, post image processing for damage detection is performed.

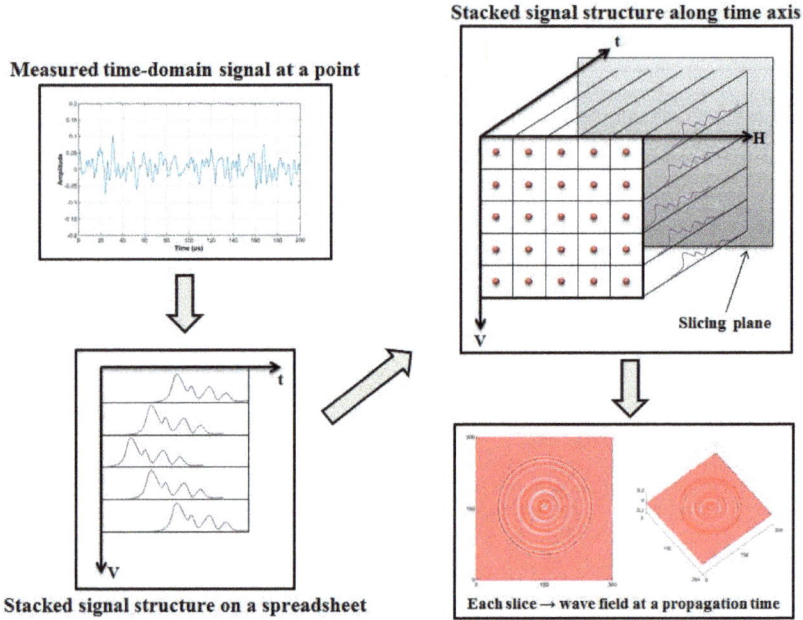

Figure 2. Ultrasonic wave propagation imaging process.

3. Imaging Process Algorithm for Damage Detection

3.1. Visualization Method Using Root Mean Square (RMS)

The propagating characteristics of ultrasonic waves can be more clearly expressed using the RMS images of the wave signals, because the RMS images can describe the energy distribution of the signals. The equation is shown in [17]:

$$w_{RMS}(x,y) = \left[\frac{1}{N} \sum_{i=1}^{N} (w_i(x,y,t))^2 \right]^{\frac{1}{2}} \tag{1}$$

where N is the number of signal samples, and $w(x, y, t)$ are the reflected signals.

Because of more frequent accumulations of the standing wave energy at the location of the sensor, the bigger RMS values will be produced in this area. This makes it hard to detect damage far from the sensor. In this case, multiplying a weighting parameter can equalize the RMS value of entire area, as follows:

$$w_{RMS}(x,y)\_W^p = \left[\frac{1}{N} \sum_{i=1}^{N} (w_i(x,y,t))^2 \cdot t^p \right]^{\frac{1}{2}} \tag{2}$$

where p is the weighting parameter and $w_{RMS}(x, y)\_W^p$ is the weighted RMS function.

In this study, the weighting parameter was 2.

3.2. Edge Detection Method

Edge detection techniques are very popular and essential image preprocessing steps, especially in the areas of feature detection and feature extraction. In an image, the quantity of data significantly reduces at an edge area, but these data still retain basic information of the objects in the area. If the value of a pixel point exceeds a designated threshold, that point is declared as an edge location.

Therefore, the edges have the higher pixel intensity values than the surrounding points. In this way, the edges can be detected by comparing the gradient value to the threshold value, and when the 1st derivative is the maximum, the 2nd derivative will be 0. This characteristic can be used for computer vision and image processing. The method has major features for a good ability to create the exact edge line. Therefore, edge detection is an active research area for better facilitating the image analysis. Nowadays, edge detection is usually used for object detection such as medical image processing, biometrics, and advanced computer imaging techniques [18]. Not only that, this method can also be used in SHM (Strutural Health Monitoring). In this study, the Sobel, Prewitt, Roberts, and Laplacian of Gaussian (LoG) operators are used to filter the images of the testing results.

3.2.1. Sobel Operator

The Sobel edge detection technique was proposed by Sobel in 1970 [19]. The method is a spatial domain gradient-based edge detector. The Sobel operator performs a 2-D spatial gradient measurement on an image which consists of two gradient masks of size 3×3, one for horizontal changes, and another for vertical changes. In general, it is used to calculate the approximate absolute gradient magnitude (edge strength) at each single pixel point. The actual Sobel masks are as follows:

$$Gx_{(sobel)} = \begin{bmatrix} -1 & 0 & +1 \\ -2 & 0 & +2 \\ -1 & 0 & +1 \end{bmatrix} \quad \text{and} \quad Gy_{(sobel)} = \begin{bmatrix} +1 & +2 & +1 \\ 0 & 0 & 0 \\ -1 & -2 & -1 \end{bmatrix} \tag{3}$$

where Gx and Gy are the gradient component at each point that contain the horizontal and vertical direction. The gradient magnitude can be calculated using the formula:

$$|G| = \sqrt{Gx^2 + Gy^2} \tag{4}$$

Then, the approximate absolute gradient magnitude can be calculated using:

$$|G| = |Gx| + |Gy| \tag{5}$$

Finally, using this information, the gradient direction θ is given by:

$$\theta = \arctan\left(\frac{Gy}{Gx}\right) \tag{6}$$

where in this case, $\theta = 0$ means the direction of maximum contrast from the color of black to white runs from left to right on the image, and other angles can be measured anti-clockwise from it.

In general, the absolute magnitude is the output that only the researchers can observe. Figure 3 shows that by using the pseudo-convolution operator, the two components of the gradient could be conveniently computed and added in a single pass over the input image.

Using this mask, the equation of approximate magnitude is given by:

$$|G| = |(P_1 + 2 \times P_2 + P_3) - (P_7 + 2 \times P_8 + P_2)| + |(P_3 + 2 \times P_6 + P_2) - (P_1 + 2 \times P_4 + P_7)| \tag{7}$$

P$_1$	P$_2$	P$_3$
P$_4$	P$_5$	P$_6$
P$_7$	P$_8$	P$_9$

Figure 3. Pseudo-convolution masks for the Sobel operator used to quickly compute the approximate gradient magnitude.

3.2.2. Roberts Cross Operator

The Roberts cross operator has a good ability to perform a simple, quick calculation and 2-D spatial gradient measurement on an image [20]. The Roberts cross operator for the input is a grayscale image, as is the output. Pixel values of each point in the output data are the estimated absolute magnitude of the spatial gradient at that point. This operator consists of a pair of 2 × 2 convolution kernels as follows:

$$Gx_{(Robert)} = \begin{bmatrix} +1 & 0 \\ 0 & -1 \end{bmatrix} \quad \text{and} \quad Gy_{(Robert)} = \begin{bmatrix} 0 & +1 \\ -1 & 0 \end{bmatrix} \tag{8}$$

where one kernel is simply rotated by 90° to the other, and this mask is very similar to the Sobel operator.

The kernels are designed to maximize the response to the edges running at 45° to the pixel grid, one kernel will correspond to each of the two perpendicular orientations. These kernels are applied separately to form gradient components in each orientation (Gx and Gy). Therefore, the gradient magnitude can be defined as:

$$\left| G_{(Robert)} \right| = \sqrt{Gx_{(Robert)}^2 + Gy_{(Robert)}^2} \tag{9}$$

The approximate magnitude can be calculated by:

$$\left| G_{(Robert)} \right| = \left| Gx_{(Robert)} \right| + \left| Gy_{(Robert)} \right| \tag{10}$$

The direction of the gradient (relative to the pixel grid orientation) is given by:

$$\theta = \arctan\left(\frac{Gy_{(Robert)}}{Gx_{(Robert)}} \right) - 3\frac{\pi}{4} \tag{11}$$

When $\theta = 0$, it has same characteristics as the Sobel operator.

Not only that, the absolute magnitude is the output that only the researchers can observe. A pseudo-convolution operator is used to computed the gradient components and add in a single pass over the input image, as shown in Figure 4.

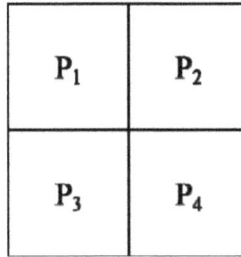

Figure 4. Pseudo-convolution masks for the Robert operator.

The approximate magnitude can be given by:

$$\left|G_{(Robert)}\right| = |(P_1 - P_4)| + |P_2 - P_3| \tag{12}$$

3.2.3. Prewitt Operator

The Prewitt operator [21] is similar to the Sobel operator. This operator can be used for detecting vertical and horizontal edges of the images. The Prewitt operator kernel is given by:

$$Gx_{(Prewitt)} = \begin{bmatrix} +1 & +1 & +1 \\ 0 & 0 & 0 \\ -1 & -1 & -1 \end{bmatrix} \quad \text{and} \quad Gy_{(Prewitt)} = \begin{bmatrix} -1 & 0 & +1 \\ -1 & 0 & +1 \\ -1 & 0 & +1 \end{bmatrix} \tag{13}$$

3.2.4. Laplacian of Gaussian (LoG) Operator

A method was proposed where finding the zero-crossings in the 2nd derivative of the image intensity can detect the edge point in an image. Unfortunately, the 2nd derivative is very sensitive to noise. In this case, the noise should be filtered before edge detection. In order to achieve that, the LoG operator performs Gaussian smoothing before applying Laplacian [22].

In this method, the image is convolved with a Gaussian filter first. This step can smoothen the image and reduces noise. Since the width of the edge increases in the smoothing process, only the point having the local maximum value should be regarded as an edge. Therefore, the 2nd derivative operator, Laplacian, is used for this purpose. In order to reduce unnecessary edge pixels, only pixels whose first-order differential values (threshold) of zero-crossings exceed a certain degree are treated as edge points.

The output of the LoG operator: $h(x, y)$ is obtained by the convolution operation:

$$h(x,y) = \Delta^2[G(x,y) \times f(x,y)] = \left[\Delta^2 G(x,y)\right] \times f(x,y) \tag{14}$$

where the following equation is normally called the Mexican hat operator.

$$\Delta^2 G_{(log)}(x,y) = \left(\frac{x^2 + y^2 - 2\sigma^2}{\sigma^4}\right)^{-(x^2+y^2)/2\sigma^2} \tag{15}$$

4. Experimental Study

4.1. Experimental Setup

In this study, the 6061-T6 aluminum plates were selected as test specimens, which had dimensions of 400×400 mm with the thickness of 3 mm. After scanning the intact specimen, the notche and the corrosion damages were artificially formed on four specimens, as shown in Figures 5 and 6. Figure 5a shows the designed condition of the first specimen; four notches were made as the same angles which are parallel to the tangent of wave front. This notch direction arrangement is used to verify the influence of the notch depth on the test results. Figure 5b shows the designed condition of the second specimen; seven artificial notches were formed on the plate, and the dimensions of each notch was 20 mm long, 1 mm long, and 2 mm deep. In addition, the direction of the notches is formed with a counter-clockwise increase of 15° for each notch, starting from the notch at the right area which is tangential to the wave front. This notch direction arrangement is used to verify the influence of the notch direction on the test results.

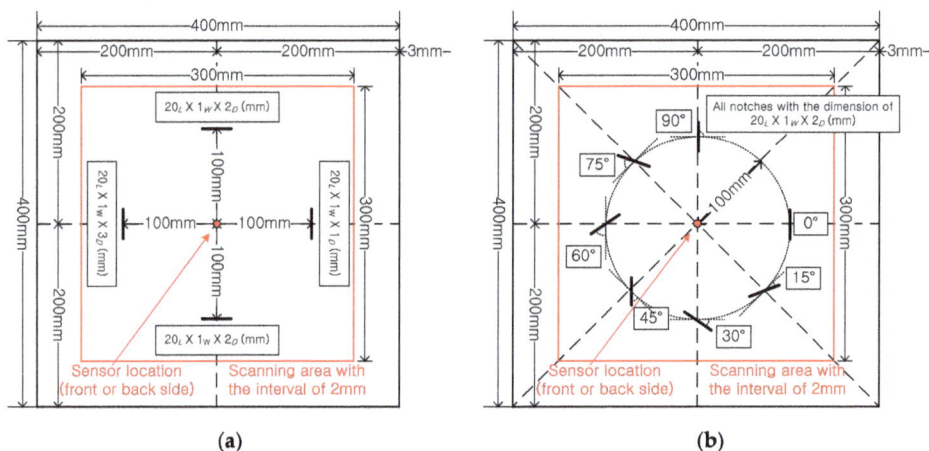

Figure 5. Configuration of notch damage on the aluminum plate (subscripts for the dimension of the notch, *L*, *W*, and *D* mean length, width, and depth, respectively), (**a**) different depth, (**b**) different angle.

Next, corrosion damages were considered. The corrosion damages on the aluminum plates were artificially formed using concentrated hydrochloric acid, as shown in Figure 6. Figure 6a shows the designed condition of the third specimen, all corroded areas had the same size of 50 × 50 mm, but they had different depths with 0.5, 1.0, 1.5, and 2.0 mm. This arrangement is used to verify the influence of the depth of the corrosion on the test results. Figure 6b shows the designed condition of the forth specimen, and the corrosion areas had the same corrosion depth but different dimensions with 5 × 5, 10 × 10, 15 × 15, and 20 × 20 mm. This arrangement was used to verify the influence of the corroded area on the test results.

Figure 7 shows that the specimen was fixed on a metal support. The bottom part of the specimen was tightly clamped with two clamps on the metal frame. In this study, the sensor was attached to the central position on the back side of the scanned surface. An amplifier-integrated acoustic emission (AE) sensor was used to measure the multiple wave signals. The AE sensor has a broadband characteristic with upper and lower cutoff frequencies of 2 MHz and 100 kHz, respectively. The resonant frequency of the sensor is 200 kHz ± 20%. The maximum sensitivity of the sensor is 120 ± 3 dB at the resonant frequency. The scanned area was 300 × 300 mm at the central part of the specimen; in this area, a 151 × 151 point grid can be generated, with the scanning interval of 2 mm. The distance between the laser mirror scanner and the target specimen was 2 m.

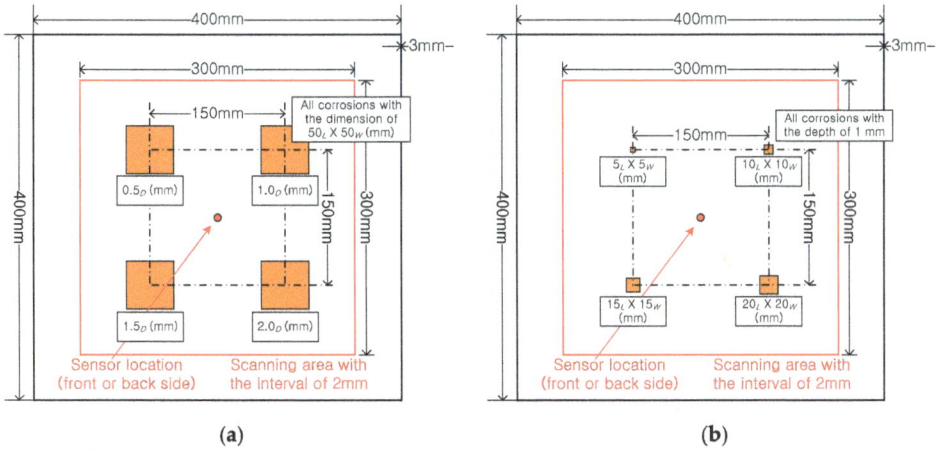

(a)   **(b)**

Figure 6. Configuration of corrosion damage on the aluminum plate (subscripts for the dimension of the notch, *L*, *W*, and *D* mean length, width, and depth, respectively), (**a**) different depth, (**b**) different size.

Figure 7. Location of the fixed target specimen during scanning.

4.2. Comparison between RMS Images and Edge Detection Results

4.2.1. Damage Case 1: Notch

Figure 8 shows the scanning results of the intact specimen at 40 μs. For the UWPI snapshots, the wave packet propagated radially in a dispersed fashion from the central location in the circumferential boundary condition as shown in Figure 8a. Figure 8b shows the estimated RMS snapshots from Figure 8a. The results showed that the color of the scanned area is uniform; this means a structural condition of the smooth plate surface. Because more wave energy was accumulated in the vicinity of the sensor location at the early stage, the color was lighter than for the other areas; this means that bigger RMS values were estimated in the sensor location.

The edge detection results for the intact condition are shown in Figure 9. In this study, four types of operators, which were Sobel, Roberts, Prewitt, and LoG operators, were applied to the RMS images. In this case, the RMS values were dramatically changed only at the vicinity of the sensor. On the other hand, the RMS values were varied smoothly at the boundary of the wave front. As a result, the edge was detected at the sensor location clearly for all operators.

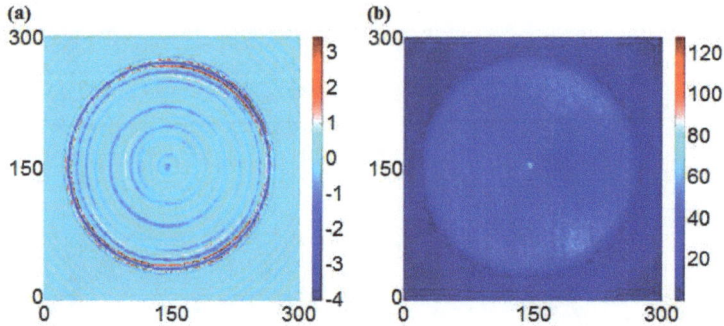

Figure 8. Snapshots of the intact specimen at 40 µs: **(a)** UWPI snapshot, **(b)** RMS (Root Mean Square) snapshot.

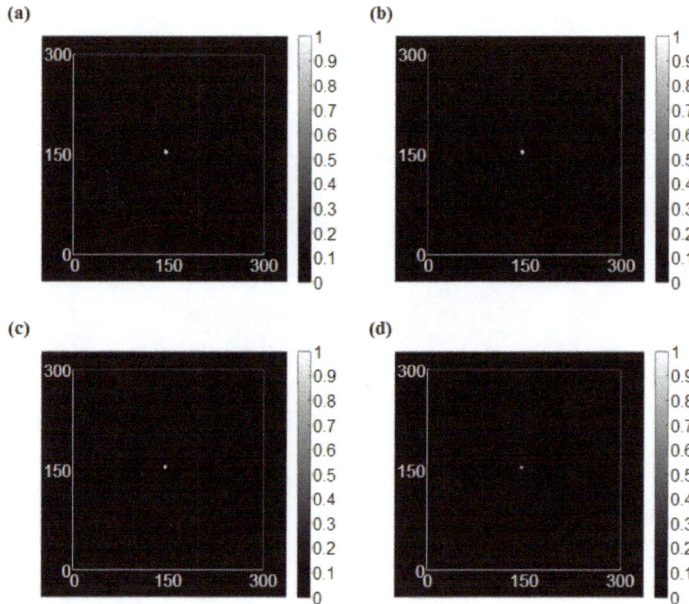

Figure 9. The edge detection results of the intact condition, **(a)** Sobel operator (threshold = 1.3), **(b)** Roberts cross operator (threshold = 0.3), **(c)** Prewitt operator (threshold = 1.4), **(d)** LoG (Laplacian of Gaussian) operator (threshold = 0.3).

Next, the variation in depth of the notch was considered. Figure 10a shows a UWPI snapshot at 40 µs, and anomalous wave due to the damage can be observed near the damage locations. Furthermore, the reflected wave became a source of new scattered waves when the propagating waves encountered the damage. The influence of the reflected waves at the right area (depth = 1 mm) was the lowest, and was also not significant. On the other hand, the influence of the reflected waves at two notches (depth = 2 mm) which are located in the upper and lower area can be more clearly observed than a notch at the right area (depth = 1 mm). Furthermore, the result at the left area (depth = 3 mm) was most significant. In Figure 10b, the results showed that the larger values occurred at reflected wave paths. Any reflected waves almost cannot be observed at the notch with the depth of

1 mm, but the other three notches were observed successfully because the RMS values are lowest at the shallowest notch. Because the two notches of the upper and lower parts have the same condition, their results were very similar, and the biggest RMS value was measured at the notch of 3 mm.

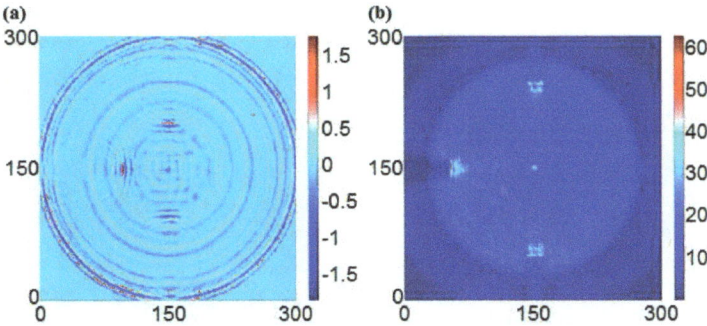

Figure 10. Snapshots of specimen 1 at 40 μs: (a) UWPI snapshot, (b) RMS snapshot.

The edge detection process was applied again in this case as shown in Figure 11. The shallowest notch at the right side can be visualized, but it is not still clear, while it is hard to identify this notch through the RMS image. Unfortunately, however, the variation in the depth of the notches is not clearly expressed in the results of the edge detection operations although the edges of the notches are clearly detected. It means that the edge detection process may be appropriate for detecting the shape of the damage. On the contrary, the method has poor quality to detect the depth of the damage.

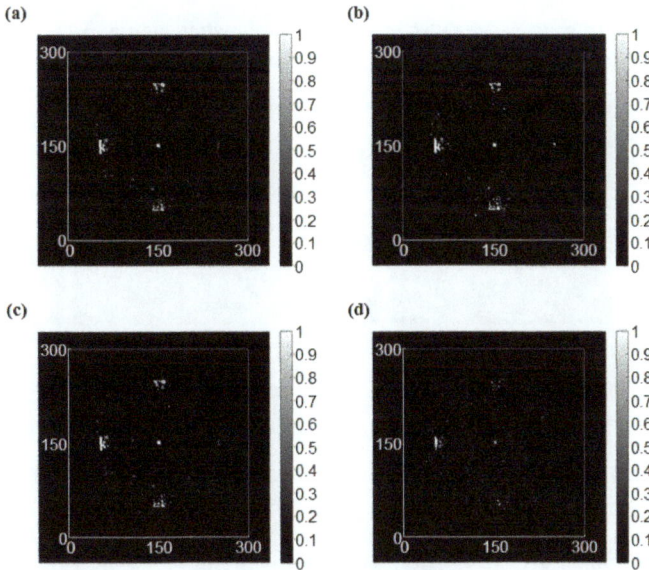

Figure 11. The edge detection results of the damaged condition with the different depth, (a) Sobel operator (threshold = 1.1), (b) Roberts cross operator (threshold = 0.2), (c) Prewitt operator (threshold = 0.8), (d) LoG operator (threshold = 0.15).

The UWPI and RMS snapshots of the notches with different angles were captured at 40 μs and are shown in Figure 12. Since the waves propagate along the radial direction, the wave portions have stress components in the vertical and horizontal directions and hence the reflected waves can be observed regardless of the angle between the wave front and the notch, as shown in Figure 12b. In this case, the energy of the reflected waves is largest when the notch is tangential to the wave front. On the other hand, the smallest value of RMS is observed when the notch is perpendicular to the wave front. This is because the wider notch can reflect the incident waves.

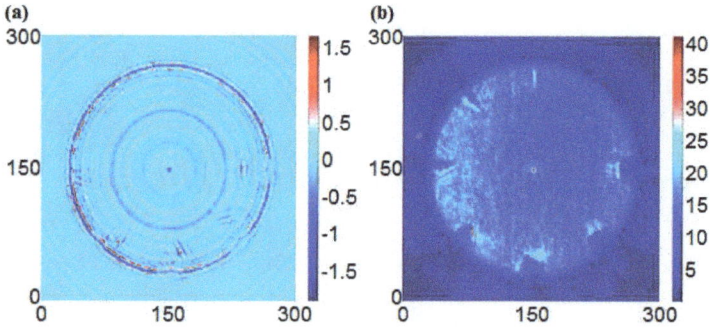

Figure 12. Snapshots of specimen 2 at 40 μs: (**a**) UWPI snapshot, (**b**) RMS snapshot.

In this case, the edges of the notches were most clearly detected compared to the other case. Additionally, the noise near the notches is hardly observed because the depths of the notches were identical and the size of the notches was enough to reflect the incident waves. As mentioned previously, the edge detection is good at detecting the shape of the damage and hence the angles of the notches are clearly visualized using the edge detection process as shown in Figure 13.

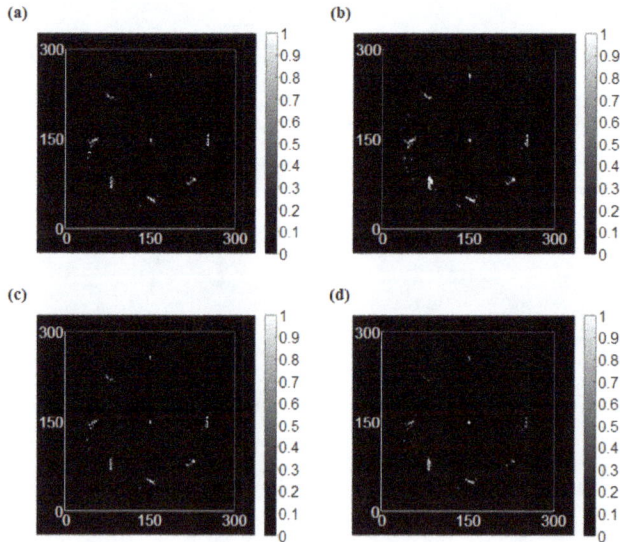

Figure 13. The edge detection results of the damaged condition with the different angle, (**a**) Sobel operator (threshold = 1.8), (**b**) Roberts cross operator (threshold = 0.3), (**c**) Prewitt operator (threshold = 1.4), (**d**) LoG operator (threshold = 0.3).

4.2.2. Damage Case 2: Corrosion

Figure 14 shows the UWPI results of the two specimens, in which one includes corrosion with different depths and the corrosion with different areas is made on the other plate, at 60 μs. Figure 14a shows a front side result of the first specimen at 60 μs, and the damage-induced anomalous wave can be observed at the damage locations. Furthermore, the reflected wave became a source of new scattering waves when the propagating waves encountered the damage. The result in Figure 14a showed that the propagating waves were obviously scattered. At the top left area, the scattering influence was the smallest, and the corrosion at the top right was more significant. Additionally, the wave reflection phenomena occurred at the bottom left and bottom right areas, which at the area with greater depth was clearer. Therefore, these features can be used for damage detection. Figure 14b shows the back side result of the first specimen. The results were almost the same as the results obtained from the front side. Figure 14c,d shows the results from both sides of the second specimen. The scattering phenomenon also occurred at the corrosion areas, and the degrees of interference at both sides were similar. Figure 15 shows the RMS result of the two specimens. The results showed that the corrosion areas were observed clearly, the damage with deeper depth showed a deeper color on the front side, and a lighter color was shown at a deeper depth on the back side. Therefore the damage was successfully detected. In addition, the results show that the RMS method has a good ability for classification of the damage depth.

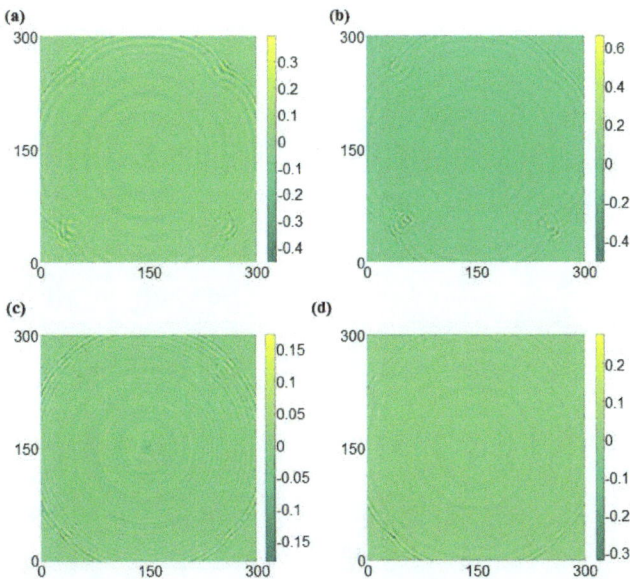

Figure 14. UWPI snapshots at 60 μs: (**a**) Front side of the first specimen, (**b**) Back side of the first specimen, (**c**) Front side of the second specimen, (**d**) Back side of the second specimen.

After edge detection processing, the grayscale images of Sobel, Roberts, Prewitt, and LoG operators are shown in Figures 16–19, respectively. The color axis showed a standard of edge strength, in which a bigger RMS value has a more dramatic change of the gray-scale (strong edge), and it also can be considered as deeper damage in this study. In this case, the threshold of the Sobel, Roberts, Prewitt, and LOG operators were 2, 0.4, 1, and 0.3, respectively.

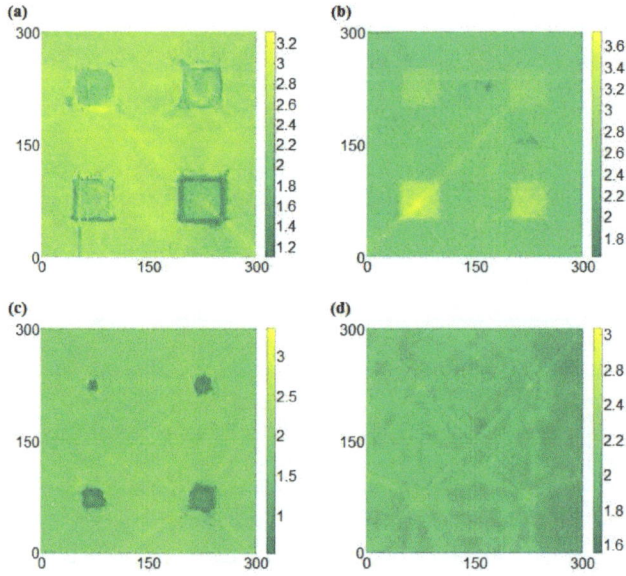

Figure 15. RMS snapshots at 400 μs: (**a**) Front side of the first specimen, (**b**) Back side of the first specimen, (**c**) Front side of the second specimen, (**d**) Back side of the second specimen.

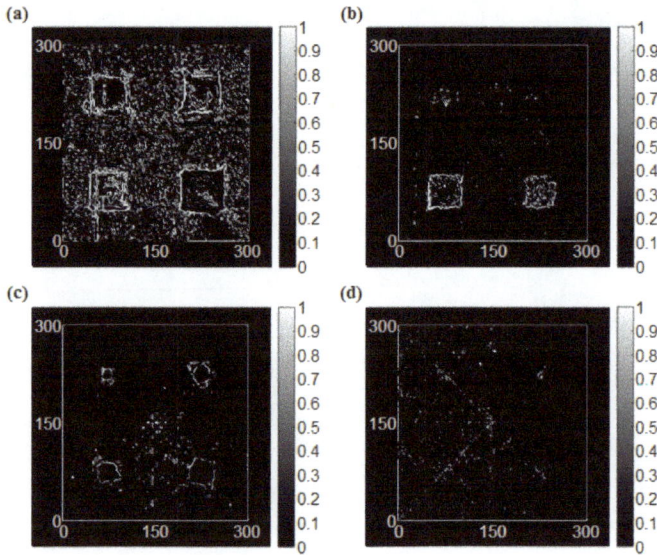

Figure 16. The edge detection results using the Sobel operator (threshold = 2); (**a**) Front side of the first specimen, (**b**) Back side of the first specimen, (**c**) Front side of the second specimen, (**d**) Back side of the second specimen.

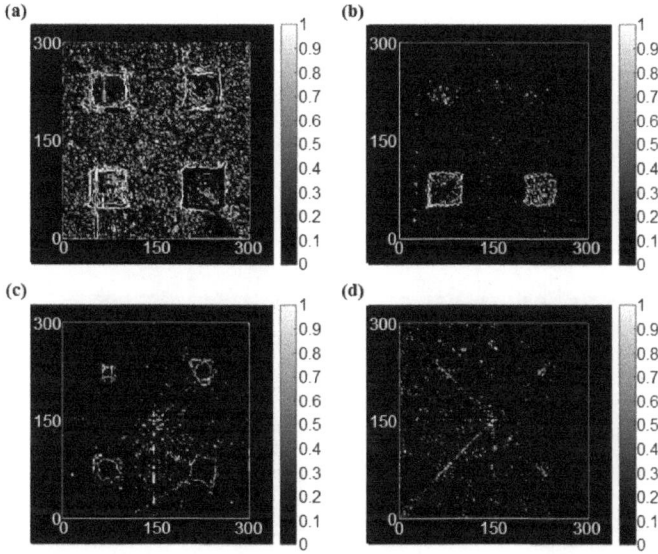

Figure 17. The edge detection results using the Roberts cross operator (threshold = 0.4); (**a**) Front side of the first specimen, (**b**) Back side of the first specimen, (**c**) Front side of the second specimen, (**d**) Back side of the second specimen.

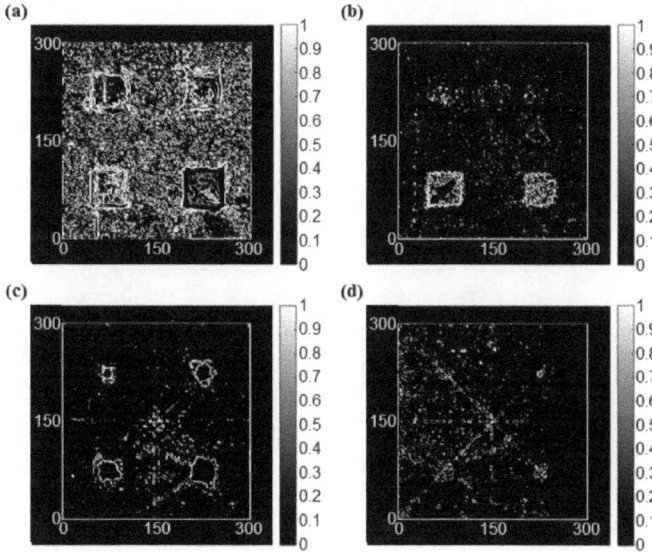

Figure 18. The edge detection results using the Prewitt operator (threshold = 1); (**a**) Front side of the first specimen, (**b**) Back side of the first specimen, (**c**) Front side of the second specimen, (**d**) Back side of the second specimen.

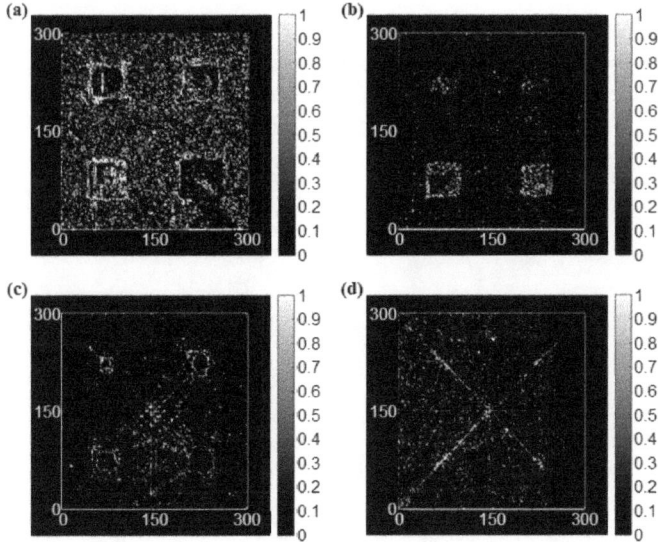

Figure 19. The edge detection results using the LoG operator (threshold = 0.3); (**a**) Front side of the first specimen, (**b**) Back side of the first specimen, (**c**) Front side of the second specimen, (**d**) Back side of the second specimen.

For the Sobel operator, Figure 16a shows a remarkable classification of the damage degrees. A deeper damage has a more dramatic change of the gray-scale (a bigger value). The result of Figure 16c shows a similar degree of edge with the same depths. For the back side, in Figure 16b the damages of 1.5 mm and 2 mm are observed clearly. However, because the depth of some damage is not deep enough (1 mm or less than 1 mm), it cannot be observed very clearly in Figure 16b,d. Compared with the Roberts and Prewitt operator, the results were very similar. It is worth noting that the effect of the Prewitt operator for a small area of corrosion damage detection was better, as shown in Figure 18c. This result means that the Prewitt operator is more sensitive to horizontal and vertical damage. However, the result of the LoG operator was not clearer than the other three operators, possibly due to the method further reducing the number of unnecessary pixels. On the other hand, the result in Figure 19d showed that the LoG operator is more sensitive to the diagonal edges. In summary, the results successfully detected the corrosion. In addition, the results from the measurement at the front side of the plate, where the corrosion damages were directly exposed to the laser beam, were better than the results from the back-side scanning. Additionally, the damage detection for the corrosion damages based on the edge detection method will not be as good as the RMS method.

5. Conclusions

In this study, notch and corrosion damage on aluminum plates were investigated using the UWPI imaging system, which utilizes a Q-switched Nd:YAG pulsed laser system and a laser scanner based on a galvanometer. First, an intact aluminum plate without any damage was scanned. In this case, the results showed continuities of the propagating waves in the snapshots at the raw UWPI, filtered UWPI, and RMS snapshots. Then, two conditions of notches were artificially formed, in which one was the notches with different depths and different angles were considered in the other condition. In the case of different depths of the notches, the reflected waves could be clearly visualized through the RMS calculation except for the shallowest notch. The energy of the reflection at the shallowest notch was

Sensors **2017**, *17*, 1224

relatively low compared to other notches because the depth which could reflect the incident waves was shallow. After the edge detection processing, the notches were clearly visualized even at the point of the shallowest notch. However, the differences in depth were hardly detected using the edge detection method. In the case of different angles of the notches, the damage was clearly detected in the RMS snapshots and in the images obtained from the edge detection process. Since the incident waves in this study propagated along the radial direction, the waves had stress components in the vertical and horizontal directions. Also, all of the notches had the same depth in this case. Therefore, the reflected waves were clearly observed regardless of the angles of the notches. However, the tangential notch to the wave front was most clearly detected while the energy of the notch perpendicular to the wave front was the lowest, because the reflected waves were affected by the width of the notch. Next, corrosion damage was considered in a similar manner with that of the notch damage. Also, in this case, the depth and the area of the corrosion damage were investigated. For both conditions, the scanning results obtained from the front side of the specimens were clearer than the back side of the specimens at the RMS and edge detection snapshots. In this case, the depth of the damage was hard to detect while the size of the damage could be clearly identified. As a result, the RMS snapshots are appropriate for observing the energy flow of the propagating wave while the image obtained from the edge detection method is good for distinguishing the shape of the damage. Unfortunately, however, the variation in the depth of the damage is hardly investigated using the edge detection method.

Acknowledgments: This work was financially supported by a grant (14SCIP-B088624-01) from the Construction Technology Research Program funded by the Ministry of Land, Infrastructure, and Transport of the Korean government, National Research Foundation of Korea (NRF) grant funded by the Korean government (MSIP) (NRF-2015R1D1A1A01059291 and 2017R1A2B3007607), and the Korea Ministry of Land, Infrastructure and Transport (MOLIT) as the 'u-City Master and Doctor Course Grant Program.'

Author Contributions: Changgil Lee, Aoqi Zhang and Byounjoon Yu conceived and designed the experiments; Aoqi Zhang and Byoungjoon Yu performed the experiments; Changgil Lee analyzed the data; Changgil Lee and Seunghee Park wrote the paper. In addition, Changgil Lee and Seunghee Park are responsible for the implementation the proposed scheme.

Conflicts of Interest: The authors declare no conflict of interest.

References

1. Lee, C.; Park, S. Flaw imaging technique for plate-like structures using scanning laser source actuation. *Shock Vib.* **2014**, *2014*, 1–14. [CrossRef]
2. Lee, J.R.; Ciang, C.C.; Jin, S.H.; Park, C.Y.; Jin, Y.D. Laser ultrasonic propagation imaging method in the frequency domain based on wavelet transformation. *Opt. Laser. Eng.* **2011**, *49*, 167–175. [CrossRef]
3. Sohn, H. Noncontact laser sensing technology for structural health monitoring and nondestructive testing. In Proceedings of the Bioinspiration, Biomimetics, and Bioreplication 2014, San Diego, CA, USA, 10–12 March 2014.
4. Green, R.E., Jr. Non-contact ultrasonic techniques. *Ultrasonics* **2004**, *42*, 9–16. [CrossRef] [PubMed]
5. Eudeline, Y.; Duflo, H.; Izbicki, J.L.; Duclos, J. Immersed narrow plate study. Lamb wave identification. In Proceedings of the 1999 IEEE Ultrasonics Symposium, Stateline, NV, USA, 17–20 October 1999.
6. Kehlenbach, M.; Kohler, B.; Cao, X.; Hanselka, H. Numerical and Experimental Investigation of Lamb Wave Interaction with Discontinuities. In Proceedings of the 4th International Workshop on Structural Health Monitoring, Stanford, CA, USA, 15–17 September 2003.
7. Staszewski, W.J.; Lee, B.C.; Mallet, L.; Scarpa, F. Structural health monitoring using scanning laser vibrometry: I. Lamb wave sensing Smart Materials and Structures. *Smart Mater. Struct.* **2004**, *13*, 251–260. [CrossRef]
8. Mallet, L.; Lee, B.C.; Staszewski, W.J.; Scarpa, F. Structural health monitoring using scanning laser vibrometry: II. Lamb waves for damage detection Smart Materials and Structures. *Smart Mater. Struct.* **2004**, *13*, 261–269. [CrossRef]
9. Leong, W.H.; Staszewski, W.J.; Lee, B.C.; Scarpa, F. Structural health monitoring using scanning laser vibrometry: III. Lamb waves for fatigue crack detection Smart Materials and Structures. *Smart Mater. Struct.* **2005**, *14*, 1387–1395. [CrossRef]

10. Johansmann, M.; Siegmund, G.; Pineda, M. Targeting the Limits of Laser Doppler Vibrometry. 2005. Available online: http://www.polytec.com/fileadmin/user_uploads/Applications/Data_Storage/Documents/LM_TP_Idema_JP_2005_E.pdf (accessed on 26 May 2017).

11. Schleyer, G.; Brebbia, C.A. *Infrastructure Risk Assessment & Management*; WIT Press: Boston, MA, USA, 2016; p. 28.

12. Lawrence, E. *Optical Measurement Techniques for Dynamic Characterization of MEMS Devices*; Polytec Inc.: Irvine, CA, USA, 2012.

13. Lee, C.; Park, S. Damage visualization of pipeline structures using laser-induced ultrasonic waves. *Struct. Health Monit.* **2015**, *14*, 475–488. [CrossRef]

14. Radzieński, M.; Doliński, L.; Krawczuk, M.; Zak, A.; Ostachowicz, W. Application of RMS for damage detection by guided elastic waves. *J. Phys. Conf. Ser.* **2011**, *305*, 012085. [CrossRef]

15. Quantel Laser CFR (200–400 mj): Lamp Pumped Solid State Laser. Available online: http://www.quantel-laser.com/en/products/item/cfr-200-400-mj--133.html (accessed on 8 June 2016).

16. Lee, J.R.; Sunuwar, N. Advances in damage visualization algorithm of ultrasonic propagation imaging system. *J. Korean Soc. Nondestruct. Test.* **2013**, *33*, 232–240.

17. Lee, C.; Kang, D.; Park, S. Visualization of fatigue cracks at structural members using a pulsed laser scanning system. *Res. Nondestruct. Eval.* **2015**, *26*, 123–132. [CrossRef]

18. Muthukrishnan, R.; Radha, M. Edge detection techniques for image segmentation. *Int. J. Comput. Sci. Inf. Technol.* **2011**, *3*, 259–267. [CrossRef]

19. Gonzalez, R.C.; Woods, R.E.; Eddins, S.L. *Digital Image Processing Using MATLAB*; Pearson Education South Asia Pte. Ltd.: Singapore, 2004.

20. Roberts, L.G. Machine Perception of 3-D Solids. Ph.D. Thesis, Massachusetts Institute of Technology, Cambridge, MA, USA, 1963.

21. Gonzalez, R.C.; Woods, R.E. *Digital Image Processing*, 2nd ed.; Prentice Hall: Upper Saddle River, NJ, USA, 2002.

22. Maini, R.; Aggarwal, H. Study and comparison of various image edge detection techniques. *Int. J. Image Proc.* **2009**, *3*, 1–11.

sensors

MDPI

Article

The Use of Flexible Ultrasound Transducers for the Detection of Laser-Induced Guided Waves on Curved Surfaces at Elevated Temperatures

Tai Chieh Wu [1],*, Makiko Kobayashi [2], Masayuki Tanabe [2] and Che Hua Yang [1]

[1] College of Mechanical and Electrical Engineering, National Taipei University of Technology, Taipei 10608, Taiwan; chyang@ntut.edu.tw

[2] Faculty of Advanced Science and Technology, Kumamoto University, Kumamoto 8608555, Japan; kobayashi@cs.kumamoto-u.ac.jp (M.K.); mtanabe@cs.kumamoto-u.ac.jp (M.T.)

* Correspondence: djwu1224@gmail.com; Tel.: +886-2-2771-2171 (ext. 4817)

Academic Editor: Xiaoning Jiang
Received: 12 May 2017; Accepted: 31 May 2017; Published: 4 June 2017

Abstract: In this study, a flexible ultrasonic transducer (FUT) was applied in a laser ultrasonic technique (LUT) for non-destructive characterization of metallic pipes at high temperatures of up to 176 °C. Compared with normal ultrasound transducers, a FUT is a piezoelectric film made of a PZT/PZT sol-gel composite which has advantages due to its high sensitivity, curved surface adaptability and high temperature durability. By operating a pulsed laser in B-scan mode along with the integration of FUT and LUT, a multi-mode dispersion spectrum of a stainless steel pipe at high temperature can be measured. In addition, dynamic wave propagation behaviors are experimentally visualized with two dimensional scanning. The images directly interpret the reflections from the interior defects and also can locate their positions. This hybrid technique shows great potential for non-destructive evaluation of structures with complex geometry, especially in high temperature environments.

Keywords: flexible ultrasonic transducer; laser ultrasonic technique; laser ultrasonic visualization; material characterization; defect detection; Non-destructive testing; high temperature measurement

1. Introduction

Structures with curved surfaces such as pipelines or pressured vessels that are required to operate at elevated temperatures are commonly seen. Defects due to corrosion, erosion, and cracks may lead to catastrophic outcomes. Nondestructive testing (NDT) techniques have been continuously developing for early detection of possible defects to ensure a structure's integrity. Among them, ultrasound techniques are widely used because of their advantages in terms of cost effectiveness, being free of radiation, and their versatility to be applied under numerous different conditions. In recent years, ultrasonic techniques (UTs) have been based on guided wave detection and were developed for the detection of subsurface and interior faults. However, UTs have some limitations such as their range of working temperatures and incomplete surface conductivity. Currently, when using ultrasonic wave detection at high temperatures, optical interferometers have overcome the temperature shortcoming, but their sensitivity is low and the preparation of a smooth specimen surface is necessary. Therefore, a more robust inspection technique, which can be applied to more complex geometry at high temperatures is very desirable.

A thorough review of various piezoelectric materials and bonding techniques was compiled by Kažys [1]. In the past decade, many researchers have developed high temperature sensors for inspection and condition monitoring [2,3]. The most often used piezoelectric materials in manufacturing ultrasonic transducers and their applications are as follows: first, lithium niobate (LiNbO$_3$) single crystals are one of the most well-known elements due to its high Curie temperature threshold of T$_c$ 1142–1210 °C and its ideal piezoelectric element performance. It has been used in high temperature transducers for ultrasonic testing at 400 °C since 1989 [4]. Since then, a series of studies have focused on the improvement of its working temperature by using the dice and fill method [5–8]. However, LiNbO$_3$ has low thermal shock durability due to its single crystal structure. In addition, LiNbO$_3$ cannot sustain high temperatures for long terms because of oxidation losses. Bismuth titanate (BIT) and modified bismuth titanate (MBIT) are also commonly used as the piezoelectric element for direct contact ultrasonic transducers. These materials exhibit a low dielectric constant, low dielectric losses and their properties are stable up to very high temperatures. In earlier applications, BIT and MBIT successfully served as a tool for ultrasonic thickness monitoring [9] and pipeline defect inspection [10] at temperatures of up to 350 °C. The most popular piezoelectric material used for manufacture of ultrasonic transducers is lead zirconate titanate (PZT). It has very good electromechanical properties, but is compromised by having a relatively low T$_c$ threshold/tolerance of 350 °C which is far below that of most other piezoelectric materials such as LiNbO$_3$, BIT, GaPO$_4$, etc. An NDT device fabricated with a combination of PZT and LiNbO$_3$ or PbTiO$_3$ (PT) and PZT as a composite piezoelectric material was successfully tested by Kobayashi et al. for uses at elevated temperatures [11,12]. In their studies, a sol-gel composite was developed to solve the problems caused by high temperatures [13]. A sol-gel composite consists of a piezoelectric powder phase and a high dielectric constant sol-gel phase. It can be applied to the piezoelectric material [14] and thick film ultrasonic transducer by using the sol-gel spray technique [15–17]. With this technique, a piezoelectric ceramic film can be easily fabricated at the desired location on the surface of molds or dies through a shadow mask. In addition, the film can be sprayed onto a thin plate substrate to be used as a flexible ultrasonic transducer (FUT) which takes advantages of its curved surface adaptability, high temperature durability, broad band frequency response and high signal-to-noise ratio [18–21]. In this process, a FUT with PZT/PZT composite film is applied to receive the laser induced guided waves propagating on the curved surface at high temperature. The acoustic performance of this sensor has been verified in previous studies [22–24].

A laser-induced ultrasonic wave is one of the most powerful techniques in the field of nondestructive evaluation (NDE) or structure health monitoring (SHM). The ability of contact-free excitation, multi-mode guided wave generation, and rapid inspection of various structures are among the main advantages of laser ultrasound [25–28]. In fact, when a pulsed laser beam is irradiated onto a solid, those waves can be considered a versatile means for the evaluation of the elastic properties of materials. In our previous studies, laser-induced ultrasound was used to measure the dispersions of guided waves and characterize material properties such as the material's hydrogen concentration [29], solid oxide full cells [30] and nickel aluminum coatings [31].

Laser ultrasound imaging (LUI) is a cutting-edge inspection technique, which employs a pulsed laser to scan over the area of interest and visualize the resulting wave propagations. It effectively shortens the analysis time and satisfies the requirements for easy interpretation of ultrasonic propagation without reference data. In previous studies [32–38], a Q-switched pulsed laser and a galvano-motorized mirror were utilized to generate guided waves, and a PZT ultrasonic transducer used as a receiver was fixed on to the specimen during scanning. The wave field visualization application included measurement of the phase and group velocities of Lamb waves, wave propagation on different structures and defect detection, and all these parameters had been successfully demonstrated. When analyzing defect detection by using LUI, the research focused on the interaction of the laser-generated Rayleigh wave on surface-breaking cracks [39,40]. However, any inner defect of a piece of equipment should be taken seriously during any industrial inspection.

The main objective of this work is material characterization and defect detection based on laser-induced ultrasound on a metallic pipe at high temperature. The dispersion curves can represent the material characterization with the change of temperature by using laser ultrasonic technique (LUT). Additionally, defect detection can be performed by using LUI to monitor the dynamic wave propagation behaviors. A PZT/PZT based FUT is applied to be a sensor due to its self-alignment to the curved surface and high temperature durability. The outline of this research is as follows: Section 2 briefly describes the fabrication process of the FUT and its ultrasonic performance testing. The experimental setup and the sketches for the LUT and LUI methods are shown in Section 3. Section 4 illustrates the results from the LUT and LUI experiments. Section 5 presents our/the conclusions of this work.

2. Flexible Ultrasound Transducer

The PZT/PZT transducer is fabricated by utilizing the sol-gel spray method which is outlined in Figure 1. In this process, a submicron fine PZT powder is first dispersed into the sol-gel solution by ball milling. The liquid mixture is sprayed directly onto a 40 mm × 40 mm, 50 μm-thick piece of stainless steel substrate (SS304) by an airbrush to form a layer of coating. The coated layer on the substrate is then dried by a plate heater at 150 °C for 5 min and then baked in a furnace at 650 °C for another 5 min. The coating and thermal processes are repeated until the sprayed film reaches the desired thickness. Later on, the coated film is electrically poled by using a corona discharging technique. For corona poling, positive high voltage power is fed into a sharp, thin needle that is above the film. After polarization through the corona discharging at room temperature, a colloidal silver is sprayed on to the sensor area of the PZT/PZT that was layered by the air brush.

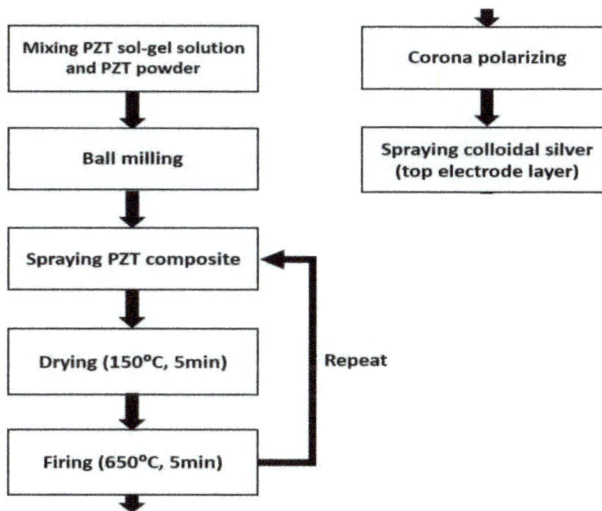

Figure 1. FUT fabrication process.

Figure 2 shows four fabricated FUTs labeled as PzPzss01, PzPzss02, PzPzss03 and PzPzss04 corresponding to sprayed film thicknesses of 80, 72, 146 and 138 μm, respectively.

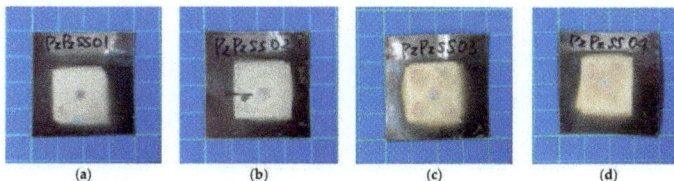

Figure 2. Fabricated FUTs: (**a**) PzPzss01; (**b**) PzPzss02; (**c**) PzPzss03; (**d**) PzPzss04.

The performance of the FUTs was further tested after the fabrication process using the experimental setup as shown in Figure 3. A pulser/receiver (Panametric 5900PR, Olympus, Waltham, MA, USA) is used to drive the FUTs in the pulse/echo mode. The testing specimen is an aluminum plate with a thickness of 8.5 mm. The detected signal was recorded with a digital oscilloscope (WR44Xi, LeCroy, Thief River Falls, MN, USA) and transferred to a computer for data processing.

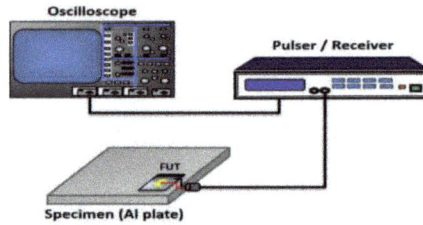

Figure 3. Schematic diagram of ultrasonic performance for FUT.

Figure 4a shows the received pulse/echo signal by using the Pzpzss01 FUT transducer with the experimental setup shown in Figure 3. In this time domain trace, S1 corresponds to the initial pulse, S2, S3, and S4 for the first, second and third reflections from the bottom surface of the plate. The signals in the window are designated as W which are multiple reflections between the top and bottom surfaces of the FUT substrate. Figure 4b shows the frequency spectrum after a Fast Fourier Transform (FFT) from the S2 signal. With a film thicknesses of 80 μm on the Pzpzss01 sensor, the central frequency is 9.18 MHz. Similar signals with good SNR for the other 3 FUTs are shown in Figures 5–7 with central frequencies of 9.18 MHz, 4.89 MHz and 6.45 MHz for the Pzpzss02, Pzpzss03 and Pzpzss04, respectively. As shown in Table 1, the central frequency of the FUT decreases as the sprayed film thickness increases. With this information, we can customize the central frequency by controlling the film thickness of the FUT.

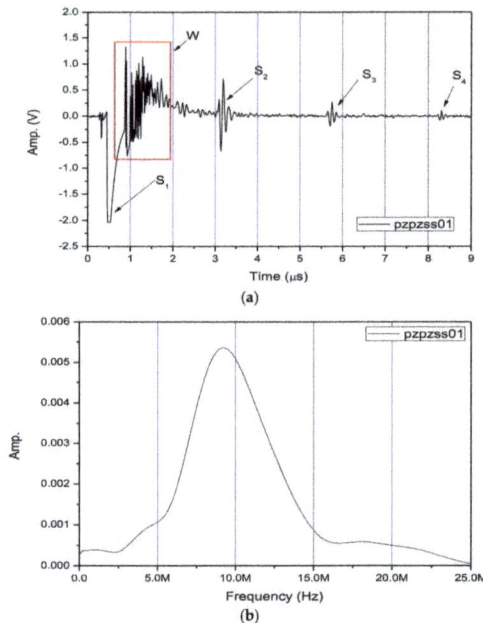

(a)

(b)

Figure 4. Performance of Pzpzss01: (**a**) Time domain signal; (**b**) spectrum.

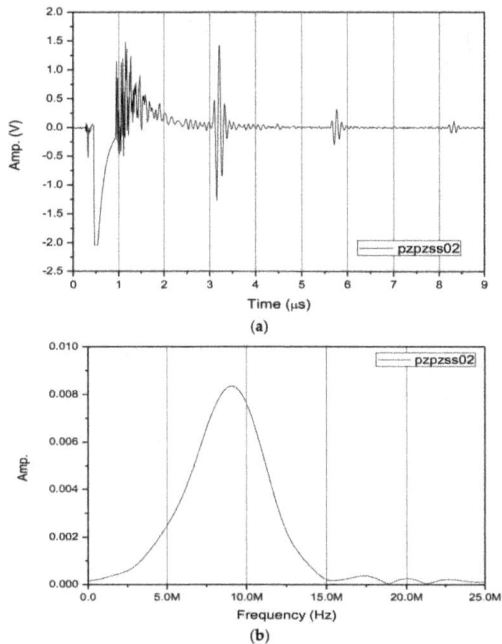

Figure 5. Performance of Pzpzss02: (**a**) Time domain signal; (**b**) spectrum.

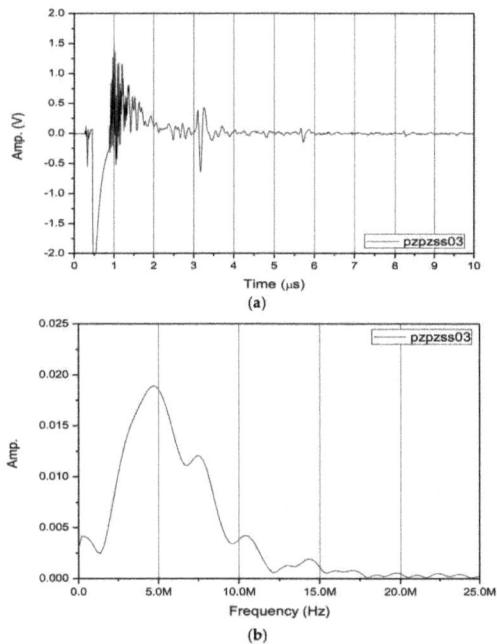

Figure 6. Performance of Pzpzss03: (**a**) Time domain signal; (**b**) spectrum.

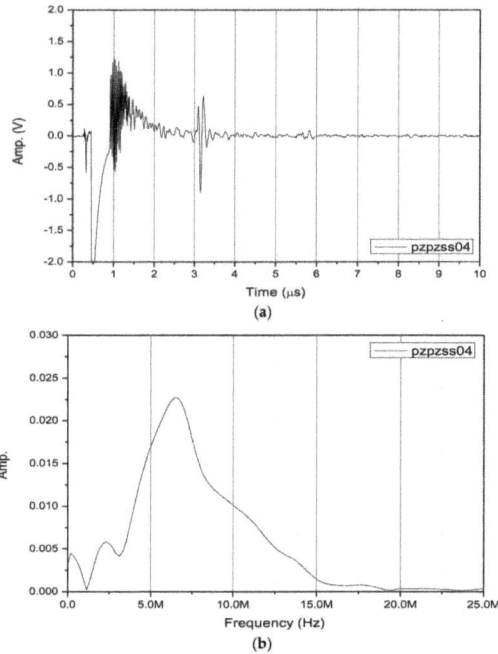

Figure 7. Performance of Pzpzss04: (**a**) Time domain signal; (**b**) spectrum.

Table 1. Film thickness and central frequency of the FUTs.

Label	Film Thickness (μm)	Central Frequency (MHz)
PzPzss01	83	9.18
PzPzss02	72	9.18
PzPzss03	146	4.89
PzPzss04	138	6.45

3. Laser Ultrasonic Technique (LUT) Tests

In the LUT test, dispersion spectra of guided acoustic waves traveling along a metal tube at elevated temperatures are measured by laser-generation and FUT detection. A stainless steel pipe with an outer diameter of 48 mm and a thickness of 2.2 mm is tested with the LUT while Pzpzss02 is used as an ultrasound detector. Figure 8 is a schematic for the experimental configuration of the LUT system at elevated temperatures. A pulsed Nd:YAG laser (Quantel, Brilliant B, Les Ulis, France) with a wavelength of 532 nm, a duration time of 6.6 ns, and an energy output of about 100 mJ is used for the ultrasound generation. The sensor is attached with a small amount of ultrasonic coupler on the surface of the pipe and fixed using a high temperature durable tape made of polytetrafluoroethylene (PTFE). Meanwhile, a thermocouple is also attached on to the interior of the pipe for monitoring the temperature. The stainless steel pipe is heated with a hot plate and covered with an asbestos cover to maintain temperatures of 25 °C, 65 °C, 92 °C, 119 °C, 148 °C and 176 °C. When the temperature reaches a steady state, the scanning stage controlled by a computer drives a mirror to scan a Nd:YAG laser beam along the axial direction of the pipe. A computer with a fast analog to digital converter (ADC) is used for controlling the scanning stage, waveform acquisition, temperature recording and further signal processing. Guided waves are generated with the pulsed laser to propagate throughout the heated stainless steel pipe.

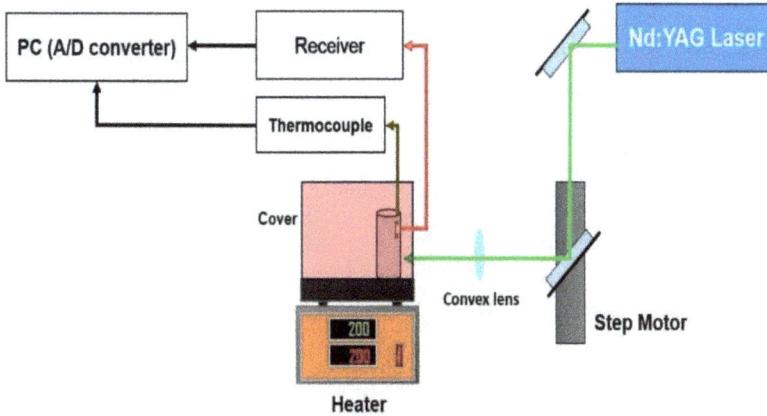

Figure 8. Experimental setup for the LUT with FUT.

By collecting the waveforms at each step, Figure 9 shows the set of B-scan data collected at room temperature with a total scanning distance of 20 mm with 200 steps. The B-scan data is further processed with a two-dimensional fast Fourier transform (2D-FFT) signal processing. During the 2D-FFT, the first FFT is taken with respect to time, and the second FFT with respect to the scanning position. The 2D-FFT transforms the B-scan data into ultrasound amplitude as a function of frequency (f) and wavenumber (k). A peak-detection routine is used to find the trajectories of peak amplitudes in the f-k space. Finally, dispersion curves in the form of ultrasound phase velocity (V) versus frequency are obtained with the aid of the relation $V = 2\pi f/k$.

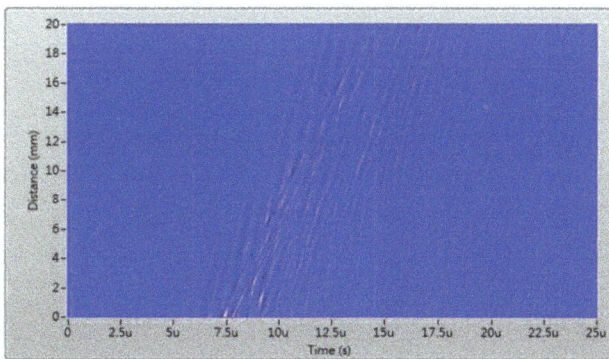

Figure 9. B-scan data for the stainless steel pipe with the LUT/FUT at room temperature.

4. Laser Ultrasonic Imaging (LUI) Tests

An aluminum pipe with an outer diameter of 50 mm and a thickness of 3 mm, which has an interior crack, is used as a specimen for LUI as shown in Figure 10. Figure 11 illustrates the experimental setup. A pulsed Nd:YLF laser (Optowave, Awave, Ronkonkoma, NY, USA) with a wavelength of 1064 nm, a maximum repetition rate of up to 20 kHz, a pulse energy of about 2 mJ, with a 0.7 mm beam diameter is employed to generate guided waves. The Pzpzss02 sensor is attached on to the surface of pipe with a small amount of ultrasonic coupler and fixed using PTFE tape. The scanning mechanism is a two axis galvano-mirror, which is controlled by the computer to scan the Nd:YLF laser beam onto the exterior surface of the pipe. The specimen is heated with an asbestos cover and operated at room temperature,

75 °C and 95 °C. A computer with a fast analog to digital converter is used for signal acquisition, scanning control and image post-processing. With the LUI system, the pulsed laser scans over the area of interest and the detected signals pile up into a data cube with the dimensions of (x, y, t) as shown in Figure 12. The data cube is time-gated at various elapsed times, so a series of pictures are created. With the aid of reciprocal theorem, these pictures represent many instantaneous frames representing wavefronts generated by the FUT that were detected at the scanning area. The measured points are arranged in a 470×150 grid with a pitch of 0.1 mm at room temperature, and a 150×100 grid with a pitch of 0.1 mm at 75 °C and 95 °C. An imaging process to superimpose all the frames is employed in order to specify the position of defect.

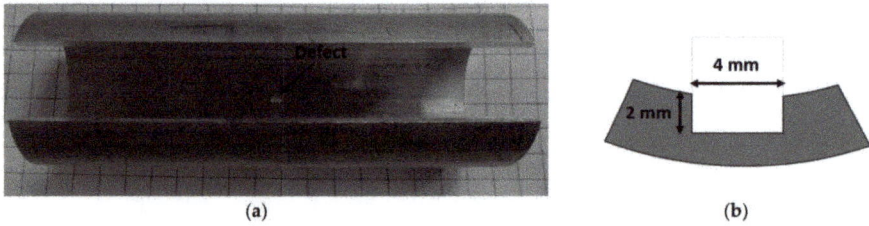

(a)   (b)

Figure 10. (**a**) Aluminum pipe with defect; and (**b**) schematic graph of defect.

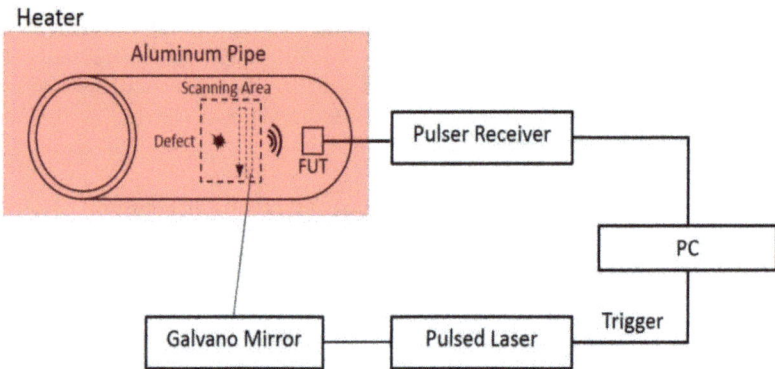

Figure 11. Schematic experiment setup graph of LUI with FUT.

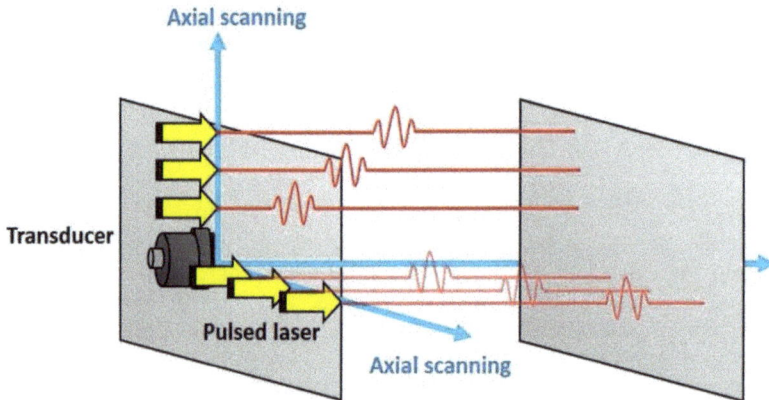

Figure 12. LUI imaging processing.

5. Results and Discussion

5.1. LUT Testing Results

Figure 13 shows waveforms generated with the pulsed Nd:YAG laser in the LUT and detected with the FUT for temperatures from 25 °C to 176 °C. With B-scan and 2D-FFT processing, Figure 14a shows the measured dispersion spectrum for the stainless steel pipe at various temperatures. Dispersion spectra with obvious multi-mode structures from the guided waves propagating through the sample at elevated temperatures are obtained. The dispersion curves shift in a downward trend towards the lower frequencies and lower phase velocities as the temperature increases. With the aid of a zoomed-in graph of the data shown in Figure 14b, the phase velocity of surface wave is noticeably reduced as the temperature increases. The measured surface wave velocity is 2900 m/s at 25 °C and 2780 m/s at 176 °C, corresponding to a reduction of 100 m/s due to the increased temperature. Compared with other ultrasound probes, a FUT mounted on a pipe for can last as long as 3 h which shows that this system has the capability to continuously inspect, monitor and gather data in an elevated temperature environment.

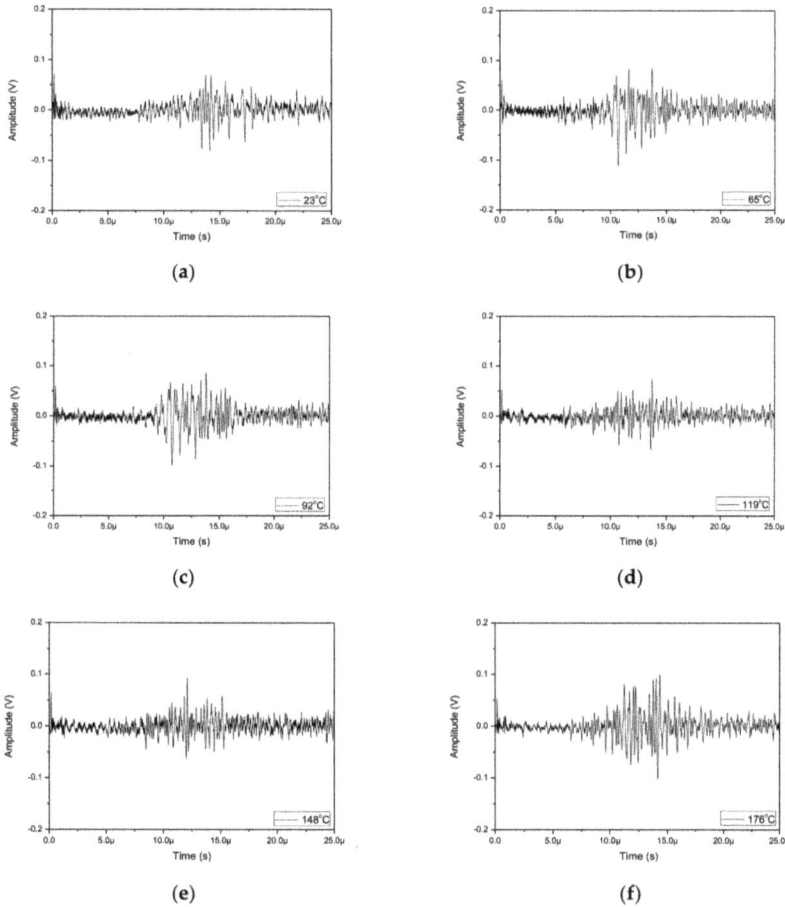

Figure 13. Measured signals in LUT/FUT at (**a**) 25 °C; (**b**) 65 °C; (**c**) 92 °C; (**d**) 119 °C; (**e**) 148 °C; and (**f**) 176 °C.

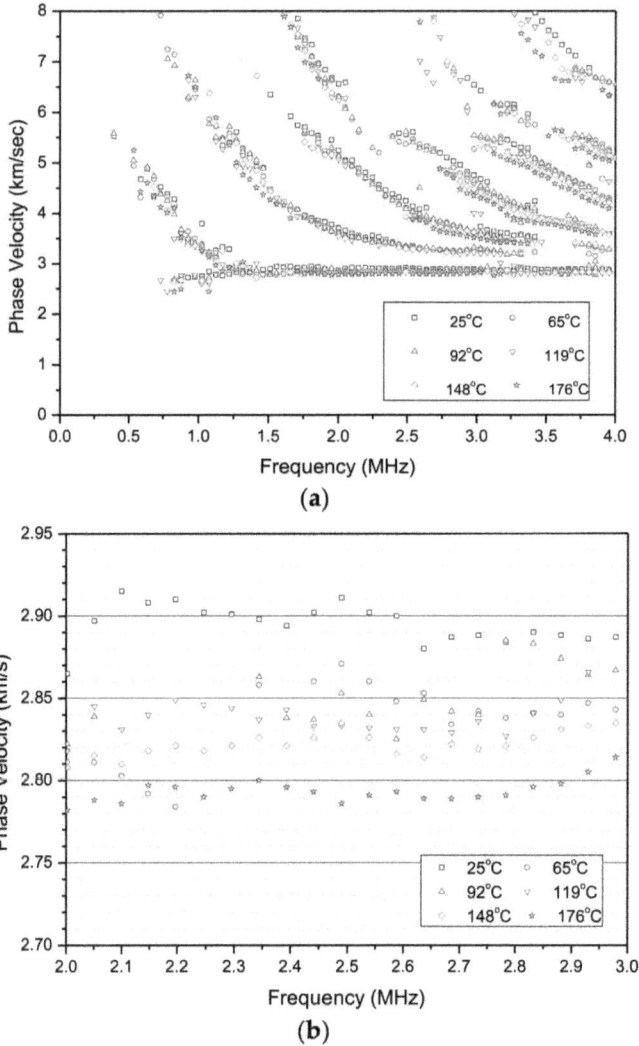

Figure 14. (a) Measured dispersions at various temperatures; and (b) a zoom-in for the measured dispersions.

5.2. LUI Testing Results

Figure 15a shows visualized images for the laser-generated/FUT-detected guided wave propagation along the pipe at room temperature. Here, at least two guided wave modes can be seen propagating along the pipe by observing different elapsed times. One is a faster mode (L(0,2)) and the other is a slower mode (L(0,1)) with a larger amplitude. The ultrasonic waves passed through a hole with a diameter of 4 mm, and are scattered in the image at 9.76 μs. The defects relative position is ensured though the dynamic imaging and our understanding of guided wave propagation behavior. Besides, the defect can be emphasized by further signal processing and can be displayed in a static image. We accumulated the amplitudes of each time domain signal at the same x and y position for every image. Figure 15b displays the processed static image obtained from the dynamic result with a total of 500 frames. By accumulating the energy of each frame, the defect region is able to be more easily seen because the reflected and scattered waves primarily originated from the defect itself.

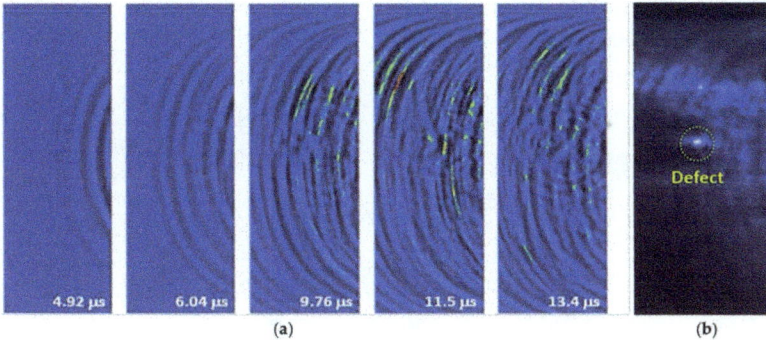

Figure 15. (**a**) Frames of guided waves propagating on the stainless steel pipe with an interior defect; and (**b**) the processed image by accumulating the intensity of each of the frames at room temperature.

Figures 16a and 17a present the wave propagation images which were extracted from the LUI at 75 °C and 95 °C. With the relative small scanning area compared with the room temperature measurements, the two guided wave modes can be observed clearly at different elapsed times. For the 75 °C LUI test, the wave propagations of (L(0,2)) mode and (L(0,1)) mode are visualized at 1.50 μs and 3.58 μs, respectively. Both of them show the reflections and the changes in the wave fronts that resulted from the interior defect. In contrast with dynamic wave propagation behavior, the change in the waveform can be observed only when the slower mode (L(0,1)), passes through the defect at the 95 °C point of the LUI test. To enhance the defects location with respect to the scanning area, the post-processed images are shown in Figures 16b and 17b. The defect comes out clearly by cumulating 500 frames of both processed static images. Some energy appeared on the top of the reconstructed image because the sensor was placed close to the edge of scanning area. There are also a few anomalies that can be seen in the reconstructed image because of the imperfect contact between the BNC cable and the FUT anode during the scanning process.

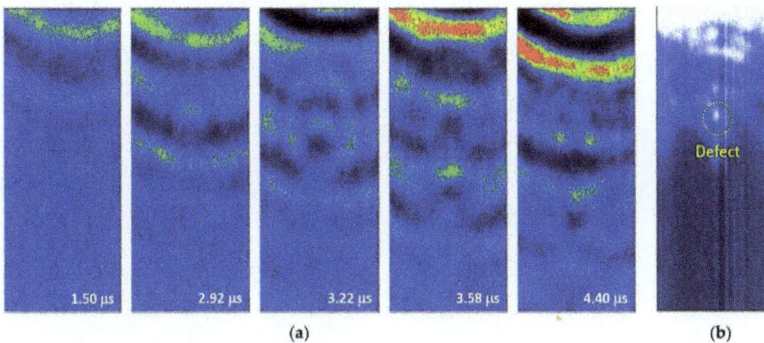

Figure 16. (**a**) Frames of guided waves propagating on the stainless steel pipe with an interior defect and (**b**) the processed image gathered by accumulating the intensity of each of the frames at 75 °C.

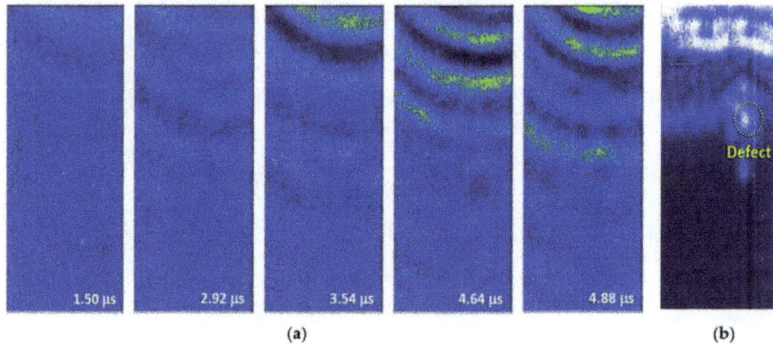

Figure 17. (a) Frames of guided waves propagating on the stainless steel pipe with an interior defect; and (b) the processed image gathered by accumulating the intensity of each of the frames at 95 °C.

6. Conclusions

This paper demonstrates that the integration of FUT's and laser-induced ultrasound applied during material characterization and defect detection for curved surfaced structures at high temperatures has merit. With FUT fabrication, the frequency response can be controlled through the sol-gel spraying process. The PZT/PZT film thickness is inversely proportional to its central frequency. In this study, a 9.18 MHz FUT with a film thickness around 80 μm was utilized to be the ultrasonic receiver in two experiments. For material characterization, the multi-mode dispersion spectrum of a stainless steel pipe can be obtained through signal processing at high temperatures of up to 176 °C. The guided wave modes shifted at a downward trend towards the lower frequencies and lower phase velocities when the temperature was increased. Furthermore, the FUT was able to continuously measure at elevated temperatures for as long as 3 h. For defect detection, although the signals were affected by the thermal noise from the heater and an inadequate connection to the FUT, the dynamic wave propagation behaviors of an aluminum pipe with an interior defect were still visualized through two-dimensional scanning. Although the reflections from the interior defect become weaker when the temperature is raised, the defect can be still highlighted by compiling each frame obtained from LUI. The performance results of LUI with the FUT were outstanding and they included curved surface analysis feasibility as well as the ability to quickly scan specific areas and easy identification of any defects from the guided wave propagation images.

Acknowledgments: This work is supported by Ministry of Science and Technology, Taiwan, under Grand Number MOST-105-221-E-027-012-MY3.

Author Contributions: Tai Chieh Wu and Che Hua Yang conceived and designed the experiments; Tai Chieh Wu performed the experiments. Tai Chieh Wu and Che Hua Yang analyzed the data; Makiko Kobayashi and Masayuki Tanabe contributed materials (PZT powder and sol-gel solution); Tai Chieh Wu wrote the paper.

Conflicts of Interest: The authors declare no conflict of interest.

References

1. Kažys, R.; Voleišis, A.; Voleišienė, B. High temperature ultrasonic transducers: Review. *Ultragarsas Ultrasound* **2008**, *63*, 7–17.

2. Damjanovic, D. Materials for high-temperature piezoelectric transducers. *Curr. Opin. Solid State Mater. Sci.* **1998**, *3*, 469–473. [CrossRef]

3. Zhang, S.; Yu, F. Piezoelectric materials for high temperature sensors. *J. Am. Ceram. Soc.* **2011**, *94*, 3153–3170. [CrossRef]

4. Fothergill, J.R.; Willis, P.; Waywell, S. Development of high temperature ultrasonic transducers for under-sodium viewing applications. *Br. J. Non Destr. Test.* **1987**, *31*, 259–264.

5. Schmarje, N.; Kirk, K.J.; Cochran, S. 1–3 Connectivity lithium niobate composites for high temperature operation. *Ultrasonics* **2007**, *47*, 15–22. [CrossRef] [PubMed]

6. Baba, A.; Searfass, C.T.; Tittmann, B.R. High temperature ultrasonic transducer up to 1000 °C using lithium niobate single crystal. *Appl. Phys. Lett.* **2010**, *97*, 232901.

7. Kirk, K.J.; Scheit, C.W.; Schmarje, N. High temperature acoustic emission tests using lithium niobate piezocomposite transducers. *Insight Non Destr. Test. Cond. Monit.* **2007**, *49*, 142–145. [CrossRef]

8. Amini, M.H.; Sinclair, A.N.; Coyle, T.W. High Temperature Ultrasonic Transducer for Real-time Inspection. *Phys. Procedia* **2015**, *70*, 343–347. [CrossRef]

9. Technologies, I.A. *Developing Permanently Installed System for Ultrasonic Thickness Monitoring Based on HotSense®Technology*; Ionix: Ipswich, UK, 2015.

10. Dhuttia, A.; Tumina, S.A.; Mohimib, A.; Kostana, M.; Gana, T.H.; Balachandrana, W.; Selcuka, C. Development of low frequency high temperature ultrasonic transducers for in-service monitoring of pipework in power plants. *Procedia Eng.* **2016**, *168*, 983–986. [CrossRef]

11. Kobayashi, M.; Ken, C.K.; Bussiere, J.F.; Wu, K.T. High-temperature integrated and flexible ultrasonic transducers for nondestructive testing. *NDT E Int.* **2009**, *42*, 157–161. [CrossRef]

12. Inoue, T.; Kobayashi, M. PbTiO$_3$/Pb(Zr,Ti)O$_3$ sol–gel composite for ultrasonic transducer applications. *Jpn. J. Appl. Phys.* **2014**, *53*, 07KC11. [CrossRef]

13. Kobayashi, M.; Ono, Y.; Jen, C.K.; Chen, C.C. High-temperature piezoelectric film ultrasonic transducers by a sol-gel spray technique and their application to process monitoring of polymer injection molding. *IEEE Sens. J.* **2006**, *6*, 55–62. [CrossRef]

14. Barrow, D.A.; Petroff, T.E.; Sayer, M. Method for Producing Thick Ceramic Films by a Sol Gel Coating Process. U.S. Patent 5,585,136, 17 December 1996.

15. Kobayashi, M.; Jen, C.K.; Ono, Y.; Kruger, S. Lead-free thick piezoelectric films as miniature high temperature ultrasonic transducers. In Proceedings of the 2004 IEEE Ultrasonics Symposium, Montreal, QC, Canada, 23–27 August 2004.

16. Kobayashi, M.; Jen, C.K.; Ono, Y.; Moisan, J.F. Integratable high temperature ultrasonic transducers for NDT of metals and industrial process monitoring. *CINDE J.* **2005**, *26*, 5–10.

17. Kobayashi, M.; Jen, C.K.; Hui, R.; Yick, S.; Wu, K.T. Fabrication and characterization of thick film piezoelectric ultrasonic transducers. In Proceedings of the 2006 IEEE Ultrasonics Symposium, Vancouver, BC, Canada, 2–6 October 2006; Volume 14, pp. 816–819.

18. Kobayashi, M.; Jen, C.K. Piezoelectric thick bismuth titanate/PZT composite film transducers for smart NDE of metals. *Smart Mater. Struct.* **2004**, *13*, 951–956. [CrossRef]

19. Shih, J.L.; Kobayashi, M.; Jen, C.K. Flexible metallic ultrasonic transducers for structural health monitoring of pipes at high temperatures. *IEEE Trans. Ultrason. Ferroelectr. Freq. Control* **2010**, *57*, 2103–2110. [CrossRef] [PubMed]

20. Searfass, C.T.; Pheil, C.; Sinding, K.; Tittmann, B.R.; Baba, A.; Agrawal, D.K. Bismuth titanate fabricated by spray-on deposition and microwave sintering for high-temperature ultrasonic transducers. *IEEE Trans. Ultrason. Ferroelectr. Freq. Control* **2016**, *63*, 139–146. [CrossRef] [PubMed]

21. Eason, T.J.; Bond, L.J.; Lozev, M.G. Ultrasonic Sol-Gel Arrays for Monitoring High-Temperature Corrosion. In Proceedings of the 19th World Conference on Non-Destructive Testing 2016, Munichm, Germany, 13–17 June 2016.

22. Kobayashi, M.; Jen, C.K.; Lévesque, D. Flexible ultrasonic transducers. *IEEE Trans. Ultrason. Ferroelectr. Freq. Control* **2006**, *53*, 1478–1486. [CrossRef] [PubMed]

23. Ouahabi, A.; Thomas, M.; Kobayashi, M.; Jen, C.K. Structural health monitoring of aerospace structures with sol-gel spray sensors. *Key Eng. Mater.* **2007**, *347*, 505–510. [CrossRef]

24. Jen, C.K.; Wu, K.T.; Kobayashi, M.; Blouin, A. NDE using laser generated ultrasound and ultrasonic transducer receivers. In Proceedings of the 2008 IEEE Ultrasonics Symposium, Beijing, China, 2–5 November 2008; pp. 1516–1519.

25. Di Scalea, F.L.; Bcrndl, T.P.; Spicer, J.U.; Djordjevic, B.B. Remote laser generation of narrow-band surface waves through optical fibers. *IEEE Trans. Ultrason. Ferroelec. Freq. Control* **1999**, *46*, 1551–1557. [CrossRef] [PubMed]

26. Aindow, A.M.; Dewhurst, R.J.; Palmer, S.B. Laser-generation of directional surface acoustic wave pulses in metals. *Opt. Commun.* **1982**, *42*, 116–120. [CrossRef]

27. Clorennec, D.; Royer, D.; Walaszek, H. Nondestructive Evaluation of Cylindrical Parts Using Laser Ultasonics. *Ultrasonics* **2002**, *40*, 783–789. [CrossRef]

28. Shi, Y.; Wooh, S.; Orwat, M. Laserultrasonic Generation of Lamb Waves in the Reaction Rorce Range. *Ultrasonics* **2003**, *41*, 623–633. [CrossRef]

29. Liu, I.-H.; Yang, C.-H. A novel procedure employing laser ultrasound technique and simplex algorism for the characterization of mechanical and geometrical properties in Zircaloy tubes with different levels of hydrogen charging. *J. Nucl. Mater.* **2011**, *408*, 96–101. [CrossRef]

30. Yang, C.-H.; Tang, S.-W. Characterization of material properties in solid oxide fuel cells using a laser ultrasound technique. In Proceedings of the Symposium on Ultrasonic Electronics, Kyoto, Japan, 18–20 November 2009; pp. 1207–1210.

31. Yeh, C.-H.; Yang, C.H.; Su, C.Y.; Hsiao, W.T. Laser ultrasound technique for material characterization of thermal sprayed nickel aluminum coatings in elevated temperature environment. *J. Acoust. Soc. Am.* **2012**, *131*, 3476. [CrossRef]

32. Yashiro, S.; Takatsubo, J.; Toyama, N. An NDT technique for composite structure using visualized Lamd-wave propagation. *Compos. Sci. Technol.* **2007**, *67*, 3202–3208. [CrossRef]

33. Lee, J.R.; Takatsubo, J.; Toyama, N.; Kang, D.H. Health monitoring of complex curved structures using an ultrasonic wavefield propagation imaging system. *Meas. Sci. Technol.* **2007**, *18*, 3816–3824. [CrossRef]

34. Lee, J.R.; Jeong, H.; Ciang, C.C.; Yoon, D.J.; Lee, S.S. Application of ultrasonic wave propagation imaging method to automatic damage visualization of nuclear power plant pipeline. *Nucl. Eng. Des.* **2010**, *240*, 3513–3520. [CrossRef]

35. Nishino, H.; Tanaka, T.; Yoshida, K.; Takasudo, J. Simultaneous measurement of the phase and group velocities of Lamb waves in a laser-generation based imaging method. *Ultrasonic* **2011**, *20*, 530–535. [CrossRef] [PubMed]

36. Yang, C.-H.; Liu, I.-H. Optical visualization of acoustic wave propagating along the wedge tip. In Proceedings of the Seventh International Symposium on Precision Engineering Measurements and Instrumentation, Lijiang, China, 7–11 August 2011.

37. Wu, C.-H.; Yang, C.-H. Laser ultrasound technique for ray tracing investigation of Lamb wave tomography. In Proceedings of the Nondestructive Characterization for Composite Materials, Aerospace Engineering, Civil Infrastructure, and Homeland Security 2011, San Diego, CA, USA, 7–11 March 2011.

38. Wu, C.-H.; Tseng, S.-P.; Yang, C.-H. A full-field mechanical property mapping reconstruction algorithm with quantitative laser ultrasound visualization system. In Proceedings of the 2012 IEEE Ultrasonics Symposium, Dresden, Germany, 7–10 October 2012.

39. Zhou, Z.G.; Zhang, K.S.; Zhou, J.H.; Sun, G.G.; Wang, J. Application of laser ultrasonic technique for non-contact detection of structural surface-breaking cracks. *Opt. Laser Technol.* **2015**, *73*, 173–178. [CrossRef]

40. Zeng, W.; Wang, H.T.; Tian, G.Y.; Wang, W. Detection of surface defects for longitudinal acoustic waves by a laser ultrasonic imaging technique. *Opt. Int. J. Light Electron Opt.* **2016**, *127*, 415–419. [CrossRef]

sensors

MDPI

Article

An Improved Scheduling Algorithm for Data Transmission in Ultrasonic Phased Arrays with Multi-Group Ultrasonic Sensors

Wenming Tang [1], Guixiong Liu [1,*], Yuzhong Li [1] and Daji Tan [2]

[1] School of Mechanical & Automotive Engineering, South China University of Technology, Guangzhou 510641, China; tang.wm@mail.scut.edu.cn (W.T.); JDXlyz@hzcollege.com (Y.L.)
[2] Guangzhou Doppler Electronic Technologies Co., Ltd., Guangzhou 510663, China; tandaji@cndoppler.cn
*   Correspondence: megxliu@scut.edu.cn; Tel.: +86-020-8711-0568

Received: 29 September 2017; Accepted: 12 October 2017; Published: 16 October 2017

Abstract: High data transmission efficiency is a key requirement for an ultrasonic phased array with multi-group ultrasonic sensors. Here, a novel FIFOs scheduling algorithm was proposed and the data transmission efficiency with hardware technology was improved. This algorithm includes FIFOs as caches for the ultrasonic scanning data obtained from the sensors with the output data in a bandwidth-sharing way, on the basis of which an optimal length ratio of all the FIFOs is achieved, allowing the reading operations to be switched among all the FIFOs without time slot waiting. Therefore, this algorithm enhances the utilization ratio of the reading bandwidth resources so as to obtain higher efficiency than the traditional scheduling algorithms. The reliability and validity of the algorithm are substantiated after its implementation in the field programmable gate array (FPGA) technology, and the bandwidth utilization ratio and the real-time performance of the ultrasonic phased array are enhanced.

Keywords: ultrasonic phased array; scheduling algorithm; FIFOs; multi-group sensors; FPGA; bandwidth utilization

1. Introduction

The technology of multi-group ultrasonic sensors that consist of lots of piezoelectric elements and various scanning patterns of an ultrasonic phased array (UPA) have recently attracted widespread attention in the non-destructive testing area [1,2]. The UPA produces a series of the ultrasonic waves controlled by the amplitudes and phases of the electrical pulses to excite a series of elements of the sensors. The waves can easily penetrate inside some materials by adjusting their radiation direction to synthesize flexible and rapidly focused scanning ultrasonic beams. The parameters of beams such as angles, focal distances, and focal spot sizes can be readily tuned with suitable software. Therefore, the beams can be used to detect defects that possibly occur at random positions of the materials [3–6].

To increase the focusing ability, a UPA instrument is often equipped with multiple ultrasonic sensors to collect the ultrasonic echo data from different directions. Each sensor can work in one or more groups so that a variety of scanning modes are generated [7–10], which can be called as a multi-group scanning, and each group scanning includes many focused beams. Hence, the number of the sensors and the scanning groups are two important factors to determine detection accuracy [11,12], such as size, location, and orientation of defects. For example, Song et al. verified that a large-aperture hemispherical phased array can restore a sharp focus and maximize acoustic energy delivery at target tissue [11]. Regardless of the orientation of individual focused beams, the multiple focused beams can change their focal depths and sweeping angles through the phase interference. As a consequence, it is possible to precisely detect the position and the size of defects by means of increasing the number of the sensors, the scanning groups, and the focused beams. However, this strategy will in turn significantly increase the amount of scanning data in the process of the defect detection, which makes these data difficult to be transmitted to a peripheral through a single (or small quantity) high-speed serial bus, and subsequently produces an ultrasound image.

Each focused beam often brings different sampling rates and sizes of data stream. During the transmission process, different data streams compete against each other to gain access to the unique high-speed serial bus. An excellent transmission scheduling algorithm should allow all the data streams to be transferred to a peripheral without any blocking in a serialized way. Otherwise, the data streams would be blocked or severely delayed. Therefore, it is very desirable to design an effective algorithm to serially transmit a great amount of the data streams. Several well-known scheduling algorithms have been proposed, such as Time Division Multiple Access (TDMA) [13] and Round Robin (RR) [14]. The verification, analysis, and comparison of the two algorithms were presented in literature [15], which proves that the TDMA strategy based on the fixed allocation of a time slot to each master process may lead to important latencies as a time slot, and the RR protocol allows any unused slots to be reallocated to a master process to provide higher bandwidth. Unfortunately, the process of the reallocation will make the time slice resources more fragmented, and increase the complexity of the scheduling algorithm. Multiple examples of implementation for the scheduling algorithm are available in the open literatures [16–23]. Srinivasan et al. designed a self-configuring scheduling protocol for ultrasonic sensor systems by using an algorithm of the timeslot allocation, which simplified the deployment of the present detection system [16]. Long et al. proposed a time-division-multiple-access-based energy consumption balancing algorithm for the general k-hop wireless sensor networks, where one data packet is collected in one cycle, and the results demonstrated the effectiveness of the algorithm in terms of energy efficiency and time slot scheduling [19]. However, although these strategies can effectively improve the efficiency of the data transmission, they increase the complexities of both hardware and software, and their application scopes are limited, which makes such strategies not suitable for UPA of the multi-group sensor scanning system because of limited hardware and software resources and high real-time request.

FPGA, which is short for the term field programmable logic gate array, has the characteristics of static system repeatable programming and dynamic system reconfiguration, so that hardware can be modified programmatically, and FPGA also is a special kind of ASIC with the advantages of parallel processing, high speed and flexibility. In this paper, we used a series of FIFOs as high-speed caches and cache times as weights to propose a novel FIFOs and bandwidth-sharing scheduling (MFBSS) algorithm of the data transmission, where the lengths of the FIFOs are achieved by a series of multivariate equations. Actually, the algorithm shows many advantages such as real-time and high efficiency when it is implemented by FPGA technology. As far as the UPA system of the array sensors is concerned, we designed a data stream transmission scheduling mode based on the MFBSS algorithm, with which reading operations among all the FIFOs shares a fast reading bus without time slot waiting when the reading bus switches between any two FIFOs. Hence, such algorithm gives the maximum bandwidth utilization ratio and improves the real-time performance of the UPA instruments with minimal consumption of time and space resources.

In Section 2 of the paper, we will describe the data transmission of ultrasonic scanning for UPA [24–26]. In Section 3, we will study scheduling mechanism of the MFBSS algorithm for the data transmission. Section 4 will describe the results of implementation for the scheduling algorithm by FPGA technology. Finally, Section 5 will summarize the research to derive the conclusion.

2. Multi-Group Sensor Scanning Ultrasonic Data Transmission

Figure 1 shows the UPA data transmission framework of the bandwidth-sharing with multiple scanning patterns [7–10]. In order to realize the optimal sampling of the UPA's echoes, different frequency echoes should be digitized with different sampling frequencies [27–29]. A sensor with a frequency of f_p Hz produces ultrasonic echoes with the same frequency after excitation, and thus the sampling frequency is $f_s = K \times f_p$ Hz (K is a scaling factor, and $K \geq 2$). Hence, N-group sensors can form N-group scanning patterns, generating N sampling frequencies ($f_{s0} \sim f_{sN-1}$, where 0 and $N-1$ represent the numbers of sampling) and forming N focusing beams with specific speeds and sizes.

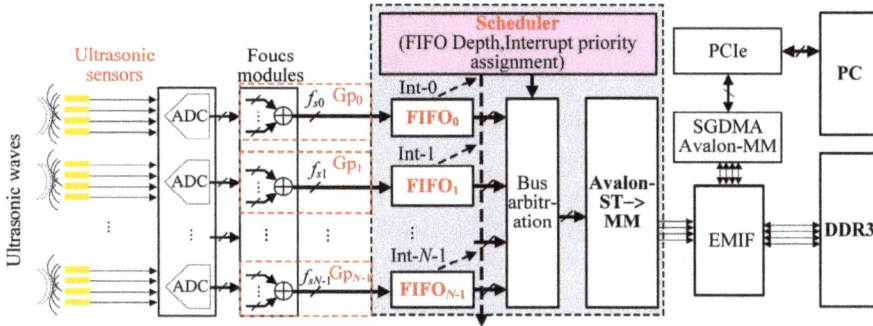

Figure 1. The diagram of the ultrasonic data transmission for the multi-sensor scanning.

As shown in Figure 1, the data of various scanning groups such as Gp_0, Gp_1, ... , and Gp_{N-1} produced from the ultrasonic sensors are written into $FIFO_0$, $FIFO_1$, ... , $FIFO_{N-1}$, respectively, which are cached by a DDR3 through the Avalon bus in the bandwidth-sharing way [30]. Then, the data from the DDR3 are transmitted to the host computer through the PCIe bus [31,32]. The entire data transmission process is controlled by a bandwidth scheduler, which is composed of a controller with all the FIFOs' lengths and a reading arbiter, and usually runs the following scheduling algorithms such as First Come First Serve (FCFS), TDMA and Equal Time Slice Polling Scheduling (ETSPS) based on the principle of the RR scheduling which will be mentioned in Section 4, and so on. This paper will adopt the MFBSS algorithm to realize reading operations from every FIFO without time slot waiting through adjusting the lengths of FIFOs, timings of the reading and writing, and priority of the interrupts. Therefore, this algorithm can not only ensure the data transmission synchronization but also maximize the bandwidth utilization in all groups, which is readily implemented by FPGA technology with parallel processing.

3. Data Transmission Scheduling Mechanism of MFBSS Algorithm

3.1. The principle of the Maximal Bandwidth Utilization

To evaluate the utilization ratio of the data transmission bandwidth of the N-group scanning in the multi-input and single-output interfaces of the UPA system, the following requirements are satisfied:

- Data transmission models $Gp(n)$, $n = 0, 1, \ldots, N-1$ are independent from each other and have identical distributions for every group.
- The sum of the data bandwidth $[\sum_{n=0}^{N-1} B_{v-Gp}(n)]$ of all the groups and the sum of the memory bandwidth ($\sum B_{v-RAM}$) and the sum of the transmission bandwidth ($\sum B_{v-Trans}$) of the peripheral need to satisfy the following inequality:

$$\sum_{n=0}^{N-1} B_{v-Gp}(n) \leq \min(\sum B_{v-RAM}, \sum B_{v-Trans}) \tag{1}$$

The defined parameters of the N-group scanning and the N FIFOs caches are listed in Table 1. The writing bandwidth and the reading bandwidth of the nth $FIFO_n$ are $V_W(n)$ [$V_W(n) = f_{sn} \times \Delta B$] and V_R

bit/s, respectively. The sum of the writing bandwidth of all the FIFOs [$\sum_{n=0}^{N-1} V_W(n)$] should equal to the sum of the transmission bandwidth of the N-group scanning data [$\sum_{n=0}^{N-1} B_{v-Gp}(n)$], i.e., $\sum_{n=0}^{N-1} V_W(n)$ = $\sum_{n=0}^{N-1} B_{v-Gp}(n)$. Likewise, the sum of the reading bandwidth (V_R) of all the FIFOs should equal to the sum of the transmission bandwidth of the DDR3 bandwidth ($\sum B_{v-RAM}$), i.e., $V_R = \sum B_{v-RAM}$. When the Equation (1) becomes an equality, the maximum bandwidth utilization ratio is achieved, i.e., the single-output bandwidth equals to the sum of the multi-input bandwidths from the FIFOs. Consequently, the mathematical principle of the maximal bandwidth utilization ratio can be written as Equation (2).

Table 1. The parameters of the N groups and the N FIFOs caches.

Group Number	Sampling Rate (Hz)	Bit Width	Cache	Length of FIFO	Input Width of FIFO (bit)	Writing Bandwidth (bit/s)	Output Width of FIFO (bit)	Reading Bandwidth (bit/s)
0	f_{s0}	ΔB	FIFO$_0$	$L(0)$	ΔB_W	$V_W(0)$	ΔB_R	V_R
1	f_{s1}	ΔB	FIFO$_1$	$L(1)$	ΔB_W	$V_W(1)$	ΔB_R	V_R
2	f_{s2}	ΔB	FIFO$_2$	$L(2)$	ΔB_W	$V_W(2)$	ΔB_R	V_R
...	...	...	...	...	...	...	...	...
$N-1$	f_{sN-1}	ΔB	FIFO$_{N-1}$	$L(N-1)$	ΔB_W	$V_W(N-1)$	ΔB_R	V_R

$$\begin{cases} \sum_{n=0}^{N-1} B_{v-Gp}(n) \leq \sum B_{v-RAM} \\ \sum_{n=0}^{N-1} B_{v-Gp}(n) = \sum_{n=0}^{N-1} V_W(n) \\ \sum B_{v-RAM} = V_R \end{cases} \Rightarrow V_R = \sum_{n=0}^{N-1} V_W(n) \tag{2}$$

3.2. Realization of the Maximal Bandwidth Utilization Ratio

According to Equation (2), the mathematical model of the N FIFOs' length functions of $L(i)$, $i = 0$, $1, \ldots, N-1$, [$L(0) \leq L(1) \leq \ldots \leq L(N-1)$, FIFO$_0$, FIFO$_1$, $\ldots$, FIFO$_{N-1}$] can be described as follows:

- Assuming that at the moment T_i^j, when the FIFO$_i$ is read until empty, the reading operation of the FIFO$_i$ will be disabled.
- At the next T_i^{j+1}, when the FIFO$_i$ is full and the amount of the data is $L(i)$ ($i = 0, 1, \ldots, N-1$), the reading operation of the FIFO$_i$ will be enabled.

When the FIFO$_i$ transfers from empty to full (where the consumed time is $\Delta T_i = T_i^{j+1} - T_i^j = \frac{L(i)}{V_W(i)}$ and a reading interrupt is produced), the FIFO$_i$ will gain access to the reading of the Avalon bus. During this process, the other FIFOs with the number of $0, 1, \ldots, i+1, i+2, \ldots, N-1$ have also transferred from full to empty with the consumed time of $\Delta T'_i = \sum_{k=0, k\neq i}^{N-1} \frac{L(k)}{V_R - V_W(k)}$. The time slot transition diagram of the N FIFOs reading operations is shown in Figure 2. Because $\Delta T_i = \Delta T'_i$, i.e., $\Delta T'_i - \Delta T_i = 0$, $i = 0, 1, \ldots, N-1$, the mathematical equations of the N FIFOs' length functions of $L(i)$, $i = 0, 1, \ldots, N-1$ can be easily described in Equation (3).

$$\begin{cases} -\dfrac{L(0)}{V_W(0)} + \dfrac{L(1)}{V_R - V_W(1)} + \dfrac{L(2)}{V_R - V_W(2)} + \cdots + \dfrac{L(N-1)}{V_R - V_W(N-1)} = 0 \\[2mm] \dfrac{L(0)}{V_R - V_W(0)} - \dfrac{L(1)}{V_W(1)} + \dfrac{L(2)}{V_R - V_W(2)} + \cdots + \dfrac{L(N-1)}{V_R - V_W(N-1)} = 0 \\[1mm] \vdots \qquad \vdots \qquad \vdots \qquad \vdots \\[1mm] \dfrac{L(0)}{V_R - V_W(0)} + \dfrac{L(1)}{V_R - V_W(1)} + \cdots + \dfrac{L(N-2)}{V_R - V_W(N-2)} - \dfrac{L(N-1)}{V_W(N-1)} = 0 \end{cases} \tag{3}$$

where $L(i) \neq 0$, $i = 0, 1, \ldots, N - 1$ in Equation (3). A series of new variables are given in Equation (4) for the simplification of Equation (3).

$$\begin{cases} K_0' = \dfrac{1}{V_W(0)}, \quad K_1' = \dfrac{1}{V_W(1)}, \cdots, K_{N-1}' = \dfrac{1}{V_W(N-1)} \\[2mm] K_0 = \dfrac{1}{V_R - V_W(0)}, K_1 = \dfrac{1}{V_R - V_W(1)}, \cdots, K_{N-1} = \dfrac{1}{V_R - V_W(N-1)} \\[2mm] V_R = \sum\limits_{i=0}^{N-1} V_W(i) \end{cases} \tag{4}$$

Equation (3) is transformed into a matrix of Equation (5):

$$\begin{pmatrix} -K_0' & K_1 & \cdots & K_{N-1} \\ K_0 & -K_1' & \cdots & K_{N-1} \\ \vdots & \vdots & \vdots & \vdots \\ K_0 & K_1 & \cdots & -K_{N-1}' \end{pmatrix} \begin{pmatrix} L(0) \\ L(1) \\ \vdots \\ L(N-1) \end{pmatrix} = \begin{pmatrix} 0 \\ 0 \\ \vdots \\ 0 \end{pmatrix}$$

$$A = \begin{pmatrix} -K_0' & K_1 & \cdots & K_{N-1} \\ K_0 & -K_1' & \cdots & K_{N-1} \\ \vdots & \vdots & \vdots & \vdots \\ K_0 & K_1 & \cdots & -K_{N-1}' \end{pmatrix}, \vec{L} = \begin{pmatrix} L(0) \\ L(1) \\ \vdots \\ L(N-1) \end{pmatrix}, \vec{0} = \begin{pmatrix} 0 \\ 0 \\ \vdots \\ 0 \end{pmatrix}$$

$$A \cdot \vec{L} = \vec{0} \tag{5}$$

The matrix A is achieved by elementary row transformation, and then the triangular array is applied:

$$A \sim \begin{pmatrix} -K_0' & K_1 & \cdots & \cdots & K_{n-1} \\ K_0 + K_0' & -(K_1 + K_1') & 0 & \cdots & 0 \\ 0 & \vdots & \vdots & \vdots & \vdots \\ \vdots & \vdots & K_{N-3} + K_{N-3}' & -(K_1 + K_{N-2}') & 0 \\ 0 & \cdots & 0 & K_{N-2} + K_{N-2}' & -(K_{N-1} + K_{N-1}') \end{pmatrix} \sim \begin{pmatrix} f_K(x_0) & K_1 & \cdots & K_{N-1} \\ 0 & f_K(x_1) & \cdots & K_{N-1} \\ \vdots & \vdots & \vdots & \vdots \\ 0 & \cdots & 0 & f_K(x_{N-1}) \end{pmatrix} \tag{6}$$

$f_K(x_{i+1}) = \dfrac{K_{i+1} + K_{i+1}'}{K_i + K_i'} \cdot f_K(x_i) + K_{i+1}$, $i = 0, 1, \ldots, N - 2$, $f_K(x_0) = -K_0'$, and $f_K(x_{i+1})$ can be done by using the following recursion:

$$\begin{aligned} f_K(x_{i+1}) &= \frac{K_{i+1} + K_{i+1}'}{K_i + K_i'} \cdot \left(\frac{K_i + K_i'}{K_{i-1} + K_{i-1}'} \cdot \left(\cdots \left(\frac{K_1 + K_1'}{K_0 + K_0'} \cdot f_K(x_0) + K_1 \right) + \cdots \right) + K_i \right) + K_{i+1} \\ &= \frac{K_{i+1} + K_{i+1}'}{K_i + K_i'} \cdot \frac{K_i + K_i'}{K_{i-1} + K_{i-1}'} \cdots \frac{K_1 + K_1'}{K_0 + K_0'} \cdot f_K(x_0) + \frac{K_{i+1} + K_{i+1}'}{K_i + K_i'} \cdot \frac{K_i + K_i'}{K_{i-1} + K_{i-1}'} \cdots \frac{K_2 + K_2'}{K_1 + K_1'} \cdot K_1 \\ &\quad + \frac{K_{i+1} + K_{i+1}'}{K_i + K_i'} \cdot \frac{K_i + K_i'}{K_{i-1} + K_{i-1}'} \cdots \frac{K_3 + K_3'}{K_2 + K_2'} \cdot K_2 + \cdots + \frac{K_{i+1} + K_{i+1}'}{K_i + K_i'} \cdot K_i + K_{i+1} \\ &= \frac{K_{i+1} + K_{i+1}'}{K_0 + K_0'} \cdot f_K(x_0) + \frac{K_{i+1} + K_{i+1}'}{K_1 + K_1'} \cdot K_1 + \frac{K_{i+1} + K_{i+1}'}{K_2 + K_2'} \cdot K_2 + \cdots + \frac{K_{i+1} + K_{i+1}'}{K_{i+1} + K_{i+1}'} \cdot K_{i+1} \\ &= (K_{i+1} + K_{i+1}') \cdot \left(\frac{K_{i+1}}{K_{i+1} + K_{i+1}'} + \frac{K_i}{K_i + K_i'} + \cdots + \frac{K_2}{K_2 + K_2'} + \frac{K_1}{K_1 + K_1'} - \frac{K_0'}{K_0 + K_0'} \right) \\ &= (K_{i+1} + K_{i+1}') \cdot \left[\left(\sum_{j=1}^{i+1} \frac{K_j}{K_j + K_j'} \right) - \frac{K_0'}{K_0 + K_0'} \right] \end{aligned}$$

According to Equation (4), $f_K(x_{i+1})$ can be described as Equation (7).

$$f_K(x_{i+1}) = \left(\frac{1}{V_R - V_W(i+1)} + \frac{1}{V_W(i+1)} \right) \cdot \left(\sum_{j=1}^{i+1} \frac{\frac{1}{V_R - V_W(j)}}{\frac{1}{V_R - V_W(j)} + \frac{1}{V_W(j)}} - \frac{\frac{1}{V_W(0)}}{\frac{1}{V_R - V_W(0)} + \frac{1}{V_W(0)}} \right)$$

$$= \frac{1}{(V_R - V_W(i+1)) \cdot V_W(i+1)} \cdot \left(\sum_{j=0}^{i+1} V_W(j) - V_R \right) \tag{7}$$

For the N-group scanning of the UPA system, when $i = N$, according to the Equation (2), $V_R = \sum_{n=0}^{N-1} V_W(n)$, and $f_K(x_{N-1}) =$

$$\frac{1}{(V_R - V_W(N-1)) \cdot V_W(N-1)} \cdot \frac{1}{V_R} \cdot \left(\sum_{j=0}^{N-1} V_W(j) - V_R \right) \Bigg|_{V_R = \sum_{j=0}^{N-1} V_W(j)} = 0. \quad \text{The matrix } A \text{ can}$$

be transformed to A' through the primary row transformation:

$$A = \begin{pmatrix} -K'_0 & K_1 & \cdots & K_{N-1} \\ K_0 & -K'_1 & \cdots & K_{N-1} \\ \vdots & \vdots & \vdots & \vdots \\ K_0 & K_1 & \cdots & -K'_{N-1} \end{pmatrix} \sim \begin{pmatrix} f_K(x_0) & K_1 - f_K(x_1) & 0 & 0 & \cdots & 0 \\ 0 & f_K(x_1) & K_2 - f_K(x_2) & 0 & \cdots & 0 \\ \vdots & \vdots & \vdots & \vdots & \vdots & 0 \\ 0 & \cdots & \cdots & 0 & f_K(x_{N-2}) & K_2 - f_K(x_{N-1}) \\ 0 & \cdots & \cdots & \cdots & 0 & f_K(x_{N-1}) \end{pmatrix}$$

$$\underline{f_K(x_{N-1}) = 0} \begin{pmatrix} f_K(x_0) & K_1 - f_K(x_1) & 0 & 0 & \cdots & 0 \\ 0 & f_K(x_1) & K_2 - f_K(x_2) & 0 & \cdots & 0 \\ \vdots & \vdots & \vdots & \vdots & \vdots & 0 \\ 0 & \cdots & \cdots & 0 & f_K(x_{N-2}) & K_2 \\ 0 & \cdots & \cdots & \cdots & 0 & 0 \end{pmatrix} = A'$$

Because the rank $R(A)$ of the matrix A and the rank $R(A')$ of the matrix A' have the following relation $R(A) = R(A') < N$, Equation (5) has an infinite number of the solutions, and because $A \cdot \vec{L} = \vec{0} \Leftrightarrow A' \cdot \vec{L} = \vec{0}$, and the solutions can be expressed as follows:

$f_K(x_i) \times L(i) + (K_{i+1} - f_K(x_{i+1})) \times L(i+1) = 0$, $(i = 0, 1, \ldots, N-2)$, and $L(i) = \dfrac{f_K(x_{i+1}) - K_{i+1}}{f_K(x_i)} \cdot L(i+1)$, $(i = 0, 1, \ldots, N-2)$, and $L(i)$ can be further deduced forward:

$$L(i) = \frac{f_K(x_{i+1}) - K_{i+1}}{f_K(x_i)} \cdot \frac{f_K(x_{i+2}) - K_{i+2}}{f_K(x_{i+1})} \cdots \cdots \frac{f_K(x_{N-1}) - K_{N-1}}{f_K(x_{N-2})} \cdot L(N-1)$$

$$= \prod_{j=i}^{N-2} \frac{f_K(x_{j+1}) - K_{j+1}}{f_K(x_j)} \cdot L(N-1) \tag{8}$$

Substituting the expression of $f_K(x_{i+1})$ from Equation (7) into Equation (8). The values of $L(i)$, $i = 0, 1, \ldots, N-1$ are obtained, as shown in Equation (9):

$$\begin{cases} L(0) = \dfrac{(V_R - V_W(0)) \cdot V_W(0)}{(V_R - V_W(N-1)) \cdot V_W(N-1)} \cdot L(N-1) \\ \vdots \\ L(i) = \dfrac{(V_R - V_W(i)) \cdot V_W(i)}{(V_R - V_W(N-1)) \cdot V_W(N-1)} \cdot L(N-1) \\ \vdots \\ L(N-2) = \dfrac{(V_R - V_W(N-2)) \cdot V_W(N-2)}{(V_R - V_W(N-1)) \cdot V_W(N-1)} \cdot L(N-1) \\ L(N-1) = L(N-1) \end{cases} \tag{9}$$

when Equation (9) is multiplied by a term of $\dfrac{(V_R - V_W(N-1)) \cdot V_W(N-1)}{L(N-1)}$, a set of fundamental solutions $\vec{\xi}$ to the equations of $A \cdot \vec{L} = \vec{0}$ will be obtained:

$$\vec{\xi} = ((V_R - V_W(0)) \cdot V_W(0),\ (V_R - V_W(1)) \cdot V_W(1),\ \cdots,\ (V_R - V_W(N-1)) \cdot V_W(N-1))^{\mathrm{T}}.$$

Therefore, the solutions to the equations of $A \cdot \vec{L} = \vec{0}$ can be expressed as $\vec{L} = \alpha \cdot \vec{\xi}$ $(\alpha \in \mathbf{R}^+)$. The length function of $L(i)$, $i = 0, 1, \ldots, N-1$ of the FIFOs has a proportional relation, as showed in Equation (10).

$$
\begin{aligned}
L(0) : L(1) : \cdots : L(N-1) =\ & (V_R - V_W(0)) \cdot V_W(0) : (V_R - V_W(1)) \cdot V_W(1) : \cdots \\
& : (V_R - V_W(N-1)) \cdot V_W(N-1)
\end{aligned}
\tag{10}
$$

Equation (10) can be used to describe the most critical conclusion to realize the MFBSS algorithm, which shares the transmission bandwidth for the N-group scanning of the UPA system. Therefore, according to the ratios of the FIFOs' lengths, i.e., the cache time of each FIFO, the reading operation can be switched among each FIFO without time slot waiting, thus maximizing the bandwidth utilization ratio.

When the algorithm is implemented by an FPGA, in order to make the consumed resources of the FIFOs minimal, the ratio of $L(0):L(1):\ldots:L(N-1)$ can often be simplified to a series of the suitable integer ratios. In the system of the N-group scanning and the N FIFOs caches, if the sampling rate f_{sn} ($n = 0, 1, \ldots, N-1$, and unit is 100 MHz) of the N groups linearly increases, $V_R = \sum\limits_{n=0}^{N-1} f_{sn} \cdot \Delta B$, and $\Delta B = \Delta B_W = \Delta B_R$. The ratios of $L(0):L(1):\ldots:L(N-1)$ of the FIFOs' lengths are calculated from Equation (10), and the results are listed in Table 2.

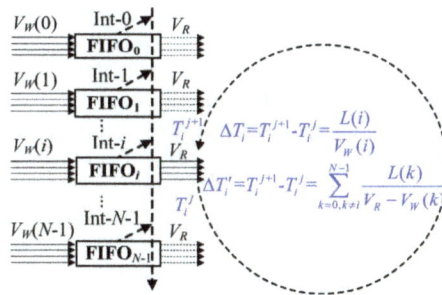

Figure 2. The time slot transition diagram of the N FIFOs reading operations.

Table 2. The N-group scanning and the N-FIFO caches depth ratios.

N	f_{s0}	f_{s1}	f_{s2}	f_{s3}	...	f_{sN-1}	$L(0):L(1):...:L(N-1)$
2	1	2	×	×	×	×	1:1
3	1	2	3	×	×	×	5:8:9
4	1	2	3	4	×	×	9:16:21:24
...	...	...	...	...	...	×	...
N	1	2	3	4	...	$N-1$	$(V_R - f_{s0}) \times f_{s0}:(V_R - f_{s1}) \times f_{s1}:...:(V_R - f_{sN-1}) \times f_{sN-1}$

Figure 3 shows the time slot switching flow chart with the sharing reading bus of the N-group scanning and the N-FIFO caches ($N = 3$ or 4). The horizontal axis represents the time (unit: s).

In the initialization phase, the $FIFO_{N-1}$ caches the maximum sampling rate beam, which is filled with the length $L(N-1)$ data. Meanwhile, the other caches $FIFO_{N-2} \sim FIFO_0$ are filled with the lengths $[L(i) - \left(\sum_{n=i+1}^{N-1} K_n \cdot L(n) \right) / K'_i](i = N-2, N-3, \dots, 1, 0)$, respectively. The working principle is described as follows:

When an FIFO is full, it will be immediately read until empty (the symbol R represents the reading state of the FIFO), and subsequently switches to the next FIFO without any time slot in the process of the data transmission. Likewise, when the next FIFO is just written fully, it will be read immediately. Therefore, the whole process is carried out in cycles without any delay, maximizing the utilization ratio of the data transmission bandwidth.

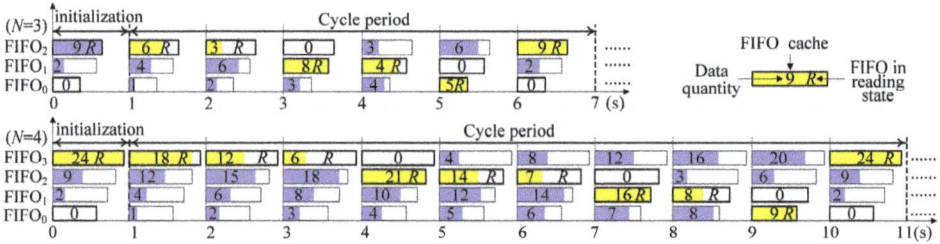

Figure 3. No time-gap switching flow chart of the N-group scanning and the N-FIFO caches shared.

4. Implementation and Performance Evaluation of the Scheduling Algorithm

The scheduling algorithm is realized by using a UPA instrument (PA2000 model), which was made by Guangzhou Doppler Electronic Technologies Co., Ltd. (Guangzhou, China), and a Cyclone V GT FPGA Development Board made by Intel Corporation (Santa Clara, CA., USA) as the PCIe communication module with the PC. The UPA data are transmitted to the PC through the PCIe interface, and the multi-group scanning images are processed.

The UPA system with a work clock frequency (f_s) of 100 MHz is mounted with four sensors with four different frequencies (f_s) of 2, 2.5, 5, or 10 MHz, and thus the system can implement 4-group scanning patterns. The echoes of all the groups are up-sampled ($f_s = 10 \times f_p$) by using digital signal processing technology, and thus the actual sampling frequencies of $f_{s0} \sim f_{s3}$ become 20, 25, 50, or 100 MHz. The bit-width (ΔB) of the echo data is 8 bits, and both widths of the input (ΔB_W) and the output (ΔB_R) ports of the FIFOs are 64 bits. Table 3 lists the parameters of the writing frequency $[V_{Wf}(n)]$ and the reading frequency (V_{Rf}) of the $FIFO_n$ caches. Obviously, $V_W(n)$ equals to $V_{Wf}(n) \times \Delta B_W$, and V_R equals to $V_{Rf} \times \Delta B_R$ for this case, hence, the scheduling algorithm can be used to allow the 4-FIFO caches to realize sharing transmission bandwidth. The capacities of the FIFOs are $L(n) \times \Delta B_W$, and the length ratios of the FIFO caches can be calculated from Equation (10), i.e., $L(0):L(1):L(2):L(3)$ = 14:17:29:38. As listed in Table 3, the value of V_{Rf} is calculated to be 24.375 MHz, but it is relatively easier to implement the value of $V_{Rf} = 25.0$ MHz ($V_{Rf} = f_s/4 = 25.0$ MHz $\approx V'_{Rf}$) by the FPGA than the value of $V_{Rf} = 24.375$ MHz, and thus we design the value of $V_{Rf} = 25.0$ MHz for the experiment.

Table 3. The parameters of the 4 groups scanning and the 4 FIFO caches.

f_p (MHz)	f_{sn} (MHz)	$V_{Wf}(n) = f_{sn} \times \Delta B/\Delta BW$ (MHz)	$V_{Rf} = \Sigma V_{Wf}(n)$ (MHz)	$L(n) \times \Delta B_W$ (bit)
2	20	2.5	24.375	14 × 64
2.5	25	3.125	24.375	17 × 64
5	50	6.25	24.375	29 × 64
10	100	12.5	24.375	38 × 64

Figure 4 shows the 4 FIFOs reading timing waves of the MFBSS algorithm from Signaltap, and a soft oscilloscope is used to observe FPGA internal signals. The signals of FIFO0_rd ~FIFO3_rd respectively control the reading operation of the 4 FIFOs, allowing it to enable output data in a time slice polling way. The times for reading the 4 FIFOs until empty are ΔT_0 ~ΔT_3. The variables of $\Delta T_0 : \Delta T_1 : \Delta T_2 : \Delta T_3$ have the following relation:

$$\Delta T_0 : \Delta T_1 : \Delta T_2 : \Delta T_3 \approx \frac{L(0)}{V_{Rf} - V_{Wf}(0)} : \frac{L(1)}{V_{Rf} - V_{Wf}(1)} : \frac{L(2)}{V_{Rf} - V_{Wf}(2)} : \frac{L(3)}{V_{Rf} - V_{Wf}(3)}$$

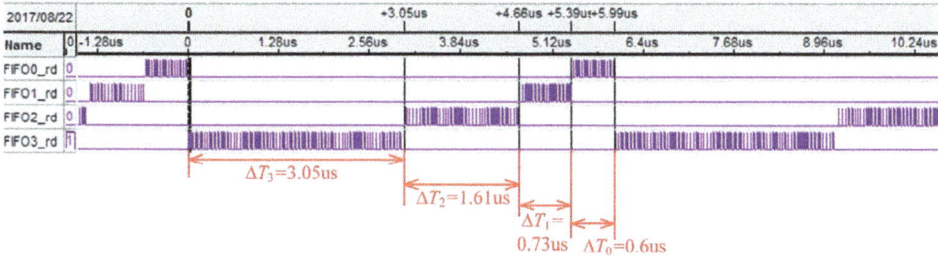

Figure 4. The 4 FIFOs read timing waves of the MFBSS algorithm from Signaltap.

All the FIFOs are readed in turn until empty in every cycle. The sum of data (D_{W-sum}) for writing into the FIFOs and the sum of data (D_{R-sum}) for reading out from the FIFOs are given by the two formulas $(\Delta t_0 \cdot V_{Wf}(0) + \Delta t_1 \cdot V_{Wf}(1) + \Delta t_2 \cdot V_{Wf}(2) + \Delta t_3 \cdot V_{Wf}(3)) \cdot \Delta B_W$ and $(\Delta t_0 + \Delta t_1 + \Delta t_2 + \Delta t_3) \cdot V_{Rf} \cdot \Delta B_W$, respectively. As a result, the experimental results show that D_{W-sum} equals to D_{R-sum}, which meets the relation $V_R = \sum_{n=0}^{N-1} V_W(n)$ of Equation (2), and also agrees well with the theoretical analysis.

In the N-group scanning system, the bandwidth utilization ratio $\eta_{bw}(N)$ of the MFBSS algorithm can be expressed by Equation (11):

$$\eta_{bw}(N) = \frac{\sum\limits_{i=0}^{3} V_{Wf}(i)}{V'_{Rf}} \times 100\%.$$ (11)

Therefore, in the experiment, when $N = 4$, the utilization ratio $\eta_{bw}(4)$ of the MFBSS algorithm used in the UPA system can be calculated by Equation (12):

$$\eta_{bw}(4) = \frac{\sum\limits_{i=0}^{3} V_{Wf}(i)}{V'_{Rf}} \times 100\% = \frac{V_{Rf}}{V'_{Rf}} = \frac{24.375}{25} \times 100\% = 97.5\%.$$ (12)

The ETSPS scheduling algorithm based on the equal allocation of a time slot to each task. As compared with the MFBSS algorithm in this work, the ETSPS scheduling algorithm has four characteristics: (i) The lengths of all the FIFO$_i$ ($i = 0, 1, 2, \ldots, N-1$) are the same as each other, i.e., $L(0) = L(1) = \ldots = L(N-1)$. (ii) All the time slice resources of the reading operation of the N FIFOs are also equal to each other. (iii) All the FIFOs have the reading speed (V'_{Rf}) which is equal to the maximum of the writing speed [$V_{Wf}(i)$], same as that of the individual FIFO, i.e., $V'_{Rf} = \max[V_{Wf}(i)]$, $i = 0, 1, \ldots, N-1$. (iv) When the FIFO$_i$ ($i = 0, 1, 2, \ldots, N-1$) is filled by writting, the reading operations of the

FIFO$_i$ will be immediately performed. Therfore, the general utilization ratio of the bandwidth-sharing transmission with N-group scanning of the UPA system can be calculated by Equation (13):

$$\eta'_{bw}(N) = \frac{\sum\limits_{j=0}^{N-1} V_{Wf}(j)}{N \cdot V'_{Rf}} \times 100\% = \frac{\sum\limits_{j=0}^{N-1} V_{Wf}(j)}{N \cdot \max\left(V_{Wf}(i)\right)} \times 100\%. \tag{13}$$
$$i = 0, 1, \cdots N - 1$$

For N-group scanning data stream with bandwidths $\{V_W(0), V_W(1), \ldots, V_W(N-1)\}$ (unit: Byte/s), we use the FPGA technology to implement the MFBSS algorithm together with the the traditional ETSPS scheduling algorithm, and analyze their bandwidth utilization ratios $\eta_{bw}(N)$ and $\eta'_{bw}(N)$. For example, the FPGA (Arria-II EP2AGX65DF29I5) with a work clock frequency of $f_{clk} = 100$ MHz. So, it is easy to produce the clock frequencies such as $F_1 = \{1, 2, 3, \ldots, f_{clk}\}$ and $F_2 = \{f_{clk}/100, f_{clk}/99, f_{clk}/98, \ldots, f_{clk}/1\}$ (unit: MHz) by using the clock f_{clk} by Digital Phase Locked Loop technology.

- The MFBSS algorithm. According to Equation (11), the theoretical value of the shared output bandwidth is V_{Rf} or ($\sum\limits_{i=0}^{3} V_{Wf}(i)$). The actual value of the shared output bandwidth is V'_{Rf}, which satisfies the following conditions: $V'_{Rf} \geq V_{Rf}$, $V'_{Rf} \in F_1$ or $V'_{Rf} \in F_2$, and the value of $(V'_{Rf} - V_{Rf})$ is minimized. For instance, when $V_{Rf} = 24.375$ HMz, and $V'_{Rf} = f_{clk}/4 = 25$ MHz, and thus the actual bandwidth utilization ratio is $\frac{V_{Rf}}{V'_{Rf}} \times 100\%$ which equals to 97.5%.

- The ETSPS algorithm. According to Equation (13), the larger the value of $\max(V_{Wf}(i))$ is, the smaller the value of $\eta'_{bw}(N)$ is. The smaller the value of $\max(V_{Wf}(i))$ is, the larger the value of $\eta'_{bw}(N)$ is. So, when the value of $\max\left(V_{Wf}(i)\right)$ equals to $\frac{1}{N} \cdot \sum\limits_{j=0}^{N-1} V_{Wf}(j)$, i.e., $V_W(0) = V_W(1) = \ldots = V_W(i) = \ldots = V_W(N-1)$, the maximum theoretical value of $\eta'_{bw}(N)$ can be expressed by Equation (14).

$$\max\left(\eta'_{bw}(N)\right) = \frac{\sum\limits_{j=0}^{N-1} V_{Wf}(j)}{N \cdot \max\left(V_{Wf}(i)\right)} \times 100\% = \eta_{bw}(N) \tag{14}$$

when the value of $\max(V_{Wf}(i))$ is close to $\sum\limits_{j=0}^{N-1} V_{Wf}(j)$, i.e., $V_{Wf}(i) \to \sum\limits_{j=0}^{N-1} V_{Wf}(j)$, the minimum theoretical value of $\eta'_{bw}(N)$ can be expressed by Equation (15).

$$\min\left(\eta'_{bw}(N)\right) \approx \frac{\sum\limits_{j=0}^{N-1} V_{Wf}(j)}{N \cdot \max\left(V_{Wf}(i)\right)} \times 100\% \approx \left(\frac{100}{N}\right)\% \tag{15}$$

Figure 5 shows the bandwidth utilization ratio curves of the two scheduling algorithms (cross axis: the theoretical value of the shared output bandwidth V_{Rf} ($N = 4$), and vertical axis: the bandwidth utilization). $\eta_{bw}(N)$ and $\eta'_{bw}(N)$ are the bandwidth utilization ratios of the MFBSS algorithm and the ETSPS algorithm, respectively.

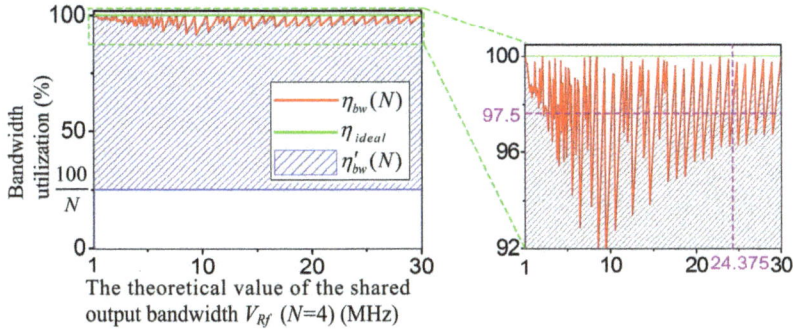

Figure 5. Comparison of the bandwidth utilization ratios of the MFBSS algorithm and the ETSPS algorithm.

The symbols $\eta_{bw}(N)$ and η_{ideal} represent the experimental and ieal values of the algorithm MFBSS, respectively. The results show that the value of $\eta_{bw}(N)$ is between 92% and 100%, for example, for the above experiment of 4-group scanning based on the MFBSS algorithm, when V_{Rf} equals to 24.375 MHz, $\eta_{bw}(N)$ equals to 97.5% and η_{ideal} equals to 100%. Whereas the value of $\eta'_{bw}(N)$ is relevant to the value of N, its value is between $(100/N)\%$ and $\eta_{bw}(N)$. For N-group scanning patterns, only when all groups have the same bandwidth, $\eta_{bw}(N)$ equals to $\eta'_{bw}(N)$. Otherwise, $\eta'_{bw}(N)$ would be much smaller than $\eta_{bw}(N)$.

Similarly, we use FPGA to implement the traditional ETSPS algorithm with the same parameters in Table 3, and collected reading timing waves of the 4 FIFOs by using Signaltap. As shown in Figure 6, the signals FIFO0_rd ~FIFO3_rd control the reading operation of the four FIFOs, and the time resources occupied by the signals are assigned by the signal FIFO_rd.

Assuming that the symbols f_{FIFO_rd}, f_{FIFO0_rd}, f_{FIFO1_rd}, f_{FIFO2_rd}, and f_{FIFO3_rd} represent the frequencies of signals FIFO_rd, FIFO0_rd, FIFO1_rd, FIFO2_rd, and FIFO3_rd, respectively, the following results can be easily obtained, as shown in Figure 6: $f_{FIFO\_rd} = \dfrac{1}{\Delta T} = 50$ MHz, $f_{FIFO0\_rd} = \dfrac{1}{\Delta T_0} = 2.5$ MHz, $f_{FIFO1\_rd} = \dfrac{1}{\Delta T_1} = 3.125$ MHz, $f_{FIFO2\_rd} = \dfrac{1}{\Delta T_2} = 6.25$ MHz, $f_{FIFO3\_rd} = \dfrac{1}{\Delta T_3} = 12.5$ MHz.

So, the utilization ratio of the data transmission with the 4-group scanning of the ETSPS algoritnm can be calculated by Equation (16):

$$
\begin{aligned}
\eta'_{bw}(4) &= \frac{f_{FIFO0\_rd} + f_{FIFO1\_rd} + f_{FIFO2\_rd} + f_{FIFO3\_rd}}{f_{FIFO\_rd}} \times 100\% = \frac{\sum\limits_{j=0}^{N-1} f_{sj}}{N \cdot \max(f_{s0}, \cdots f_{s3})} \times 100\% \\
&= \frac{2.5 + 3.125 + 6.25 + 12.5}{50} \times 100\% \\
&= 48.75\%
\end{aligned}
\tag{16}
$$

As a consequence, the bandwidth utilization ratio of the MFBSS algorithm $\eta_{bw}(4)$ reaches to 97.5% as shown in the inset of Figure 5, while the bandwidth utilization of the ETSPS algorithm $\eta'_{bw}(4)$ is only 48.75%. The experimental results demonstrate that the MFBSS algorithm is efficient when used in the multi-group sensors scanning UPA system.

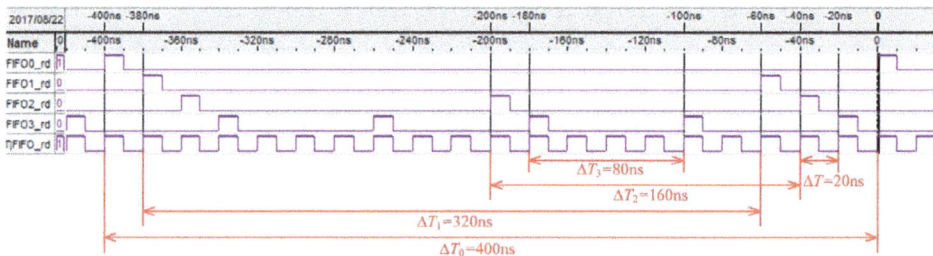

Figure 6. The 4 FIFOs reading timing waves of the ETSPS algorithm from Signaltap.

5. Conclusions

The novel MFBSS algorithm was proposed on the basis of the FIFOs variable lengths by FPGA technology, and was used for the multi-sensor scanning UPA system to maximize the bandwidth utilization ratio. The mathematical modeling of the MFBSS algorithm was established, and the formula $V_R = \sum_{n=0}^{N-1} V_W(n)$ of maximizing bandwidth transmission utilization ratio in the N-group scanning patterns was successfully deduced. The lengths of the N-group FIFOs were achieved by using the designed equations, from which the length ratios were readily calculated. The algorithm was realized by FPGA technology, which made the reading operation of one FIFO switch to another FIFO without any time slot waiting, and thus it obtained the data transmission bandwidth utilization of no less than 92% hence allowing the UPA system to have the bandwidth utilization higher than that of the traditional ETSPS algorithm. In order to improve transmission efficiency of the large data generated by the sensor systems and the real-time performance of the algorithm through the multi-FPGA technology, the MFBSS scheduling algorithm based on data transmission has important applications in the multi-sensor systems, and the future research is likely to focus on designing some special scheduling algorithm module for different sensor systems.

Acknowledgments: This work was financially supported by the National Key Foundation for Exploring Scientific Instrument (2013YQ230575) and Guangzhou Science and Technology Plan Project (201509010008).

Author Contributions: Wenming Tang and Guixiong Liu conceived the idea of the paper; Wenming Tang and Daji Tan performed the experiments, and Yuzhong Li carried out the system model; Wenming Tang and Cuixiong Liu wrote the paper.

Conflicts of Interest: The authors declare no conflict of interest.

References

1. Walter, S.; Hersog, T.; Schubert, F.; Heuer, H. Investigations of PMN-PT composites for high sensitive ultrasonic phased array probes in NDE. Proceeding of the 2015 IEEE Sensors, Busan, Korea, 1–4 November 2015; pp. 1–4.
2. Yuan, C.; Xie, C.; Li, L.; Zhang, F.; Gubanski, S.M. Ultrasonic phased array detection of internal defects in composite insulators. *IEEE Trans. Dielectr. Electr. Insul.* **2016**, *23*, 525–531. [CrossRef]
3. Rubtsov, V.; Tarasov, S.; Kolubaev, E.; Psakhie, S. Ultrasonic phase array and eddy current methods for diagnostics of flaws in friction stir welds. In Proceedings of the International Conference on Physical Mesomechanics of Multilevel Systems, Tomsk, Russia, 3–5 September 2014; pp. 539–542.
4. Hynynen, K.; Clement, G.N.; Vykhodtseva, N.; King, R.; White, P.J.; Vitek, S. 500-element ultrasound phased array system for noninvasive focal surgery of the brain: A preliminary rabbit study with ex vivo human skulls. *Magn. Reson. Med.* **2004**, *52*, 100–107. [CrossRef] [PubMed]
5. Qiu, Y.; Gigliotti, J.V.; Wallace, M.; Griggio, F.; Demore, C.E.M.; Cochran, S. Piezoelectric micromachined ultrasound transducer (PMUT) arrays for integrated sensing, actuation and imaging. *Sensors* **2015**, *15*, 8020–8041. [CrossRef] [PubMed]

6.   An, J.; Song, K.; Zhang, S.; Yang, J.; Cao, P. Design of a broadband electrical impedance matching network for piezoelectric ultrasound transducers based on a genetic algorithm. *Sensors* **2014**, *14*, 6828–6843. [CrossRef] [PubMed]

7.   Taylor, K.J.; Milan, J. Differential diagnosis of chronic splenomegaly by grey-scale ultrasonography: Clinical observations and digital A-scan analysis. *Br. J. Radiol.* **1976**, *49*, 519–525. [CrossRef] [PubMed]

8.   Dutt, V.; Greenleaf, J.F. Adaptive speckle reduction filter for log-compressed B-scan images. *IEEE Trans. Med. Imaging* **1996**, *15*, 802–813. [CrossRef] [PubMed]

9.   Li, Y.; Blalock, T.N.; Hossack, J.A. Synthetic axial acquisition-full resolution, low-cost C-scan ultrasonic imaging. *IEEE Trans. Ultrason. Ferroelectr. Freq. Control* **2008**, *55*, 236–239. [PubMed]

10. Lin, R.B.; Liu, G.X.; Tang, W.M. FPGA implementation of ultrasonic s-scan coordinate conversion based on radix-4 CORDIC algorithm. *Zeitschrift Gemeine Mikrobiologie* **2012**, *15*, 505–512. [CrossRef]

11. Song, J.; Pulkkinen, A.; Huang, Y.; Hynynen, K. Investigation of standing-wave formation in a human skull for a clinical prototype of a large-aperture, transcranial MR-guided focused ultrasound (MRgFUS) phased array: An experimental and simulation study. *IEEE Trans. Biomed. Eng.* **2012**, *59*, 435–444. [CrossRef] [PubMed]

12. ASTM F2491-13. *Standard Guide for Evaluating Performance Characteristics of Phased-Array Ultrasonic Testing Instruments and Systems*; ASTM: West Conshohocken, PA, USA, 2013.

13. Choi, H.K.; Choi, J.D.; Jang, Y.S. Numerical analysis of queuing delay in cyclic bandwidth allocation TDMA system. *Electr. Lett.* **2014**, *50*, 1204–1205. [CrossRef]

14. Park, H.; Choi, K. Adaptively weighted round-robin arbitration for equality of service in a many-core network-on-chip. *IET Comput. Digit. Tech.* **2016**, *10*, 37–44. [CrossRef]

15. Slimane, M.B.; Hafaiedh, I.B.; Robbana, R. Formal-based design and verification of SoC arbitration protocols: A comparative analysis of TDMA and round-robin. *IEEE Des. Test* **2017**, *34*, 54–62. [CrossRef]

16. Srinivasan, S.; Pandharipande, A. Self-configuring scheduling protocol for ultrasonic sensor systems. *IEEE Sens. J.* **2013**, *13*, 2517–2518. [CrossRef]

17. Patil, S.; Kulkarni, R.A.; Patil, S.H.; Balaji, N. Performance improvement in cloud computing through dynamic task scheduling algorithm. In Proceedings of the International Conference on Next Generation Computing Technologies, Dehradun, India, 4–5 September 2015; pp. 96–100.

18. Chronaki, K.; Rico, A.; Casas, M.; Moreto, M.; Badia, R.; Ayguade, E. Task scheduling techniques for asymmetric multi-core systems. *IEEE Trans. Parallel Distrib. Syst.* **2017**, *28*, 2074–2087. [CrossRef]

19. Long, J.; Dong, M.; Ota, K.; Liu, A. A Green TDMA Scheduling algorithm for prolonging lifetime in wireless sensor networks. *IEEE Syst. J.* **2017**, *11*, 868–877. [CrossRef]

20. Yaashuwanth, C.; Ramesh, R. *A New Scheduling Algorithm for Real Time System*; Auto-Ordnance Corporation: West Hurley, NY, USA, 2010; pp. 1104–1106.

21. Liu, J.; Soleimanifar, M.; Lu, M. Resource-loaded piping spool fabrication scheduling: Material-supply-driven optimization. *Visual. Eng.* **2017**, *5*, 1–14. [CrossRef]

22. Fischetti, M.; Monaci, M. Using a general-purpose mixed-integer linear programming solver for the practical solution of real-time train rescheduling. *Eur. J. Oper. Res.* **2017**, *263*, 258–264. [CrossRef]

23. Ou, X.; Chang, Q.; Chakraborty, N.; Wang, J. Gantry scheduling for multi-gantry production system by online task allocation method. *IEEE Robot. Autom. Lett.* **2017**, *99*, 1848–1855. [CrossRef]

24. SCFIFO and DCFIFO IP Cores User Guide. Available online: https://www.altera.com/content/dam/altera-www/global/en_US/pdfs/literature/ug/ug_fifo.pdf (accessed on 2 January 2017).

25. Pan, D.; Yang, Y. FIFO-based multicast scheduling algorithm for virtual output queued packet switches. *IEEE Trans. Comput.* **2005**, *54*, 1283–1297.

26. Fernandez, G.; Jalle, J.; Abella, J.; Quinones, E.; Vardanega, T.; Cazorla, F.J. Computing safe contention bounds for multicore resources with round-robin and FIFO arbitration. *IEEE Trans. Comput.* **2017**, *66*, 586–600. [CrossRef]

27. Tassart, S. Time-invariant context for sample rate conversion systems. *IEEE Trans. Signal Proc.* **2012**, *60*, 1098–1107. [CrossRef]

28. Castiglioni, P.; Piccini, L.; Rienzo, M.D. Interpolation technique for extracting features from ECG signals sampled at low sampling rates. In Proceedings of the Computers in Cardiology, Thessaloniki Chalkidiki, Greece, 21–24 September 2003; pp. 481–484.

29. Samson, C.A.; Bezanson, A.; Brown, J.A. A sub-nyquist, variable sampling, high-frequency phased array beamformer. *IEEE Trans. Ultrason. Ferroelectr. Freq. Control* **2017**, *64*, 568–576. [CrossRef] [PubMed]
30. Avalon Interface Specifications. Available online: https://www.altera.com/content/dam/altera-www/global/en_US/pdfs/literature/manual/mnl_avalon_spec.pdf (accessed on 10 February2017).
31. Cyclone V Avalon-ST Interface for PCIe Solutions User Guide. Available online: https://www.altera.com/content/dam/altera-www/global/en_US/pdfs/literature/ug/ug_c5_pcie_avst.pdf (accessed on 15 January 2017).
32. Durante, P.; Neufeld, N.; Schwemmer, R.; Balbi, G.; Marconi, U. 100 Gbps PCI-express readout for the LHCb upgrade. *IEEE Trans. Nuclear Sci.* **2015**, *62*, 1752–1757. [CrossRef]

sensors

MDPI

Article

Imaging of Subsurface Corrosion Using Gradient-Field Pulsed Eddy Current Probes with Uniform Field Excitation

Yong Li [1,*] (iD), Shuting Ren [1], Bei Yan [1], Ilham Mukriz Zainal Abidin [2] and Yi Wang [1]

1   State Key Laboratory for Strength and Vibration of Mechanical Structures,
    Shaanxi Engineering Research Center of NDT and Structural Integrity Evaluation,
    Xi'an Jiaotong University, Xi'an 710049, China; renshuting1@stu.xjtu.edu.cn (S.R.);
    yanbei@stu.xjtu.edu.cn (B.Y.); wybit2008@stu.xjtu.edu.cn (Y.W.)
2   Leading Edge NDT Technology (LENDT) Group, Malaysian Nuclear Agency, Bangi 43000, Kajang, Selangor,
    Malaysia; mukriz@nuclearmalaysia.gov.my
*   Correspondence: yong.li@mail.xjtu.edu.cn; Tel.: +86-029-8266-5721

Received: 28 June 2017; Accepted: 27 July 2017; Published: 31 July 2017

Abstract: A corrosive environment leaves in-service conductive structures prone to subsurface corrosion which poses a severe threat to the structural integrity. It is indispensable to detect and quantitatively evaluate subsurface corrosion via non-destructive evaluation techniques. Although the gradient-field pulsed eddy current technique (GPEC) has been found to be superior in the evaluation of corrosion in conductors, it suffers from a technical drawback resulting from the non-uniform field excited by the conventional pancake coil. In light of this, a new GPEC probe with uniform field excitation for the imaging of subsurface corrosion is proposed in this paper. The excited uniform field makes the GPEC signal correspond only to the field perturbation due to the presence of subsurface corrosion, which benefits the corrosion profiling and sizing. A 3D analytical model of GPEC is established to analyze the characteristics of the uniform field induced within a conductor. Following this, experiments regarding the imaging of subsurface corrosion via GPEC have been carried out. It has been found from the results that the proposed GPEC probe with uniform field excitation not only applies to the imaging of subsurface corrosion in conductive structures, but provides high-sensitivity imaging results regarding the corrosion profile and opening size.

Keywords: electromagnetic nondestructive evaluation; gradient-field pulsed eddy current inspection; subsurface corrosion; analytical modeling; corrosion imaging; uniform field excitation

1. Introduction

Conductive structures of nonmagnetic materials such as aluminum and copper are widely employed in engineering fields involving energy, transportation, as well as aerospace. Despite anti-corrosion measures, in-service conductive structures are still vulnerable to corrosion due to hostile and particularly corrosive environments [1,2]. Among various types of corrosion, subsurface corrosion poses the most severe threat to structural integrity. The reason lies in the fact that it normally occurs either within the conductor body or on the back surface of the conductor. Such conventional non-destructive evaluation (NDE) techniques as visual testing (VT) [3], penetrant testing (PT) [4], single-frequency eddy current testing (ECT) [5] and magnetic particle inspection (MPI) [6], etc., which target cracks in industry may leave subsurface corrosion undetected. Therefore, it is highly desirable to noninvasively evaluate and in particular visualize subsurface corrosion in conductive structures via appropriate NDE methods before catastrophic accidents happen.

Gradient-field pulsed eddy current technique (GPEC) is an extension of the pulsed eddy current technique (PEC), which is capable of evaluating the integrity of a conductive structure with a thickness up to 10 mm [7]. It has been found to be superior in the high-sensitivity evaluation of hidden material degradation and corrosion in conductors [8,9]. Even though subsurface corrosion could be visualized using GPEC probes, which consist of pancake coils for the excitation of the incident/primary magnetic field and magnetic sensors for quantifying the gradient of the magnetic field (namely the gradient field), there is a large discrepancy between the true corrosion profile and imaging result [8]. It could be mostly because of the fact that the incident magnetic field excited by pancake coils is non-uniform and thus, the acquired signals from magnetic sensors detect the gradient field resulting from not only the distortion of eddy currents due to anomalies in the conductor under inspection, but also the original incident magnetic field. This technical drawback arising from the previous probe configuration opens up the optimization of GPEC probes by realizing uniform field excitation, which gives localized uniform distributions of the eddy current and incident magnetic field. A schematic illustration exhibiting the interaction of the uniform eddy current induced in a conductor with subsurface corrosion is presented in Figure 1. It can be noticed from Figure 1 that in corrosion-free region the gradient field is null due to uniform distributions of the eddy currents and the incident magnetic field. In contrast, in the defect area and particularly at the edges of the subsurface corrosion, the eddy current is significantly disturbed due to material discontinuity, thus leading to the gradient field. In such case, the gradient-field signal is independent of the incident magnetic field, but only relies on the presence of subsurface corrosion. In view of this, a GPEC probe with uniform field excitation could be beneficial to high-sensitivity imaging for the profiling and sizing of subsurface corrosion.

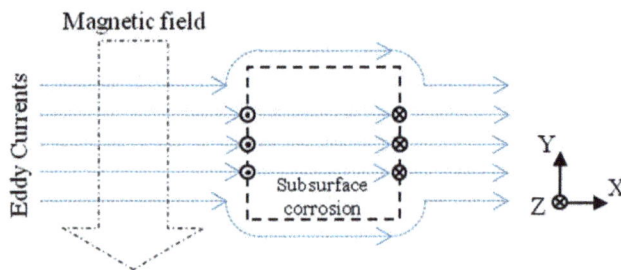

Figure 1. Schematic illustration regarding the interaction of the uniform eddy current with subsurface corrosion (top view).

As the key technology, uniform field excitation plays a vital role in alternating current field measurement (ACFM), which uses the sinusoidal excitation current, and is usually used for the detection and sizing of cracks in conductive structures [10]. In order to perform crack sizing, LeTessier et al. adopted an induction coil with its lateral surface facing the conductor surface to induce uniform eddy currents in the conductive areas of contact heaters and tanks [11]. By using a probe with the similar configuration, Knight et al. intensively investigated the influence of residual stress on ACFM for the inspection of cracks in drill-collar threaded connections [12]. Li et al. used an encircling coil for the generation of a uniform field in the external surfaces of pipes for the detection of cracks [13]. For arbitrary-angle cracks in planar conductors, Li et al. also proposed a double U-shaped orthogonal inducer to induce the uniform eddy current whose main axis could rotate at each scanning point [14]. It is believed that along with uniform field excitation, GPEC responses to subsurface corrosion, especially to its profile, could be enhanced. However, to the authors' knowledge the application of uniform field excitation to GPEC in the detection and imaging of subsurface corrosion has barely been investigated.

In this paper, a new GPEC probe together with uniform field excitation is proposed for the profiling and sizing of subsurface corrosion in nonmagnetic planar conductors through corrosion imaging. The uniformities of the eddy current and magnetic field are investigated through simulations based on their closed-form expressions formulated via the analytical modeling i.e., extended truncated region eigenfunction expansion (ETREE) modeling [15]. Following this, experiments were carried out in order to assess the capability of the proposed GPEC probe in high-sensitivity imaging of subsurface corrosion in conductors. The imaging accuracy in terms of the profile identification and estimation of the opening size of subsurface corrosion is evaluated.

2. Field Formulation and Investigation of Uniform Field Characteristics

2.1. Field Formulation

In difference to the previous probe configuration [8,9], the proposed GPEC probe consists of: (1) a rectangular coil (in lieu of the pancake coil) for generating the incident magnetic field; and (2) a magnetic sensor for measuring the gradient field. During inspection, it is placed over the upper surface of a layered conductor. It is noteworthy that in a bid to implement uniform field excitation, the rectangular coil is perpendicularly placed on the conductor with its lateral winding facing the conductor's upper surface. The model is shown in Figure 2.

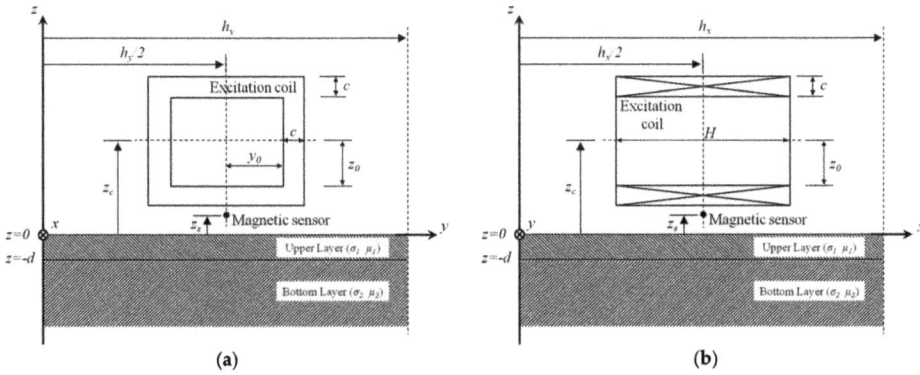

Figure 2. Model of the proposed GPEC probe placed over a two-layer conductor. (**a**) Side view in *x*-direction; (**b**) side view in *y*-direction.

Based on ETREE modeling [15–18], the net magnetic field in the air gap between the rectangular coil and conductor is written as:

$$\begin{cases} \overrightarrow{B}(t) = \frac{16\mu_0 NI(t)}{Hch_xh_y} \otimes \sum_{m=1}^{\infty}\sum_{n=1}^{\infty} C_{mn}\left\{[e^{\kappa_{mn}z} - e^{-\kappa_{mn}z}\zeta_{mn}(t)]\left(u_m\eta_1\overrightarrow{x}_0 + v_n\eta_2\overrightarrow{y}_0\right) + [e^{\kappa_{mn}z} + e^{-\kappa_{mn}z}\zeta_{mn}(t)]\kappa_{mn}\eta_3\overrightarrow{z}_0\right\} \\ \eta_1 = \cos(u_mx)\sin(v_ny); \ \eta_2 = \sin(u_mx)\cos(v_ny); \ \eta_3 = \sin(u_mx)\sin(v_ny) \end{cases} \quad (1)$$

Following Equation (1), the closed-form expression of the signal regarding each component of the gradient field $g(\overrightarrow{B})$, $g(\overrightarrow{B}) = \nabla\overrightarrow{B}$ measured by the magnetic sensor at an arbitrary location (x, y, z_s) between the rectangular coil and conductor is formulated as:

$$\begin{cases} g_x(\overrightarrow{B}) = \frac{16\mu_0 NI(t)}{Hch_xh_y} \otimes \sum_{m=1}^{\infty}\sum_{n=1}^{\infty} u_m C_{mn}\left\{[e^{-\kappa_{mn}z_s}\zeta_{mn}(t) - e^{\kappa_{mn}z_s}]\left(u_m\eta_{1x}\overrightarrow{x}_0 - v_n\eta_{2x}\overrightarrow{y}_0\right) + [e^{\kappa_{mn}z_s} + e^{-\kappa_{mn}z_s}\zeta_{mn}(t)]\kappa_{mn}\eta_{3x}\overrightarrow{z}_0\right\} \\ g_y(\overrightarrow{B}) = \frac{16\mu_0 NI(t)}{Hch_xh_y} \otimes \sum_{m=1}^{\infty}\sum_{n=1}^{\infty} v_n C_{mn}\left\{[e^{\kappa_{mn}z_s} - e^{-\kappa_{mn}z_s}\zeta_{mn}(t)]\left(u_m\eta_{1y}\overrightarrow{x}_0 - v_n\eta_{2y}\overrightarrow{y}_0\right) + [e^{\kappa_{mn}z_s} + e^{-\kappa_{mn}z_s}\zeta_{mn}(t)]\kappa_{mn}\eta_{3y}\overrightarrow{z}_0\right\} \\ g_z(\overrightarrow{B}) = \frac{16\mu_0 NI(t)}{Hch_xh_y} \otimes \sum_{m=1}^{\infty}\sum_{n=1}^{\infty} \kappa_{mn} C_{mn}\left\{[e^{\kappa_{mn}z_s} + e^{-\kappa_{mn}z_s}\zeta_{mn}(t)]\left(u_m\eta_1\overrightarrow{x}_0 + v_n\eta_2\overrightarrow{y}_0\right) + [e^{\kappa_{mn}z_s} - e^{-\kappa_{mn}z_s}\zeta_{mn}(t)]\kappa_{mn}\eta_3\overrightarrow{z}_0\right\} \end{cases} \quad (2)$$

where,

$$
\begin{cases}
\eta_{1x} = \sin(u_m x)\sin(v_n y); \ \eta_{2x} = \cos(u_m x)\cos(v_n y); \ \eta_{3x} = \cos(u_m x)\sin(v_n y) \\
\eta_{1y} = \cos(u_m x)\cos(v_n y); \ \eta_{2y} = \sin(u_m x)\sin(v_n y); \ \eta_{3y} = \sin(u_m x)\cos(v_n y)
\end{cases}
\tag{3}
$$

In Equations (1) and (2), $\otimes$ denotes convolution. $\vec{x}_0$, $\vec{y}_0$ and $\vec{z}_0$ are unit vectors. μ_0 is the permeability of vacuum. $I(t)$ and N stand for the excitation current in an arbitrary waveform and the number of turns of the rectangular coil, respectively. $\kappa_{mn} = \sqrt{u_m^2 + v_n^2}$, where, $u_m = m\pi/h_x$ and $v_n = n\pi/h_y$ (m and n are integers). The other terms include [19,20]:

$$
\begin{cases}
C_{mn} = \dfrac{\Phi_{mn}\cos(u_m h_x/2)\sin(u_m H/2)\sin(v_n h_y/2)e^{-\kappa_{mn} z_c}}{\kappa_{mn}^2 v_n} \\[2mm]
\Phi_{mn} = \dfrac{1}{\kappa_{mn}^2 + v_n^2}\left\{
\begin{array}{l}
\kappa_{mn}\sin[v_n(y_0+c)]\cosh[\kappa_{mn}(z_0+c)] - \kappa_{mn}\sin(v_n y_0)\cosh(\kappa_{mn} z_0) \\
-v_n\cos[v_n(y_0+c)]\sinh[\kappa_{mn}(z_0+c)] + v_n\cos(v_n y_0)\sinh(\kappa_{mn} z_0)
\end{array}\right\}
\end{cases}
\tag{4}
$$

It is noteworthy that in Equations (1) and (2) $\zeta_{mn}(t)$ denotes the time-domain expression of the conductor reflection coefficient [8,21]. It can be readily computed via inverse Fourier transform of its time-harmonic form $\zeta_{mn}(\omega)$, where, ω denotes the angular frequency of each harmonic in the spectrum of the excitation current [22]. For a two-layer conductor comprising an upper layer with the finite thickness d and a bottom layer with infinite thickness (as shown in Figure 2), $\zeta_{mn}(\omega)$ can be written as:

$$
\begin{cases}
\zeta_{mn}(\omega) = \dfrac{1}{\rho_{mn}}\left[(\lambda_1\mu_2 + \lambda_2\mu_1)(\kappa_{mn}\mu_1 - \lambda_1) + e^{-2\lambda_1 d}(\lambda_1\mu_2 - \lambda_2\mu_1)(\kappa_{mn}\mu_1 + \lambda_1)\right] \\[2mm]
\rho_{mn} = (\lambda_1\mu_2 + \lambda_2\mu_1)(\kappa_{mn}\mu_1 + \lambda_1) + e^{-2\lambda_1 d}(\lambda_1\mu_2 - \lambda_2\mu_1)(\kappa_{mn}\mu_1 - \lambda_1)
\end{cases}
\tag{5}
$$

where, $\lambda_i = \sqrt{\kappa_{mn}^2 + j\omega\sigma_i\mu_i\mu_0}$, $i = 1, 2$. σ_i and μ_i denote the conductivity and relative permeability of each layer, respectively.

Following Equation (1), the density of eddy currents induced at an arbitrary position within the upper layer is formulated as:

$$
\vec{J}_{ec}(t) = \frac{16\mu_0\sigma_1 N\{\partial[I(t)]/\partial t\}}{Hch_x h_y} \otimes \sum_{m=1}^{\infty}\sum_{n=1}^{\infty} C_{mn}\left(-v_n\eta_2\vec{x}_0 + u_m\eta_1\vec{y}_0\right)\left[e^{\kappa_{mn} z}\alpha_{mn}(t) + e^{-\kappa_{mn} z}\beta_{mn}(t)\right]
\tag{6}
$$

It is noted that due to the characteristics of eddy currents in flawless conductors, the z-component of $\vec{J}_{ec}$ vanishes [23]. In Equation (6), $\alpha_{mn}(t)$ and $\beta_{mn}(t)$ can be readily recovered through the inverse Fourier transform of their time-harmonic forms $\alpha_{mn}(\omega)$ and $\beta_{mn}(\omega)$, respectively. $\alpha_{mn}(\omega)$ and $\beta_{mn}(\omega)$ are written as:

$$
\begin{cases}
\alpha_{mn}(\omega) = [2\mu_1\kappa_{mn}(\lambda_1\mu_2 + \lambda_2\mu_1)]/\rho_{mn} \\[2mm]
\beta_{mn}(\omega) = \left[2\mu_1\kappa_{mn}(\lambda_1\mu_2 - \lambda_2\mu_1)e^{-2\lambda_1 d}\right]/\rho_{mn}
\end{cases}
\tag{7}
$$

Considering a conductive plate under inspection, $\sigma_2 = 0$ MS/m and $\mu_2 = 1$. Therefore, Equations (5) and (7) can further be simplified into:

$$
\begin{cases}
\zeta_{mn}(\omega) = \left\{\left[(\kappa_{mn}\mu_1)^2 - \lambda_1^2\right]\left(1 - e^{-2\lambda_1 d}\right)\right\}/\rho_{mn} \\[2mm]
\rho_{mn} = (\lambda_1 + \kappa_{mn}\mu_1)^2 - e^{-2\lambda_1 d}(\lambda_1 - \kappa_{mn}\mu_1)^2
\end{cases}
\tag{8}
$$

$$
\begin{cases}
\alpha_{mn}(\omega) = [2\mu_1\kappa_{mn}(\lambda_1 + \kappa_{mn}\mu_1)]/\rho_{mn} \\[2mm]
\beta_{mn}(\omega) = \left[2\mu_1\kappa_{mn}(\lambda_1 - \kappa_{mn}\mu_1)e^{-2\lambda_1 d}\right]/\rho_{mn}
\end{cases}
\tag{9}
$$

It should be pointed out that since GPEC normally utilizes the excitation current in quasi-rectangular waveform, $I(t)$ in Equations (1), (2) and (6) can thus be analytically formulated in the form of a Fourier series as:

$$I(t) = I_0 \left\{ \left[v + \frac{1}{T} \tau e^{-\frac{vT}{\tau}} \left(1 - e^{\frac{T}{\tau}(2v-1)} \right) \right] + \frac{1}{\pi} \sum_{l=1}^{\infty} \left[a_l \cos\left(\frac{2l\pi t}{T} \right) + b_l \sin\left(\frac{2l\pi t}{T} \right) \right] \right\} \tag{10}$$

where I_0, T, v and τ are the maximum amplitude, period, duty cycle and rising/falling time constant of the current signal, respectively. a_l and b_l are written as:

$$a_l = \frac{\sin(2l\pi v)}{l} - \frac{2\pi}{T} \left[\frac{1}{\tau^2} + \left(\frac{2l\pi}{T} \right)^2 \right]^{-1} \left\{ \frac{1}{\tau} \left[1 + e^{\frac{T(v-1)}{\tau}} \right] + \left[\frac{2l\pi \sin(2l\pi v)}{T} - \frac{\cos(2l\pi v)}{\tau} \right] \left(1 + e^{-\frac{vT}{\tau}} \right) \right\} \tag{11}$$

$$b_l = \frac{1-\cos(2l\pi v)}{l} - \frac{2\pi}{T} \left[\frac{1}{\tau^2} + \left(\frac{2l\pi}{T} \right)^2 \right]^{-1} \left\{ \frac{2l\pi}{T} \left[1 + e^{\frac{T(v-1)}{\tau}} \right] - \left[\frac{2l\pi \cos(2l\pi v)}{T} + \frac{\sin(2l\pi v)}{\tau} \right] \left(1 + e^{-\frac{vT}{\tau}} \right) \right\} \tag{12}$$

Equations (1) and (6) facilitate the computation of the electromagnetic field excited by the rectangular coil and subsequent analysis of its characteristics involving the uniformities of: (1) magnetic field over the upper surface of the conductor; and (2) eddy currents within the conductor.

2.2. Characteristics of the Uniform Field

Since the uniform field is of great importance for the proposed GPEC probe, it is essential to investigate the characteristics of the excited electromagnetic field and identify the area where the effective uniform field distributes. Simulations based on Equations (1) and (6) are consequently carried out to analyze the field characteristics with the proposed probe whose configuration is exhibited in Figure 2. It is assumed that subsurface corrosion occurs in the back surface of a conductive plate. Therefore, the uniformity of the eddy current in the plate back surface is intensively analyzed whilst the net magnetic field over the plate upper surface is computed in an effort to investigate its uniformity.

In light of the fact that the excitation current in the quasi-rectangular waveform is employed to drive the rectangular coil, special attention is given to the selection of the time instant when the field response is picked up by a magnetic sensor deployed at the location $(h_x/2, h_y/2, z_s)$, to the subsurface corrosion is the highest. It is noted that in such a case, y- and z-components of the net magnetic field (B_y and B_z) vanish. In a sense of the conductor size, subsurface corrosion with a dimension considerably larger than that of the excitation coil is analogous to the wall-thinning defect. Consequently, in the presence of the subsurface corrosion the plate thickness decreases with Δd from the back surface. By referring to [8,21,22], the response of x-component of the net magnetic field B_x to the initial subsurface corrosion with $\Delta d \rightarrow 0$ is written as:

$$\lim_{\Delta d \to 0} \frac{\Delta B_x(t)}{\Delta d} = \frac{\partial [B_x(t)]}{\partial d} = \frac{16\mu_0 N I(t)}{H c h_x h_y} \otimes \sum_{m=1}^{\infty} \sum_{n=1}^{\infty} u_m C_{mn} e^{-\kappa_{mn} z} \zeta'_{mn}(t) \cos\left(\frac{m\pi}{2} \right) \sin\left(\frac{n\pi}{2} \right) \tag{13}$$

where, $\zeta'_{mn}(t)$ can be readily recovered via the inverse Fourier transform of $\partial[\zeta_{mn}(\omega)]/\partial d$ which is formulated as:

$$\frac{\partial [\zeta(\omega)]}{\partial d} = \frac{2\kappa_{mn}\mu_1 \lambda_1^2 \left[(\kappa_{mn}\mu_1)^2 - \lambda_1^2 \right]}{\left\{ 2\kappa_{mn}\mu_1\lambda_1 \cosh(\lambda_1 d) + \left[(\kappa_{mn}\mu_1)^2 + \lambda_1^2 \right] \sinh(\lambda_1 d) \right\}^2} \tag{14}$$

Equation (13) is subsequently adopted for the computation of the field response to the initial subsurface corrosion in an effort to choose the time instant when the uniformity of the electromagnetic field involving the eddy current and net magnetic field is intensively analyzed.

Tables 1 and 2 list the parameters employed in simulations. The material of the conductive plate is Aluminum. As illustrated in Figure 3, the fundamental frequency, duty cycle, rising time and maximum amplitude of the excitation current $I(t)$ are 100 Hz, 50%, 50 μs and 0.5 A, respectively. By applying Equation (13), the field response to the initial subsurface corrosion is calculated and presented in Figure 4.

Table 1. Parameters of the probe.

Coil Parameter	Value
Inner length, $2y_0$ (mm)	24.0
Inner width, $2z_0$ (mm)	12.0
Height, H (mm)	20.3
Winding thickness, c (mm)	1.2
Lift-off, z_c (mm)	1.0
Number of turns, N	289
Sensor stand-off, z_s (mm)	0.5

Table 2. Parameters of the conductive plate.

Plate Parameter	Value
Thickness, d (mm)	6.0
Conductivity, σ_1 (MS/m)	34.2
Relative permeability μ_1	1.0
Length, h_y (mm)	300
Width, h_x (mm)	300

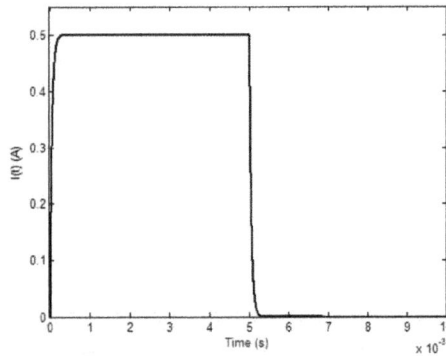

Figure 3. The excitation current $I(t)$.

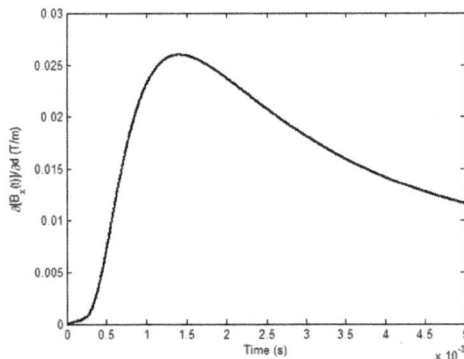

Figure 4. The computed response of B_x to the initial subsurface corrosion ($0 \leq t \leq 5$ ms).

It can be seen from Figure 4 that the field response to the initial subsurface corrosion is the highest at the time of approximately 1.4 ms. This indicates that due to the presence of the initial subsurface corrosion the perturbation of the eddy current, especially over the back surface of the conductive plate, reaches the maximum at the same time. Further analysis provides the precise temporal value, which is 1406.7 µs, and is used for analysis regarding the uniformities of the eddy current and magnetic field. The calculated distrubtion of the eddy current over the plate back surface and the magentic field above the plate upper surface are exhibited in Figures 5 and 6, respectively. It is noted that the eddy current and magnetic field are invesitigated within the XY plane which covers the lateral surface of the excitation coil (facing the plate upper surface). The coordinate of the plane centre is (0, 0), which corresponds to $(h_x/2, h_y/2)$ in Figure 2.

It can be observed from Figures 5 and 6 that the distributions of the eddy current and net magnetic field in the central region, particularly in the region of interest (ROI) (2 mm × 2 mm area with the center at (0, 0)), where the gradient-field sensor is deployed in experiments are relatively uniform. In a bid to evaluate the uniformity of the uniform field, including the eddy current and net magnetic field within the ROI, an algorithm for the uniformity evaluation regarding the magnetic field [24] is utilized. The computed degrees of field uniformity (DFU) are: 20.6 ppm for the eddy current (averaged value over DFUs of J_x and J_y) and 5.9 ppm for the net magnetic field (averaged value over DFUs of B_x, B_y and B_z), which indicates that in the ROI the eddy current on the plate back surface and the net magnetic field over the plate upper surface are highly uniform. This benefits the high-sensitivity detection and imaging of subsurface corrosion that breaks the field uniformity and thus results in the non-zero gradient-field signal from the GPEC probe.

It is also noticeable from Figures 5 and 6, that in the ROI—compared with the averaged value of J_x which is approximately zero—J_y is over 7×10^6 A/mm^2, whilst B_x has a considerably larger magnitude (over 1.54×10^{-3} Tesla) than B_y and B_z. This implies that: (1) J_x, B_y and B_z are barely sensitive subsurface corrosion on the conductive plate; and (2) the material discontinuity, which is introduced by subsurface corrosion and especially transverse to the direction of J_y, significantly perturbs the distribution of J_y, and thus breaks the uniformity of J_y. As a result, the gradient of the resultant B_x over the plate upper surface, which was originally null for the flawless scenario, is non-zero in the ROI. Consequently, it provides a good implication regarding the presence of subsurface corrosion, and is beneficial to the imaging of subsurface corrosion, particularly its opening profile. In light of this, in the following experimental investigation regarding the GPEC imaging of subsurface corrosion, the x-direction gradient of x-component of the net magnetic field $g_x(B_x)$ was measured by using a magnetic sensor which is placed right under the lateral winding of a rectangular coil.

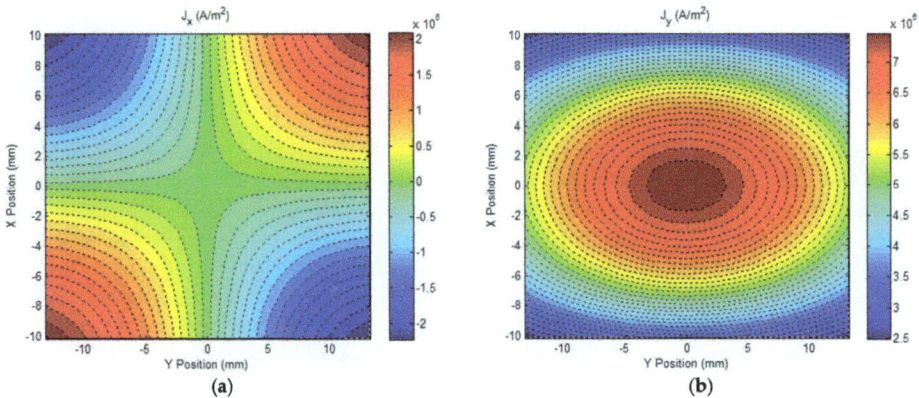

Figure 5. The computed density of eddy currents over the plate back surface. (**a**) x-component of the eddy current density J_x; (**b**) y-component of the eddy current density J_y.

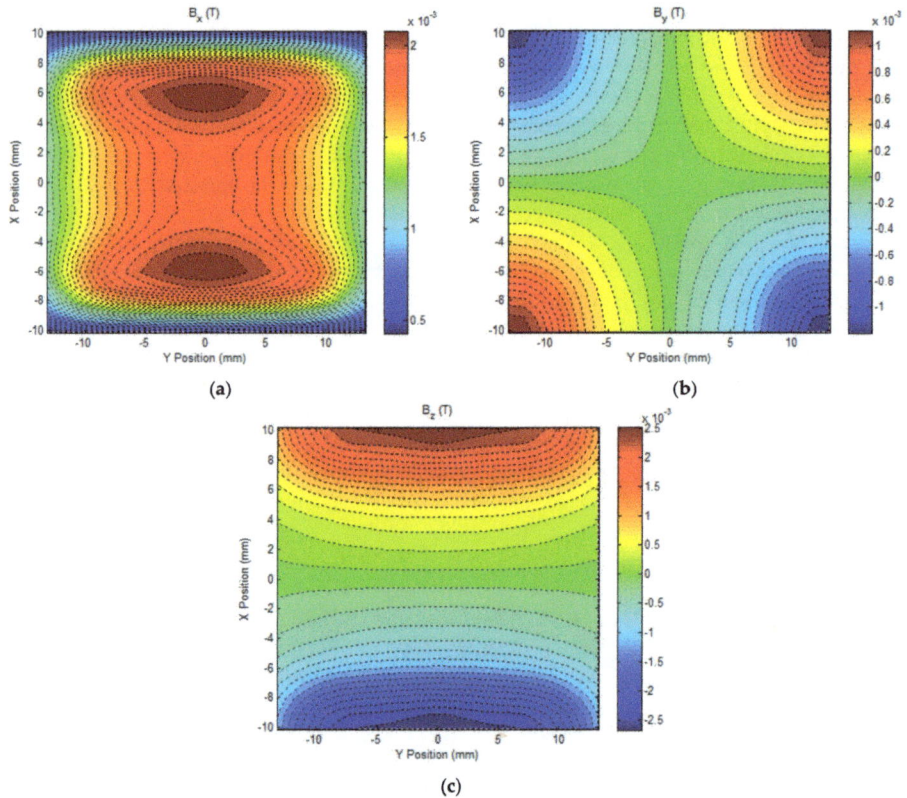

Figure 6. The computed net magnetic field above the plate upper surface. (a) B_x; (b) B_y; (c) B_z.

3. Experiments

3.1. System Setup

A corrosion imaging system was set up to further assess the applicability of the proposed GPEC probe in the imaging of subsurface corrosion in nonmagnetic planar conductors. The system setup is shown in Figure 7. The parameters of the rectangular coil were the same as those tabulated in Table 1. The pulse repetition frequency and duty cycle of the excitation current driving the rectangular coil were set as 100 Hz and 50%, respectively, whilst it had a maximum amplitude of 0.3 A and a rising time of 58.7 µs, which were directly measured from the acquired current signal. In a bid to acquire signals of the gradient field with high sensitivity, a tunnel magneto-resistance sensor (MultiDimension TMR-4002) was adopted. It was deployed right under the lateral winding of the rectangular coil. It is noted that the component of the net magnetic field and direction of its gradient field, which is sensed by the sensor, were both parallel to the axis of the rectangular coil. For example, for the case shown in Figure 7, the x-direction gradient field of the x-component of the net magnetic field $g_x(B_x)$ was the main component measured.

A plate of aluminum alloy with subsurface corrosion in different profiles and sizes was fabricated in a bid to simulate the nonmagnetic planar conductor with subsurface corrosion. It was taken as the sample under inspection. The conductivity and thickness of the plate were 33.6 MS/m and 6 mm. It is noted that the plate conductivity was measured via the direct current potential drop method [25,26]. The profile and opening size of each corrosion are tabulated in Table 3.

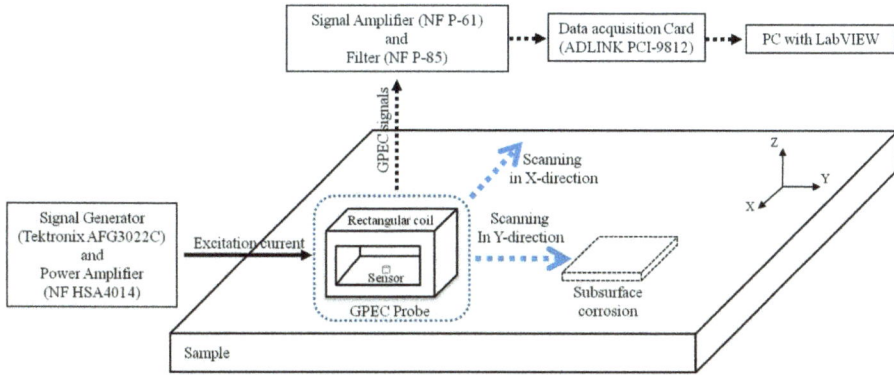

Figure 7. Schematic illustration of the corrosion imaging system with the proposed GPEC probe and sample.

Table 3. Profiles and sizes of subsurface corrosion.

Corrosion Number	Corrosion Profile (in XY Plane)	Sizes
#1		30 mm × 5 mm (diameter × depth)
#2		20 mm × 5 mm (diameter × depth)
#3		30 mm × 4 mm (diameter × depth)
#4		20 mm × 4 mm (diameter × depth)
#5		20 mm × 20 mm × 4 mm (length × width × depth)
#6	C1 / C2	C1: 20 mm × 4 mm (diameter × depth) C2: 30 mm × 4 mm (diameter × depth)

During experiments, the 2D probe scanning over the plate surface was carried out with a spatial resolution of 0.5 mm. It is noteworthy that in order for the corrosion imaging to be carried out, the 2D probe scanning was conducted twice. After completing the first-round probe scanning with the axis of the rectangular coil parallel to the X axis alongside the signal acquisition of $g_x(B_x)$ at each scanning position, the probe was rotated by 90° in the XY plane, and the second-round probe scanning was carried out along with the measurement of the y-direction gradient field of the y-component of the net magnetic field $g_y(B_y)$. At each scanning position, particularly within the corrosion region, the peak values (PVs) of the acquired gradient-field signals regarding $g_x(B_x)$ and $g_y(B_y)$—which are null when the probe is placed over a flawless area of the sample—are extracted in an effort to construct the corrosion images. The final image of every corrosion is derived from the superposition of images regarding $g_x(B_x)$ and $g_y(B_y)$.

3.2. Imaging Results and Discussion

Prior to the corrosion imaging, the system was further calibrated by placing the probe above the defect-free area of the sample, and setting the magnitude of the acquired signal as zero. The gradient-field signals from the GPEC probe deployed right over the edges of the subsurface corrosion involving Corrosion #1, Corrosion #2 and Corrosion #4 were firstly investigated, and are exhibited in Figure 8.

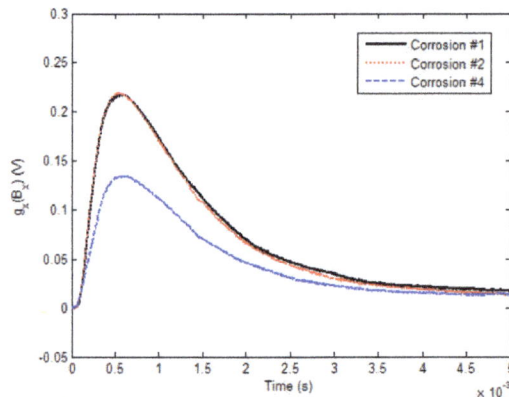

Figure 8. Gradient-field signals with respect to Corrosion #1, Corrosion #2 and Corrosion #4.

It can be seen from Figure 8 that the magnitude of the gradient-field signal changes with the corrosion volume. The PV of the signal rises as the corrosion depth increases, whilst it is insensitive to the corrosion diameter. This indicates that the variation in the gradient-field signal, as well as its PV, mostly relies on the interface thickness at the boundary of a material discontinuity, and could be exploited for approximating the depth of the subsurface corrosion.

Following the probe scanning procedure, multiple scanning curves (PV of the GPEC signal against the probe location) were acquired, and subsequently adopted for deriving the corrosion images. It is noted that apart from Corrosion #6, the coordinate of the corrosion center was set as X = 0, Y = 0. The center of Corrosion #6 was set at the joining point of two circle-shaped regions (C1 and C2 in Table 3). Figure 9 presents the scanning curves (PVs of $g_x(B_x)$ vs. probe positions) when the probe scans over Corrosion #1. The scanning curves along the central and offset scanning lines, which are illustrated in Figure 9c, are shown in Figure 9b along with the locations of the corrosion edges implicated by the dotted lines. Note that the scanning curves are normalized with the maximum amplitude corresponding to 1. It is noticeable from Figure 9 that the corrosion edges can be localized without much loss in accuracy by finding the peaks of the acquired scanning curves. This benefits the subsequent identification of the corrosion profile and the estimation of the corrosion size via GPEC imaging of subsurface corrosion.

After the probe scanning and data processing, the final image for each corrosion was constructed and shown in Figure 10, along with the true corrosion profile indicated by the dashed line. It is noteworthy that all the data i.e., PVs in corrosion images were normalized with the maximum PV corresponding to 1, following which the hue-saturation value (HSV) color was utilized to map the normalized PVs for the production of corrosion images.

It can be qualitatively observed from Figure 10 that: (1) each subsurface corrosion can be detected; and (2) the profile/shape of each subsurface corrosion can be directly visualized and identified by using the proposed GPEC probe with uniform field excitation. The corrosion profile implied by the image is in good agreement with the true profile of each corrosion. This is beneficial to the determination regarding the profile of subsurface corrosion in the conductor. Further analysis has revealed that the difference in HSV components, i.e., image contrast of the corrosion image, is highly dependent on the corrosion depth. This is because the distortion of the eddy currents due to subsurface corrosion—which results in the gradient-field signal—depends mostly on the interface thickness at the boundary of the material discontinuity. The image contrast is enhanced as the corrosion depth increases, whilst the image has lower contrast when the detected corrosion is shallow. This implies that: (1) the image quality for corrosion detection and sizing is dependent on the corrosion dimension, particularly its depth; and (2) the image contrast could be used for further evaluation of the corrosion depth. It is also noticeable from Figure 10f that the corrosion image for Corrosion #6 with the complex profile is seemingly "noisy". The reasoning could lie in the fact that the gradient field resulting from the small circle-shaped area (C1 in Table 3) interferes with that of the large section (C2 in Table 3), which leads to a more complicated distribution of the gradient field for Corrosion #6 than the other corrosion scenarios. Whereas the profiles of Corrosion #6 and particularly two corrosion sections (C1 and C2) can still be identified. Therefore, by using the proposed GPEC probe with uniform field excitation, multiple subsurface corrosions close to each other could be individually detected together with their profiles visualized by corrosion imaging.

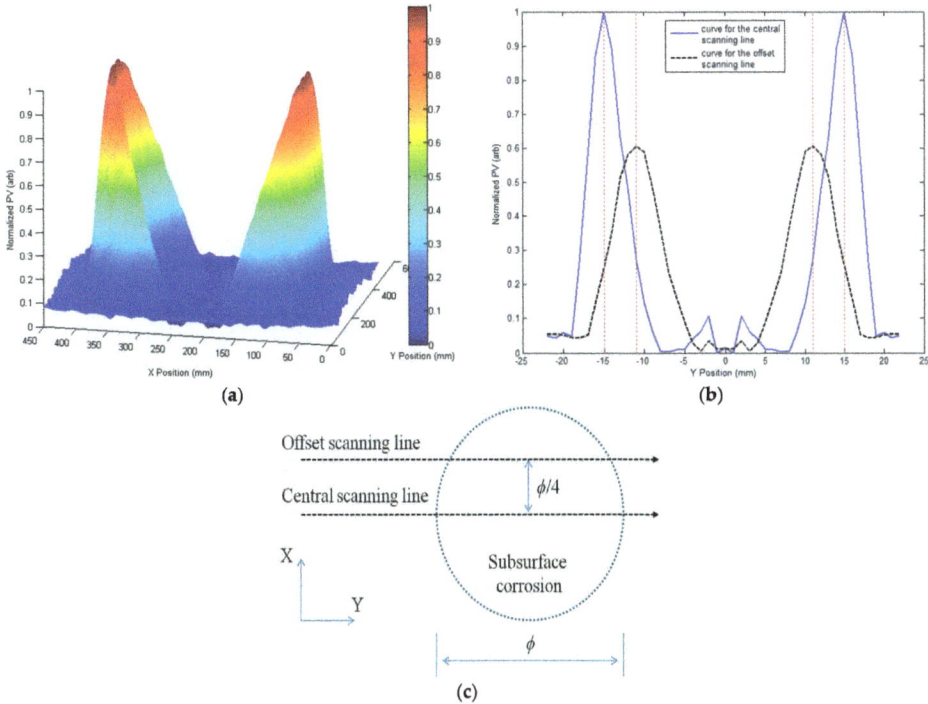

Figure 9. Peak values (PVs) of $g_x(B_x)$ vs. probe positions. (**a**) Scanning curves for Corrosion #1; (**b**) scanning curves along the central and offset scanning lines; (**c**) schematic illustration of the central and offset scanning lines.

Following the identification of the corrosion profile, the corrosion sizing based on acquired images was further investigated. The maximum magnitude in each image was extracted in an effort to quantitatively assess the diameter/length of each subsurface corrosion. The comparison regarding the corrosion size between the estimated and true values is presented in Table 4. It can be seen from Table 4 that for the given subsurface corrosion the approximated diameter/length of each subsurface corrosion is in good agreement with the true value. The maximum relative error is less than 10%. It is also noticeable that the discrepancy between the estimated and true values increases when either the depth or diameter/length of the corrosion has decreased, which is because of the drop in evaluation sensitivity due to less perturbation of eddy currents at the corrosion boundary. This indicates that the detection and sizing of subsurface corrosion depends on its dimension. Additional image processing techniques should be employed in an effort to enhance the accuracy in sizing of either shallow or small-volume corrosion. Nevertheless, it is noteworthy from Figure 10 and Table 4 that the corrosion imaging based on GPEC with uniform field excitation is promising in not only the identification of the corrosion profile but also the quantitative evaluation of the opening size of subsurface corrosion in nonmagnetic planar conductors without much loss in accuracy.

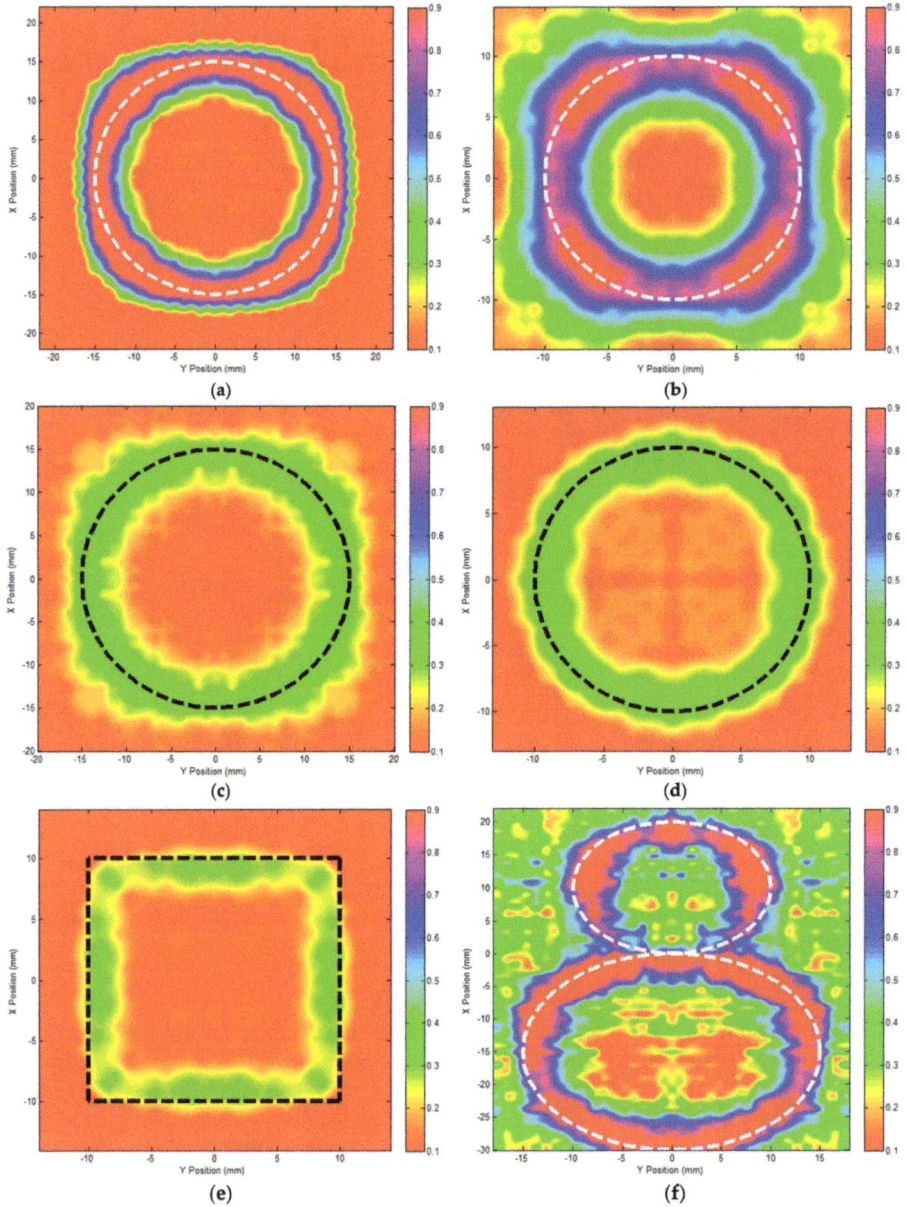

Figure 10. Images of subsurface corrosion via GPEC with uniform field excitation. (**a**) Corrosion #1; (**b**) Corrosion #2; (**c**) Corrosion #3; (**d**) Corrosion #4; (**e**) Corrosion #5; (**f**) Corrosion #6.

Table 4. Comparison between the estimated corrosion opening size and true values.

Corrosion Number	True Diameter/Length	Estimated Diameter/Length	Relative Error
#1	30 mm	30.3 mm	1.0%
#2	20 mm	19.7 mm	1.5%
#3	30 mm	28.6 mm	4.7%
#4	20 mm	21.8 mm	9.0%
#5	20 mm	18.5 mm	7.5%
#6	C1: 20 mm	C1: 21.0 mm	C1: 5.0%
	C2: 30 mm	C2: 29.1 mm	C2: 3.0%

4. Conclusions

In this paper, a GPEC probe with uniform field excitation was proposed for the imaging of subsurface corrosion within nonmagnetic conductors. A 3D analytical model of the proposed GPEC probe was established, and the closed-form expression of the magnetic field, gradient field and eddy current density were formulated. The characteristic uniformities of the electromagnetic field, including the magnetic field above the conductor and the eddy current density on the conductor back surface, were investigated via a series of simulations based on analytical modeling. Following this, experiments regarding the imaging of subsurface corrosion by using the proposed GPEC probe were conducted. It can be found from the experimental results that: (1) the magnitude of the GPEC signal, image contrast and sizing accuracy regarding subsurface corrosion are dependent on the corrosion dimension, especially its depth; and (2) the proposed GPEC probe with uniform field excitation is capable of not only identifying the corrosion profile, but also of providing an estimation regarding the opening size of subsurface corrosion in nonmagnetic planar conductors without much loss in accuracy.

Following the current work, further research regarding the proposed probe includes: (1) quantitative evaluation of subsurface corrosion in terms of corrosion sizing involving the depth and volume; (2) assessment regarding the capability of the proposed probe in detection, sizing and shape determination of the practical/natural corrosion; and (3) signal/image processing for detection and sizing of superficial and small-volume corrosion.

Acknowledgments: The work is supported by the National Magnetic Confinement Fusion Program of China (Grant No. 2013GB113005), Natural Science Foundation of China (Grant No. 51477127), Natural Science Basic Research Plan in Shaanxi Province of China (Program No. 2016JM5075) and STI Joint Committee Malaysia-China.

Author Contributions: Yong Li, Shuting Ren and Bei Yan designed and carried out the experiments; Yong Li, Ilham Mukriz Zainal Abidin and Yi Wang established the analytical model and formulated the closed-form expressions; Shuting Ren and Bei Yan analyzed the data; Yong Li wrote the paper.

Conflicts of Interest: The authors declare no conflict of interest.

References

1. Calabrese, L.; Proverbio, E.; Galtieri, G.; Borsellino, C. Effect of Corrosion Degradation on Failure Mechanisms of Aluminium/Steel Clinched Joints. *Mater. Des.* **2015**, *87*, 473–481. [CrossRef]
2. Saha, D.; Pandya, A.; Singh, J.K.; Paswan, S.; Singh, D.D.N. Role of Environmental Particulate Matters on Corrosion of Copper. *Atmos. Pollut. Res.* **2016**, *7*, 1037–1042. [CrossRef]
3. Trinity NDT. Available online: http://trinityndt.com/ (accessed on 19 July 2017).
4. Xu, G.R.; Guan, X.S.; Qiao, Y.L.; Gao, Y. Analysis and Innovation for Penetrant Testing for Airplane Parts. In Proceedings of the 2014 Asia-Pacific International Symposium on Aerospace Technology, Shanghai, China, 24–26 September 2014; pp. 1438–1442.
5. Javier, M.; Jaime, G.; Ernesto, S. Non-Destructive Techniques Based on Eddy Current Testing. *Sensors* **2011**, *11*, 2525–2565.
6. Singh, R. *Applied Welding Engineering: Processes, Codes and Standards*, 1st ed.; Butterworth-Heinemann: Waltham, MA, USA, 2012; ISBN 978-0123919168.

7. Zhou, D.Q.; Li, Y.; Yan, X.; You, L.; Zhang, Q.; Liu, X.B.; Qi, Y. The Investigation on the Optimal Design of Rectangular PECT Probes for Evaluation of Defects in Conductive Structures. *Int. J. Appl. Electromagn. Mech.* **2013**, *42*, 319–326.

8. Li, Y.; Yan, B.; Li, D.; Li, Y.L.; Zhou, D.Q. Gradient-field Pulsed Eddy Current Probes for Imaging of Hidden Corrosion in Conductive Structures. *Sens. Actuators A Phys.* **2016**, *238*, 251–265. [CrossRef]

9. Li, Y.; Jing, H.Q.; Abidin, I.; Yan, B. A Gradient-Field Pulsed Eddy Current Probe for Evaluation of Hidden Material Degradation in Conductive Structures Based on Lift-Off Invariance. *Sensors* **2017**, *17*, 943. [CrossRef] [PubMed]

10. Dover, W.D.; Collins, R.; Michael, D.H. The Use of Ac-Field Measurements for Crack Detection and Sizing in Air and Underwater. *Philos. Trans. R. Soc. A Math. Phys. Eng. Sci.* **1986**, *320*, 271–283. [CrossRef]

11. LeTessier, R.; Coade, R.W.; Geneve, B. Sizing of Cracks Using the Alternating Current Field Measurement Technique. *Int. J. Press. Vessel. Pip.* **2002**, *79*, 549–554. [CrossRef]

12. Knight, M.J.; Brennan, F.P.; Dover, W.D. Effect of Residual Stress on ACFM Crack Measurements in Drill Collar Threaded Connections. *NDT E Int.* **2004**, *37*, 337–343. [CrossRef]

13. Li, W.; Yuan, X.A.; Chen, G.M.; Yin, X.K.; Ge, J.H. A Feed-Through ACFM Probe with Sensor Array for Pipe String Cracks Inspection. *NDT E Int.* **2014**, *67*, 17–23. [CrossRef]

14. Li, W.; Yuan, X.A.; Chen, G.M.; Ge, J.H.; Yin, X.K.; Li, K.J. High Sensitivity Rotating Alternating Current Field Measurement for Arbitrary-Angle Underwater Cracks. *NDT E Int.* **2016**, *79*, 123–131. [CrossRef]

15. Zhang, Y.; Li, Y. Magnetic-Field-Based 3D ETREE Modelling for Multi-Frequency Eddy Current Inspection. In Proceedings of the IFIP Advances in Information and Communication Technology, Nanchang, China, 22–25 October 2010; pp. 221–230.

16. Theodoulidis, T.; Pichenot, G. Integration of Tilted Coil Models in a Volume Integral Method for Realistic Simulations of Eddy Current Inspections. In Proceedings of the 12th International Workshop on Electromagnetic Nondestructive Evaluation, Cardiff, UK, 19–21 June 2007.

17. Reboud, C.; Theodoulidis, T. Field Computations of Inductive Sensors with Various Shapes for Semi-Analytical ECT Simulation. In Proceedings of the 16th International Workshop on Electromagnetic Nondestructive Evaluation, Chennai, India, 10–12 March 2011.

18. Theodoulidis, T.; Kriezis, E.E. *Eddy Current Canonical Problems (with Applications to Nondestructive Evaluation)*; Tech Science Press: Forsyth, GA, USA, 2006; ISBN 978-0971788015.

19. Theodoulidis, T.P.; Kriezis, E.E. Impedance Evaluation of Rectangular Coils for Eddy Current Testing of Planar Media. *NDT E Int.* **2002**, *35*, 407–414. [CrossRef]

20. Bond, L.J.; Clark, R. Response of Horizontal-Axis Eddy-Current Coils to Layered Media: A Theoretical and Experimental Study. *IEE Proc. A Sci. Meas. Technol.* **1990**, *137*, 141–146. [CrossRef]

21. Li, Y.; Yan, B.; Li, D.; Jing, H.Q.; Li, Y.L.; Chen, Z.M. Pulse-Modulation Eddy Current Inspection of Subsurface Corrosion in Conductive Structures. *NDT E Int.* **2016**, *79*, 142–149. [CrossRef]

22. Li, Y.; Yan, B.; Li, W.J.; Jing, H.Q.; Chen, Z.M.; Li, D. Pulse-Modulation Eddy Current Probes for Imaging of External Corrosion in Nonmagnetic Pipes. *NDT E Int.* **2017**, *88*, 51–58. [CrossRef]

23. Udpa, S.S. *Nondestructive Testing Handbook Volume 5: Electromagnetic Testing*, 3rd ed.; American Society for Nondestructive Testing: Columbus, OH, USA, 2014; ISBN 1-57117-046-4.

24. Kędzia, P.; Czechowski, T.; Baranowski, M.; Jurga, J.; Szcześniak, E. Analysis of Uniformity of Magnetic Field Generated by the Two-Pair Coil System. *Appl. Magn. Reson.* **2013**, *44*, 605–618. [CrossRef] [PubMed]

25. Brinnel, V.; Dobereiner, B.; Munstermann, S. Characterizing Ductile Damage and Failure: Application of the Direct Current Potential Drop Method to Uncracked Tensile Specimens. *Procedia Mater. Sci.* **2014**, *3*, 1161–1166. [CrossRef]

26. Ali, M.R.; Saka, M.; Tohmyoh, H. A Time-dependent Direct Current Potential Drop Method to Evaluate Thickness of an Oxide Layer Formed Naturally and Thermally on a Large Surface of Carbon Steel. *Thin Solid Films* **2012**, *525*, 77–83. [CrossRef]

sensors

MDPI

Article

A Smart Eddy Current Sensor Dedicated to the Nondestructive Evaluation of Carbon Fibers Reinforced Polymers

Mohammed Naidjate [1,2,*], Bachir Helifa [1], Mouloud Feliachi [2], Iben-Khaldoun Lefkaier [1], Henning Heuer [3] and Martin Schulze [3]

[1] Laboratoire de Physique des Matériaux, Université de Laghouat, Laghouat 03000, Algeria; helifa@yahoo.fr (B.H.); lefkaier_ik@yahoo.fr (I.-K.L.)
[2] IREENA-IUT, Université de Nantes, 44602 Saint-Nazaire, France; mouloud.feliachi@univ-nantes.fr
[3] Fraunhofer Institute for Ceramic Technologies and Systems IKTS, 01109 Dresden, Germany; henning.heuer@ikts.fraunhofer.de (H.H.); martin.schulze@ikts.fraunhofer.de (M.S.)
* Correspondence: m.naidjate@lagh-univ.dz; Tel.: +213-668-132-376

Received: 6 July 2017; Accepted: 25 August 2017; Published: 31 August 2017

Abstract: This paper propose a new concept of an eddy current (EC) multi-element sensor for the characterization of carbon fiber-reinforced polymers (CFRP) to evaluate the orientations of plies in CFRP and the order of their stacking. The main advantage of the new sensors is the flexible parametrization by electronical switching that reduces the effort for mechanical manipulation. The sensor response was calculated and proved by 3D finite element (FE) modeling. This sensor is dedicated to nondestructive testing (NDT) and can be an alternative for conventional mechanical rotating and rectangular sensors.

Keywords: multi-element sensor; eddy current; CFRP characterization; nondestructive testing (NDT); electromagnetic field computation; FEM modeling

1. Introduction

Non-destructive testing (NDT) is one of the most common and powerful techniques employed in the inspection of materials during their manufacture or use. When dealing with an electrically conductive body, eddy current non-destructive testing (EC-NDT) is the most efficient way to test the state of health of materials (low cost, readily implemented...). The efficiency of EC-NDT depends directly on the performance of the sensor used. During the last few years, researchers have focused their work to a new generation of EC probes consisting of miniaturized sensors forming a rectangular sensor array that could detect defects with high accuracy, even in complex materials such as carbon fiber-reinforced polymers (CFRP), the material studied as an application in this article. Nevertheless, the highly anisotropic and complex structure of CFRP may require, depending on the type of control to be performed, specific EC sensor configurations. For the detection of defects or the fibers' orientation, actual solutions are based on high frequency sensors [1], rotating sensors [2,3] or rectangular sensors [4,5] These types of sensors are used to draw a polar diagram giving the intensity of the measured signal in terms of the rotation angle of the sensor. The angles of the obtained lobes determine the different fibers orientations whereas their amplitude indicates the position of the ply in the sample.

In this work, a new multi-element sensor array design is suggested with the aim of evaluating CFRP materials. This multi-element sensor is presented as an alternative to the rotating and the rectangular sensors in order to increase the sensitivity and resolution and to reduce the mechanical effort for rotation.

2. Conception

The proposed sensor consists of flat triangular coils combined into an array. Figure 1 displays the design of the sensor's elements as well as the simplified geometry introduced as input to the computational code for simulation.

The triangular form allows a higher flexibility on the desired shapes of the generated electromagnetic (EM) field. The elements are arranged that they can give a large number of possible EM configurations, thereby, avoid the mechanical swiveling of the sensor occurring in conventional characterization. In addition to this, such a structure can cover a relatively wide inspection area which saves the number of manual scanning operations. The size of the sensor elements can be selected to optimize precision in the defect detection.

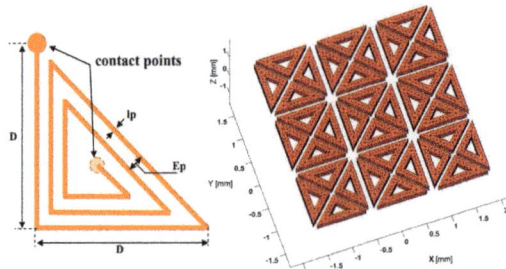

Figure 1. (**Left**) Design of a single element and (**Right**) the simplified array configuration.

The basic idea rests on the fact that the EM fields generated by two parallel wires traversed by currents with the same amplitude and in opposite direction cancel each other. This property is exploited to generate different field forms, acting only on the excitation currents' distribution. Thereby, Figure 2 represents four triangular coils excited in a way that the resulting field is similar to that obtained by a square coil: the fields generated by currents flowing in the "diagonal" conductors oppose and cancel each other. Then, the resulting field is practically that due to the currents flowing in the "external quadrature" conductors.

The mode of calculation is described hereafter.

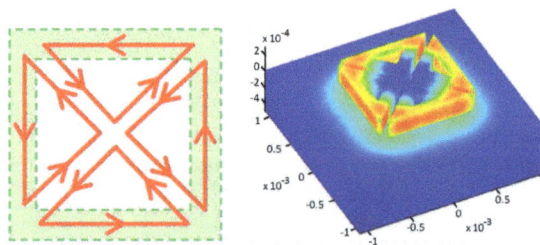

Figure 2. Top view of the current flow upon jointly exciting the four coils (**left**) and perspective view of the three-dimensional distribution of the magnetic potential vector for the four coils (**right**).

3. Sensor Characteristics

3.1. Geometrical Characterization

The proposed sensor is an array assembly of 36 identical coils in the shape of isosceles triangles whose angles at the base worth 45°. In each coil, the developed length or the total length of the wire l_{total} and the total effective surface S_{total} are given by the Equations (1) and (2) respectively (see Appendix A):

$$l_{totale} \approx n\left[\left(2+\sqrt{2}\right)D - (n-1)(l_p + E_p)\left(1+\frac{2}{\tan(\pi/8)}\right)\right] \tag{1}$$

$$S_{totale} \approx \frac{1}{2}\sum_{k=1}^{n}\left[D - (k-1)(l_p + E_p)\left(1+\frac{1}{\tan(\pi/8)}\right)\right]^2 \tag{2}$$

where D is the external rib of the coil (see Figure 1), l_p is the line width, E_p is the inter-lines distance and n is the number of turns. These geometrical parameters are indispensable to calculate the electrical parameters as their influence is direct.

3.2. Electrical Characterization

The theoretical model of a coil is given in Figure 3 [6]. To determine the coil inductance L, and for the sake of accuracy, an evaluation of the stored magnetic energy (Equation (5)) was provided via the FE model developed in Section 4. However, to determine the resistance R and the capacitance C, basic models have been adopted to simplify the calculation. The relations through which the electrical parameters have been estimated are given below:

$$R = \frac{\rho \times l_{totale}}{l_p \times h_p}, \tag{3}$$

$$C = \left[\sum_{k=2}^{n}\left(1/\varepsilon\frac{h_p \times l_k}{E_p}\right)\right]^{-1}, \tag{4}$$

$$L = \frac{\omega}{I^2}\iiint_{\Omega}\frac{1}{\mu}\left|\vec{B}\right|^2 d\Omega, \tag{5}$$

where h_p is the height of the line, ρ is its electrical resistivity of the wire, ε is the electric permittivity, ω is the angular frequency, Ω is the whole computation area (sensor and air box), μ is the magnetic permeability and B is the magnetic flux density.

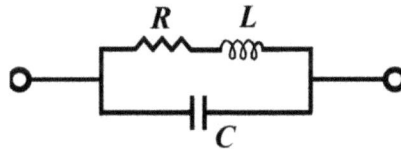

Figure 3. Electrical model of a coil.

3.3. Physical Characterization

Knowledge of its geometrical and electrical characteristics is necessary but insufficient to fully qualify the electromagnetic behaviour of a coil. As an EM sensor, the coil needs to meet other requirements depending on the intended mode of its use. As an emitter, its emissive ability must be calculated. If it's used as receiver, it is necessary to determine its sensitivity and its electrical noise signal. In the proposed sensor, coils have the versatility to work in emission and reception simultaneously or separately, which implies a complete and rigorous study of the sensitive element. Ravat, C. [6] exposes in his work these parameters as follows:

- According to Faraday-Lenz's law, at a frequency f, the sensitivity of a coil is:

$$S = \left|\frac{dV}{dB}\right| = 2\pi f S_{totale}, \tag{6}$$

where dV is the voltage variation provoked by a variation in the received magnetic induction dB.

- The noise of a coil when it is not carrying current is only a thermal agitation noise. This effective voltage v_b at a temperature T and in measuring frequency range Δf is given by;

$$v_b = \sqrt{4K \times T \times R \times \Delta f},\tag{7}$$

where K is Boltzmann's constant.

- The emissive ability P_e is the ratio between the emitted field "B" and the current "I" necessary for its emission:

$$p_e = \frac{B}{I} = \frac{L}{S_{totale}},\tag{8}$$

3.4. Optimization of the Coil

The relationship between the geometrical, electrical and physical characteristics developed previously allows us to study the influence of each parameter and thus to determine the optimum dimensions of the coil appropriate for a desired application. Table 1 provides the characteristics of the selected coil to non-destructively evaluate a CFRP. According to the theoretical model (Figure 3), Figure 4 shows the frequency response of a triangular coil using data given in Table 1. It can be seen that the coil can be used as EM field sensor above 800 kHz where it shows a strong inductive behavior with a phase greater than 60°. The cut-off frequency is much higher than 100 MHz.

Table 1. Numerical values of the coil characteristics calculated at 1 MHz.

	Parameter	Numerical Value	Unit
Coil dimensions	external length D	1	[mm]
	line width l_p	6	[μm]
	inter-line space E_p	3	[μm]
	number of turns n	33	
Electrical parameters	resistance R	4.24	[ohm]
	inductance L	1.44	[μH]
	capacity C	3.5	[fF]
	sensitivity S	35	[V/T]
	noise voltage v_b	0.83	[μV]
	emissive ability P_e	254	[mT/A]

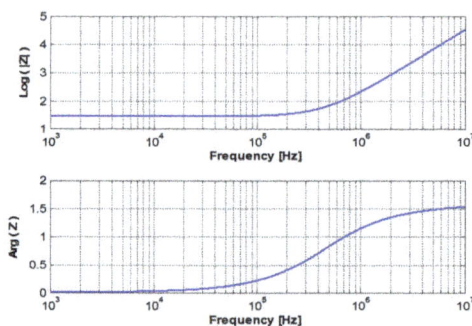

Figure 4. Frequency response of the sensor.

4. Modeling

After the construction of the geometry and the mesh generation using the open-source software GMSH, the problem data are sent to our 3D finite element solver in which was implemented the magneto-dynamics formulation AV-A (Equation (9)); mathematical model chosen to describe the EM behaviour of our problem. The calculations are carried out in the harmonic regime. A penalty term is introduced to ensure the uniqueness of the solution [7]:

$$
\begin{cases}
\vec{\nabla} \times \frac{1}{\mu} \vec{\nabla} \times \vec{A} - \vec{\nabla}\left(\frac{1}{\mu}\vec{\nabla}\cdot\vec{A}\right) + \bar{\bar{\sigma}}\left(j\omega\vec{A} + \vec{\nabla}V\right) = \vec{J}_s \\
\vec{\nabla}\cdot\left(j\omega\bar{\bar{\sigma}}\left(\vec{A} + \vec{\nabla}V\right)\right) = 0
\end{cases}
\tag{9}
$$

where $\vec{A}$ and V are respectively the magnetic vector potential and electric scalar potential, μ is the magnetic permeability and $\bar{\bar{\sigma}}$ is the electrical conductivity tensor given according to the ply orientation by [2]:

$$
\bar{\bar{\sigma}} = \begin{pmatrix}
\sigma_{//}\cos^2(\theta) + \sigma_{\perp}\sin^2(\theta) & \frac{\sigma_{//}-\sigma_{\perp}}{2}\sin(2\theta) & 0 \\
\frac{\sigma_{//}-\sigma_{\perp}}{2}\sin(2\theta) & \sigma_{//}\cos^2(\theta) + \sigma_{\perp}\sin^2(\theta) & 0 \\
0 & 0 & \sigma_{zz}
\end{pmatrix},
\tag{10}
$$

where $\sigma_{//}$ is the electrical conductivity in the fibers direction, $\sigma_{\perp}$ is the conductivity in the transverse direction of the fibers and σ_{zz} is the conductivity in the direction of the plies stacking.

5. Results

The EC-NDT concept is based on the distribution and circulation of the induced currents in the component being inspected. This distribution is strongly linked to the profile of the excitation EM field. With this in mind, numerical experiments were carried out to discern the ability of our sensor to emulate the EM field configurations obtained by conventional sensors such as a rectangular coil.

5.1. Sensor and EM Field

Figure 5 shows that the electromagnetic field generated by a set of triangular elements excited simultaneously is similar to the EM field created by a conventional rectangular coil. The electric field calculated at the front surface of the load illustrated in Figure 6 confirms this equivalence of the global EM behaviour for the two systems. Nevertheless, we note some electric field irregularities in the case of the multi-element sensor, which is due to the discontinuities in the geometry of the inductor and to the current singularities in the bends of triangles.

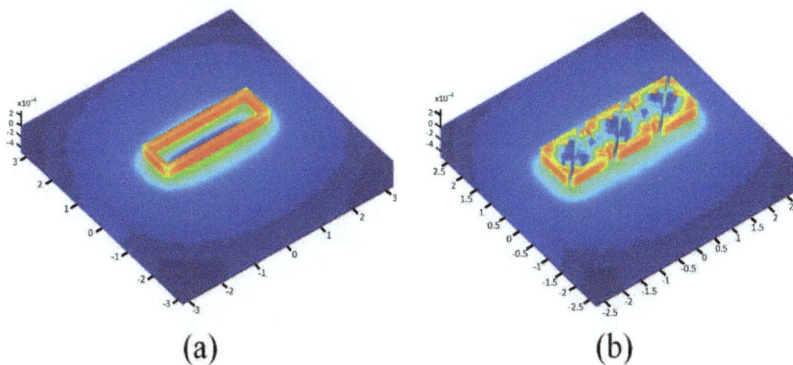

(a)   (b)

Figure 5. Magnetic vector potential calculated for a system of: (**a**) rectangular coil; (**b**) proposed multi-element sensor.

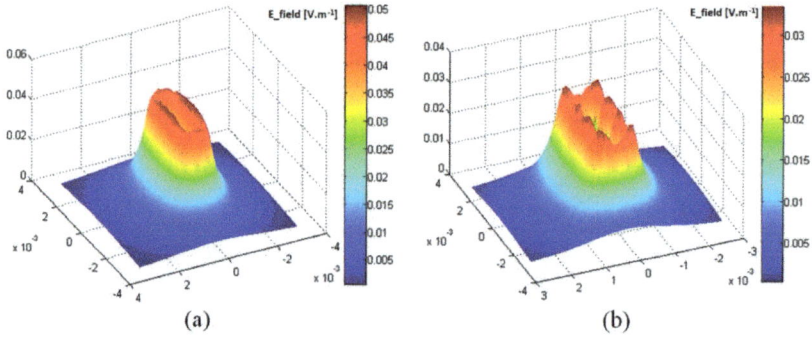

Figure 6. Electric field magnitude at the front surface of the load (z = 0): (**a**) rectangular coil; (**b**) multi-element sensor.

The interaction between the excited elements and those "at rest" was studied too with the aim of evaluating the coupling effect. The obtained results displayed in Figure 7 show that the non-excited adjacent elements are without a slightest action on the configuration of the field or on its amplitude at the operating frequency of 1 MHz.

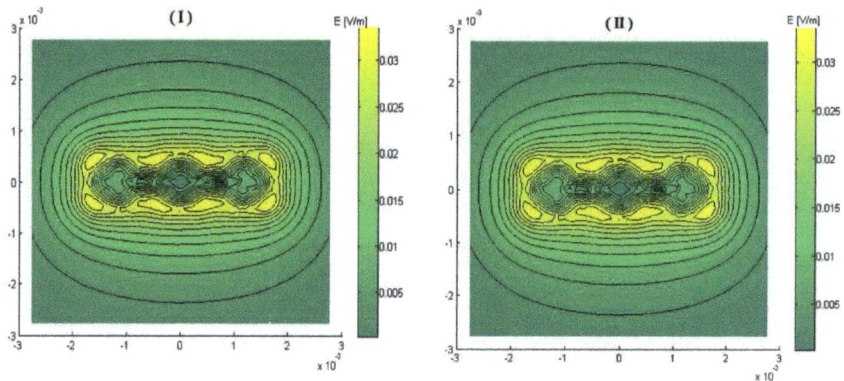

Figure 7. Electric field magnitude at the front surface of the load (z = 0): (**I**) without regard to non-excited elements (σ = 0); (**II**) non-excited elements are physically represented (σ = 30, 6 × 10^6 S/m).

In addition to the foregoing, the proposed sensor allows a high flexibility in terms of modes of excitation and measurement. The figure below illustrates different field configurations obtained for different excitations. It can be seen in Figure 8 that the sensor can substitute rectangular coil oriented at $0°, 45°, 90°$ and $-45°$ without recurring to mechanical rotation. This property will be exploited and applied to a laminate of CFRP.

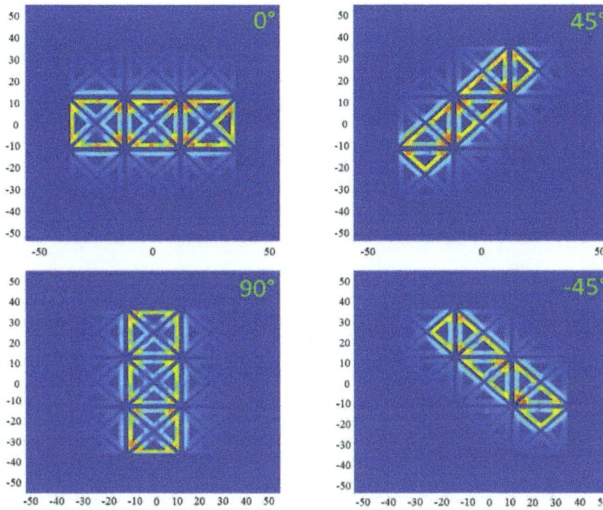

Figure 8. Field configurations generated by the multi-element sensor equivalent to a rectangular coil oriented at 0°, 45°, 90° and −45°.

5.2. Application to CFRP

The modeled system is a stack of four plies oriented at [0°, 45°, 90°, −45°]. The physical and geometrical characteristics are given by Table 2. Figure 9 illustrates the distribution of eddy currents produced by a rectangular inductor oriented at 0° and its equivalent generated by the multi-element sensor. It is noted that the distributions of eddy currents generated by the two systems, in each ply of the laminate, is typically identical. This leads to expect, consequently, an analogue dissipated power, hence an identical response in terms of impedance.

Table 2. Numerical values of the system characteristics

	Parameter	Numerical Values	Unit
	Number of plies	4	
Laminate	Fibers orientation	0°, 45°, 90° and −45°	[°]
	Conductivity ($\sigma_{//}, \sigma_{\perp}, \sigma_{zz}$)	(10^4, 2×10^2, 10)	[S/m]
	Ply thickness	125	[μm]
	Number of coils	36	
	Gap inter-coils	0.08	[mm]
Sensor	Lift-off	0.125	[mm]
	Current intensity	20	[mA]
	Frequency	1	[MHz]

Furthermore, the results presented in Figure 10 prove that the proposed sensor can detect the orientation of the plies and their order of stacking. The comparison between the amplitudes of peaks shows that they are decreasing according to the stacking order of the plies. However, our values do not coincide with those calculated by [3] due to the mismatch of the two systems (number of turns and dimensions of coils). We note also that there is no large difference between the peaks at 45° and 90°; this can be explained by the change of sizes (length to width ratio) of the equivalent rectangular coil generated at 45° and 90° (see Figure 8).

1st Ply 0° →

2nd Ply 45° ╱

3rd Ply 90° ↑

4th Ply -45° ╲

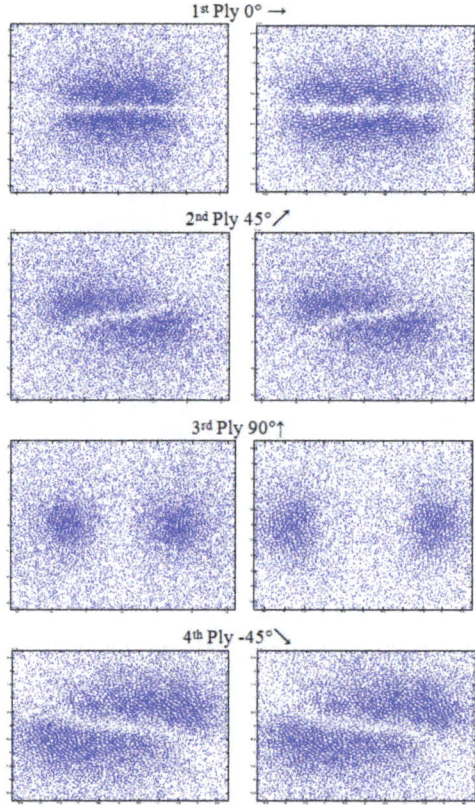

Figure 9. Eddy current distribution in different plies of laminate ($0°$, $45°$, $90°$, $-45°$); at the right caused by the multi-element sensor and at left by a rectangular coil (3.2 mm $\times$ 1 mm).

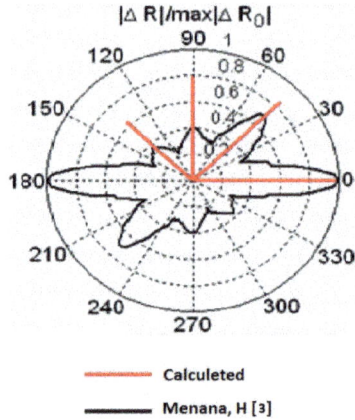

|Δ R|/max|Δ R_0|

——— Calculeted

——— Menana, H [3]

Figure 10. The normalized resistance of the coil as function of its rotation angle above a laminate of four plies [$0°$, $45°$, $90°$, $-45°$].

6. Conclusions

A new design of an eddy current multi-element sensor is proposed. The EM field computation results reveal that this sensor is able to control the EM field shape and generate the plurality of configurations often needed in EC–NDT of carbon fiber-reinforced polymers. The application of the proposed sensor on a sample of CFRP shows its capability to detect the plies' orientations and their stacking order by acting only on the excitation currents. This sensor can thus be an alternative to the mechanical rotating sensors and the rectangular sensors.

Acknowledgments: This research was partwise funded by the European Union Regional Development Fund (EFRE) and the Free State of Saxony (grant "3D-Fast", No. 100224749). The authors would like to thank the mentioned institutions for providing the funding.

Author Contributions: M.N. and B.H. modeled the sensor, wrote the manuscript, contributed to discussion, and reviewed; M.F. and I.-K.L. presented the research subject, provided guidance in modeling, contributed to discussion, and reviewed; H.H. and M.S. conceived and designed the experiments.

Conflicts of Interest: The authors declare no conflict of interest.

Appendix A

Appendix A.1. Calculation of the Developed Length of the Coil

The coil is composed of n triangular turns (see Figure 1) numbered from the outside to the inside. These triangular turns are isosceles whose angles at the base worth 45°. The lengths of the three edges of the *k-th* turn are:

$$d_1(k) \approx d_2(k) = D - (k-1)(l_p + E_p)\left(1 + \frac{1}{tg\frac{\pi}{8}}\right), \tag{A1}$$

$$d_3(k) \approx D\sqrt{2} - 2(k-1)\left(\frac{l_p + E_p}{tg\frac{\pi}{8}}\right), \tag{A2}$$

where d_1 and d_2 are the identical edges, d_3 is the hypotenuse.

Therefore, the developed length of a turn number k is:

$$l_k \approx d_1(k) + d_2(k) + d_3(k), \tag{A3}$$

Thus, the total developed length of the entire coil is simply the sum of l_k:

$$l_{totale} \approx \sum_{k=1}^{n} l_k, \tag{A4}$$

$$l_{totale} \approx \sum_{k=1}^{n}\left(2\left[D - (k-1)(l_p + E_p)\left(1 + \frac{1}{tg\frac{\pi}{8}}\right)\right] + \left[D\sqrt{2} - 2(k-1)\left(\frac{l_p + E_p}{tg\frac{\pi}{8}}\right)\right]\right), \tag{A5}$$

Or more simply:

$$l_{totale} \approx n\left[\left(2 + \sqrt{2}\right)D - (n-1)(l_p + E_p)\left(1 + \frac{2}{\tan(\pi/8)}\right)\right], \tag{A6}$$

Appendix A.2. Calculation of the Total Area

The surface enclosed by the turn number k is:

$$S_k \approx \frac{1}{2}\left[D - (k-1)(l_p + E_p)\left(1 + \frac{1}{\tan(\pi/8)}\right)\right]^2, \tag{A7}$$

Then, the total effective surface of the coil is:

$$S_{totale} \approx \frac{1}{2}\sum_{k=1}^{n}\left[D - (k-1)\left(l_p + E_p\right)\left(1 + \frac{1}{\tan(\pi/8)}\right)\right]^2,\qquad (A8)$$

References

1. Heuer, H.; Schulze, M.; Pooch, M.; Gäbler, S.; Nocke, A.; Bardl, G.; Cherif, C.; Klein, M.; Kupke, R.; Vetter, R.; et al. Review on quality assurance along the CFRP value chain—Non-destructive testing of fabrics, preforms and CFRP by HF radio wave techniques. *Compos. Part B Eng.* **2015**, *77*, 494–501. [CrossRef]
2. Mook, G.; Lange, R.; Koeser, O. Non-destructive characterization of carbon-fiber reinforced plastics by means of eddy currents. *Compos. Sci. Technol.* **2001**, *61*, 865–873. [CrossRef]
3. Menana, H.; Feliachi, M. Non destructive evaluation of the conductivity tensor of a CFRP plate using a rotating eddy current sensor. In Proceedings of the XIV International Symposium on Electromagnetic Fields, Arras, France, 10–12 September 2009.
4. Savin, A.R.; Grimberg, R.; Chifan, S. Evaluation of delamination in Carbon Fiber Composites Using the Eddy Current Method. In Proceedings of the 15th World Conference on Non-Destructive Testing, Roma, Italy, 15–21 October 2000.
5. Grimberg, R.; Savin, A.; Steigmann, R.; Bruma, A. Eddy current examination of carbon fibres in carbon-epoxy composites and Kevlar. In Proceedings of the 8th International Conference of the Slovenian Society for Non-Destructive Testing, Portorož, Slovenia, 1–3 September 2005; pp. 223–228.
6. Ravat, C. Conception de Multicapteurs à Courants de Foucault et Inversion des Signaux Associés Pour le Contrôle non Destructif. Ph.D. Thesis, Sciences and Technologys of Information and telecommunications Systems, University of Paris-Sud, Paris, France, 2008.
7. Helifa, B. Contribution à la Simulation du CND par Courants de Foucault en vue de la Caractérisation des Fissures Débouchantes. Ph.D. Thesis, Computer and Mathematical Sciences and Technologies, Nantes University, Nantes, France, 2012.

sensors

MDPI

Letter

A Study of Applying Pulsed Remote Field Eddy Current in Ferromagnetic Pipes Testing

Qingwang Luo [1,2], Yibing Shi [1,2,*], Zhigang Wang [2], Wei Zhang [1,2] and Yanjun Li [1,2]

1   Center for Information Geoscience, University of Electronic Science and Technology of China,
     Chengdu 611731, China; qwluo@std.uestc.edu.cn (Q.L.);
     weizhang@uestc.edu.cn (W.Z.); yjli@uestc.edu.cn (Y.L.)
2   School of Automation Engineering, University of Electronic Science and Technology of China,
     Chengdu 611731, China; wangzhigang@uestc.edu.cn
*   Correspondence: ybshi@uestc.edu.cn

Academic Editor: Vittorio M. N. Passaro
Received: 24 March 2017; Accepted: 3 May 2017; Published: 5 May 2017

Abstract: Pulsed Remote Field Eddy Current Testing (PRFECT) attracts the attention in the testing of ferromagnetic pipes because of its continuous spectrum. This paper simulated the practical PRFECT of pipes by using ANSYS software and employed Least Squares Support Vector Regression (LSSVR) to extract the zero-crossing time to analyze the pipe thickness. As a result, a secondary peak is found in zero-crossing time when transmitter passed by a defect. The secondary peak will lead to wrong quantification and the localization of defects, especially when defects are found only at the transmitter location. Aiming to eliminate the secondary peaks, double sensing coils are set in the transition zone and Wiener deconvolution filter is applied. In the proposed method, position dependent response of the differential signals from the double sensing coils is calibrated by employing zero-mean normalization. The methods proposed in this paper are validated by analyzing the simulation signals and can improve the practicality of PRFECT of ferromagnetic pipes.

Keywords: PRFECT; ANSYS software; LSSVR; zero-crossing time; double sensing coils; Wiener deconvolution filter; zero-mean normalization

1. Introduction

Ferromagnetic pipes are largely used in the transportation and exploration of oil and gas, and their monitoring and prevention of corrosion is highly important [1–3]. Remote Field Eddy Current Testing (RFECT) is primarily interesting because of its advantages, such as equal sensitivity to inner and outer defects, insensitivity to lift-off effects and contactless testing [4,5]. In order to overcome some inherent drawbacks of RFECT [4,5], such as the single spectrum, long probes, weak testing signal, and so on, studies are further focused on applying Pulsed Remote Field Eddy Current Testing (PRFECT) in ferromagnetic pipes [6–10]. A majority of these studies are performed in time domain [6–10], and the voltage magnitude and the zero-crossing time are extracted to analyze the effects of pipe inner diameter and pipe thickness, respectively.

In order to study the features of applying PRFECT in ferromagnetic pipes, the authors simulated the practical PRFECT of pipes by using ANSYS software. In the simulation, the transmitter and sensing coil moved simultaneously with the step of 10 mm, only one defect was built, and the simulation was performed 100 times. Because the pipe thickness affects the value of the zero-crossing time [6–10], in the signal processing, the zero-crossing time of every testing signal was extracted by employing Least Squares Support Vector Regression (LSSVR) [11,12] to analyze the pipe thickness, and came to the conclusion that, in the zero-crossing time, a secondary peak appeared when the transmitter passed by the defect. The secondary peak appeared in practical PRFECT of pipes is nearly not studied all over the world, even though the secondary peak leads to a wrong assessment of the defect, including the quantification and the localization. This paper studies the secondary peaks in PRFECT of ferromagnetic pipes, and provides methods to eliminate the secondary peaks. According to the authors' studies on RFECT in ferromagnetic pipes [13–15], double sensing coils are set in the transition zone and the Wiener deconvolution filter is used to remove the secondary peak. In the paper, the errors caused by the double sensing coils are also discussed and calibrated. This paper provides methods to process practical problems of PRFECT of ferromagnetic pipes and they are verified by analyzing the simulation signals.

2. Methods and Models

2.1. LSSVR

Because of the good approximation accuracy and generalization ability of Least Squares Support Vector Regression (LSSVR) [11,12], this paper used LSSVR to extract the zero-crossing time from simulation signals to analyze the wall thickness feature of practical PRFECT in ferromagnetic pipes. In order to ensure the uniqueness of the zero-crossing time extracted from a simulation signal and to simplify the inversion model, only parts of a simulation signal are used to fit the inverse function, as shown in Figure 1.

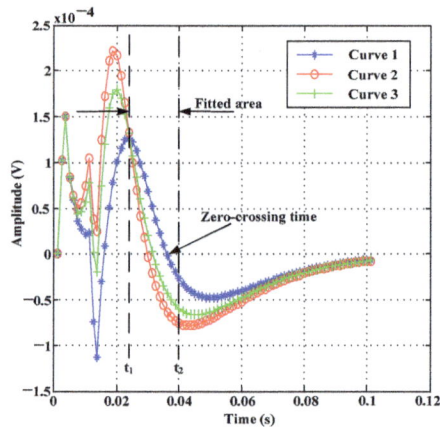

Figure 1. Schematic layout of the fitted area nearby zero-crossing time: Curve 1 is the induced signal when both sensing coil and transmitter are placed non-defect area; Curve 2 is the induced signal when only sensing coil is placed defect area; and Curve 3 is the induced signal when only the transmitter is placed defect area.

As shown in Figure 1, the zero-crossing time is between t_1 and t_2. In order to obtain the zero-crossing time of every simulation signal, the inversion model built with LSSVR [11,12] can be described as follows,

$$f(x) = \sum_{i=1}^{m} \alpha_i \kappa(x, x_i) + b \tag{1}$$

where $f(x)$ is the inversion time, $\kappa(x, x_i)$ is the Radial Basis Function (RBF), x_i is the amplitude between t_1 and t_2, m is the number of amplitudes, and α_i and b are the coefficients.

The procedures of obtaining $f(x)$ are realized in MATLAB with the LSSVR toolkit, and $f(0)$ is used to inverse the zero-crossing time of every simulation signal. The inversion procedures are operated 200 times, and it takes 40 s in total.

2.2. Testing Model

The zero-crossing time is extracted to analyze the features of pipe thickness in PRFECT of pipes by employing LSSVR, and a secondary peak is found when the transmitter passed by defects. To remove the secondary peak, double sensing coils are set in transition zone, as shown in Figure 2.

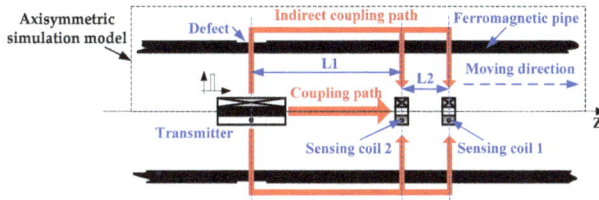

Figure 2. The model used for removing secondary peaks in Pulsed Remote Field Eddy Current Testing (PRFECT) of ferromagnetic pipes.

In Figure 2, sensing coil 1 and sensing coil 2 are both set in the transition zone, and they are made of the same material and are the same sizes. The transmitter is employed to excite a pulsed signal at a low frequency (0–100 Hz). Pulsed signal is regarded as a superposition of sinusoidal signals with different frequencies. Wall thickness affects these sinusoidal signals following the well-known "skin-effect" Equation [6],

$$A = A_0 e^{-d\sqrt{\pi f \mu \sigma}} \sin(2\pi f t - d\sqrt{\pi f \mu \sigma}) \tag{2}$$

where A is the induced signal at any depth of the sample, A_0 is the amplitude of excited signal, f is the frequency of excited signal, d is the distance of the signal penetrated, t is signal time, and μ and σ are the permeability and conductivity, respectively.

According to Equation (2), at any time (t), the influence of a defect on the amplitude of a sample can be given by,

$$A_c = A_0 e^{-(d-\Delta d)\sqrt{\pi f \mu \sigma}} \tag{3}$$

where Δd is the thickness of a defect.

Taking the logarithm of both sides of Equation (3), and transforming,

$$\ln A_c = \ln A_0 e^{-d\sqrt{\pi f \mu \sigma}} + \Delta d\sqrt{\pi f \mu \sigma} \tag{4}$$

According to Equation (4), the influence of a defect on the amplitude of a sample exhibits a superimposed effect. Based on Equation (4), the influence of defects nearby the transmitter on the testing signals from double sensing coils can be described as follows,

$$\begin{cases} A_1(T) = A_1{}'(T) + \Delta A(T) & (5a) \\ A_2(T) = A_2{}'(T) + \Delta A(T) & (5b) \end{cases}$$

where $A_1(T)$ and $A_2(T)$ are the testing signal from sensing coil 1 and sensing coil 2, respectively, $\Delta A(T)$ is the influence of defects nearby the transmitter, and T indicates the testing time.

Because at the testing place $T + \Delta T$, the sensing coil 2 has moved to where sensing coil 1 was ΔT ago, Equation (6) is obtained,

$$A_2'(T + \Delta T) = A_1'(T) \tag{6}$$

where $\Delta T = L2/v$, and v is the moving speed of the transmitter and sensing coils.

After substituting Equation (6) into Equation (5b) and subtracting Equation (5a) from Equation (5b), Equation (7) is obtained.

$$A_2(T + \Delta T) - A_1(T) = \Delta A(T) * (h(T + \Delta T) - h(T)) \tag{7}$$

where $*$ is the convolution operation, and $h(T)$ is the impulse function.

Based on Equation (7), Wiener deconvolution filter is employed to obtain $\Delta A(T)$ (details in [11–13]). After these processing, the influence of defects nearby the transmitter ($\Delta A(T)$) can be removed from the testing signals.

2.3. Calibrations

Two calibrations should be made in Section 2.2. One calibration is made because the distance between the sensing coil 1 and transmitter is unequal to that between sensing coil 2 and transmitter. These unequal distances make Equation (6) invalid, and the zero-mean normalization method is used to eliminate this inequality, as shown in Equation (8).

$$A_i'(T) = [A_i(T) - E(A_i(T))] / \sqrt{D(A_i(T))} \tag{8}$$

where $A_i(T)$ is the zero-crossing time extracted from the sensing coils, and $i = 1, 2$ indicate sensing coil 1 and sensing 2, respectively; $E(A_i(T))$ is the mean value of $A_i(T)$; and $D(A_i(T))$ is the variance of $A_i(T)$.

Another calibration is made because of the factor $h(T + \Delta T) - h(T)$ in Equation (7). Aimnig to explain this, $\Delta A(T)$ is decomposed as in Equation (9).

$$\Delta A(T) = \Delta A'(T) + N \tag{9}$$

where N can be any real constant and is irrelevant to testing time (T).

By substituting Equation (9) into Equation (7), Equation (10) is obtained as follows:

$$
\begin{aligned}
A_2(T + \Delta T) - A_1(T) &= (\Delta A'(T) + N) * (h(T + \Delta T) - h(T)) \\
&= \Delta A'(T) * (h(T + \Delta T) - h(T)).
\end{aligned} \tag{10}
$$

Equation (10) indicates that $\Delta A'(T)$ is the results obtained from Equation (7) and there exist a real constant (N) between $\Delta A'(T)$ and $\Delta A(T)$. The real constant (N) can be compensated as follows:

1. By subtracting Equation (5b) from Equation (5a), Equation (11) is obtained as,

$$A_1(T) - A_2(T) = A_1'(T) - A_2'(T) \tag{11}$$

2. By substituting Equation (6) into Equation (11), Equation (12) is obtained as,

$$A_1(T) - A_2(T) = A_1'(T) * (h(T) - h(T - \Delta T)) \tag{12}$$

   $A_1'(T)$ can be used to indicate the defects nearby sensing coil 1.
3. Based on Equation (12), the Wiener deconvolution filter is applied to obtain $A_1'(T)$.
4. Then N is computed by Equation (13),

$$N = -\Delta A'(T_j) \tag{13}$$

T_j satisfies $A_1'(T_j) \leq |\delta|$, and δ is a constant approaches to zero.

The Steps 1 to 4 are realized in MATLAB, and several N are saved to obtain the mean value.

2.4. Simulation Sets

The model used to simulate practical PRFECT of ferromagnetic is built using ANSYS software. The simulation model is axisymmetric and 2D, as indicated in Figure 2 (the dotted portion). The amplitude of exciting pulse is 80 V, the repetition rate of excitation is 10 Hz, and the pulse duration is 10 ms. There is only one defect built on pipe wall, and its length and depth are 50 mm and

6 mm, respectively. The relative permeability of all coils is 1, and the relative permeability of pipe is 80. The other parameters of coils and ferromagnetic pipe set in ANSYS model are given in Table 1.

Table 1. Parameters of the coils and pipe set in ANSYS.

Name	Length (mm)	Inner Diameter (mm)	Outer Diameter (mm)	Turns	Resistivity (ohm/m)	Wire Diameter (mm)
Transmitter	167	28.4	44.4	3775	4.247×10^{-8}	0.58
Sensing coil 1	19.1	26.3	32.7	9275	3.083×10^{-7}	0.051
Sensing coil 2	19.1	26.3	32.7	9275	3.083×10^{-7}	0.051
Pipe	2050	153.7	177.1		3.083×10^{-7}	

The range of the transition zone is usually within one times the inner diameter of testing pipe, and the distance between the two sensing coils should be less than one times the inner diameter when the length of every sensing coil is taken into consideration. The larger distance between the two sensing coils the better resolution for large area defects. However, large distance between the two sensing coils also increase the distance from the transmitter to the farther sensing coil, and this will enhance the power consumption.

The distance between middle of sensing coil 2 and middle of transmitter is 2.0 times the pipe inner diameter. Several distances between the middle of two sensing coils are simulated (0.2, 0.4 and 0.6 times the pipe inner diameter), and 0.4 times is chosen to present the theoretical validation of the proposed methods. However, in practical PRFECT of pipes, the distance between the middle of the two sensing coils should be chosen with the consideration of defect resolution, power consumption and pipe diameter range etc. The parameter values of the coils in this paper are obtained by optimizing the parameters of practical RFECT tool. In simulations, the transmitter and sensing coils moved simultaneously with the step of 10 mm. It takes nearly 25 min of each simulation on a computer with Intel(R) Core(TM) i5-4690 CPU@3.5GHz, 8.00 GB RAM, and simulations are operated 100 times.

3. Results and Discussion

The results of using LSSVR to extract zero-crossing times from the induced signals on sensing coils are shown in Figure 3a. The calibrations of distance effect between the signals extracted from sensing coil 1 and sensing coil 2 by using zero-mean normalization are shown in Figure 3b.

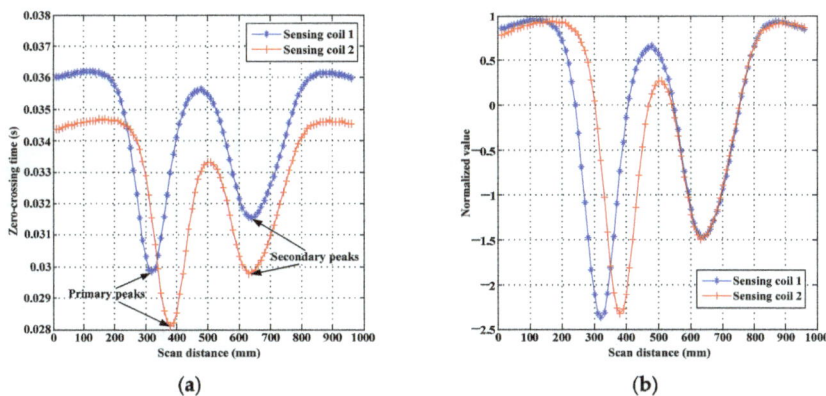

(a)   (b)

Figure 3. Demonstration of the extracted zero-crossing times: (**a**) the zero-crossing times directly extracted from the sensing coils by employing Least Squares Support Vector Regression (LSSVR); and (**b**) the calibrations made between zero-crossing times extracted from sensing coils 1 and sensing coils 2 using zero-mean normalization.

As shown in Figure 3a, the zero-crossing time curve extracted from each sensing coil appears two peaks, even there is only one defect built in simulations. One peak appears when the sensing coil passed by the defect (primary peak), and the secondary peak arises when the transmitter passed by the defect. Figure 3a indicates that PRFECT has a same feature as the RFECT, and the feature reveals the secondary peak caused by the transmitter passing by defects. When the defects only occurred near the transmitter, wrong localization and quantification of defects will obtained because testing place (sensing coil place) has no defects. When the defects occurred at both locations of the transmitter and the sensing coils, the value of zero-crossing time (primary peaks) related to pipe thickness near the sensing coil is largely affected by the defects near the transmitter (which means the secondary peak are superimposed in the primary peak, please refer to [11–13] for more discussions about secondary peaks). The secondary peak that appears in Figure 3 will lead to a wrong localization and quantification of a defect, and its removal is necessary.

In order to remove the secondary peak from zero-crossing time, two sensing coils are used in the transition zone, and the calibration of the independent distances from sensing coils to transmitter is shown in Figure 3b. According to Figure 3b, the influence of the distance between the sensing coils and the transmitter is eliminated by applying zero-mean normalization method, and the scales of zero-crossing times extracted from the sensing coils are unified.

This paper takes the signals from sensing coil 1 for instance. The results of removing the secondary peak are shown in Figure 4a, and the results of calibrating N are shown in Figure 4b. As shown in Figure 4a, the secondary peak is removed efficiently while the primary peak is maintained. Figure 4a validates the correctness of proposed method for removing the secondary peaks in PRFECT of pipes. It is apparent that there is a difference quantity between the two curves shown in Figure 4a. The difference quantity is the real constant (N) discussed in Section 2.3. By employing the method provided in Equations (11)–(13), the difference quantity (N) can be eliminated, as shown in Figure 4b.

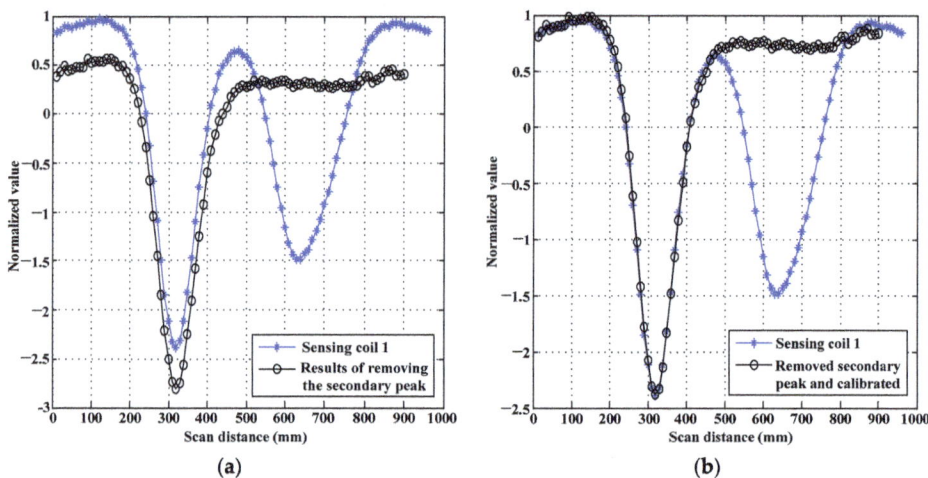

Figure 4. Schematic layout of the processing results by proposed method: (**a**) the comparison between the zero-crossing time and its removal of the secondary peak; and (**b**) the calibration of N.

4. Conclusions

This paper investigates the features of applying Pulsed Remote Field Eddy Current Testing (PRFECT) in ferromagnetic pipes. In the investigation, the zero-crossing time is extracted using Least Squares Support Vector Regression (LSSVR) to analyze pipe thickness. As a result, the zero-crossing time exhibits the same feature as the signal in Remote Field Eddy Current Testing (RFECT) of pipes.

The feature indicates that the secondary peaks are found when the transmitter passes by defects. The secondary peak appearing in the PRFECT of ferromagnetic pipes will lead to wrong localization and quantification of a defect, and its removal is essential. Further, double sensing coils are set in the transition zone to remove the secondary peaks, and the essential calibrations are also discussed and worked out. The methods proposed in this paper are validated by simulating in ANSYS software. Because the studies provided in this paper are an extension of authors' research on RFECT of pipes, and the circuits and coils used to test the pipe are fixed in RFECT, it is a shortcoming that the practical validation of the discussed method is not provided. The parameters set in the simulation model of PRFECT are obtained by optimizing the parameters of practical RFECT tool. The future work will focus on the modification of the RFECT tool to implement the validation of proposed method, and it would be a huge work. This paper has already provided a good theoretical guidance of preprocessing the signals in practical PRFECT of ferromagnetic pipes.

Acknowledgments: This research is supported by the Major National Science and Technology Projects During the 12th Five-Year Plan under the Grant number 2011ZX05020-006-005, National Natural Science Foundation of China under the Grant number 61201131 and the Fundamental Research Funds for the Central Universities under the Grant number ZYGX2012J092.

Author Contributions: Qingwang Luo and Yibing Shi conceived and designed the testing model and methods; Zhigang Wang built the simulation models; Wei Zhang contributed analysis tools; Qingwang Luo wrote the paper and analyzed the data, and Yanjun Li revised the paper.

Conflicts of Interest: The authors declare no conflict of interest.

References

1. Teitsma, A.; Takach, S.; Maupin, J.; Fox, J.; Shuttleworth, P.; Seger, P. Small diameter remote field eddy current inspection for unpiggable pipelines. *J. Press. Vessel Technol.* **2005**, *27*, 269–273. [CrossRef]
2. Schempf, H.; Mutschler, E.; Gavaert, A.; Skoptsov, G.; Crowley, W. Visual and nondestructive evaluation inspection of live gas mains using the explore family of pipe robots. *J. Field Robot.* **2010**, *27*, 217–249.
3. Gantala, G.; Krishnamurthy, C.V.; Balasubramaniam, K. Location and sizing of defects in coated metallic pipes using limited view scattered data in frequency domain. *J. Nondestruct. Eval.* **2016**, *35*, 1–13. [CrossRef]
4. Xu, X.J.; Liu, M.; Zhang, Z.B.; Jia, Y.L. A novel high sensitivity sensor for remote field eddy current non-destructive testing based on orthogonal magnetic field. *Sensors* **2014**, *14*, 24098–24115. [CrossRef] [PubMed]
5. Xue, X.J.; Peng, W.L. Rapid defect reconstruction based on genetic algorithm and similar model in remote field eddy current non-destructive testing. *Appl. Mech. Mater.* **2014**, *1*, 269–274. [CrossRef]
6. Vasić, D.; Bilas, V.; Ambruškim, C. Measurement of ferromagnetic tube wall thickness using pulsed remote field technique. In Proceedings of the 12th IMEKO TC4 International Symposium Electrical Measurements and Instrumentation, Zagreb, Croatia, 25–27 September 2002.
7. Vasić, D.; Bilas, V.; Ambruškim, C. Pulsed eddy-current nondestructive testing of ferromagnetic tubes. *IEEE Trans. Instrum. Meas.* **2004**, *53*, 1289–1294. [CrossRef]
8. Yang, B.F.; Li, X.C. Pulsed remote field technique used for nondestructive inspection of ferromagnetic tube. *NDT E Int.* **2013**, *53*, 47–52. [CrossRef]
9. Vasić, D.; Bilas, V.; Šnajder, B. Analytical modelling in low-frequency electromagnetic measurements of steel casing properties. *NDT E Int.* **2007**, *40*, 103–111. [CrossRef]
10. Yang, B.F.; Li, X.C. Pulsed remote eddy current field array technique for nondestructive inspection of ferromagnetic tube. *Nondestruct. Test. Eval.* **2010**, *25*, 3–12. [CrossRef]
11. Liu, Z.J.; Li, Q.; Liu, X.H.; Mu, C.D. A hybrid LSSVR/HMM-based prognostic approach. *Sensors* **2013**, *13*, 5542–5560. [CrossRef] [PubMed]
12. Xiong, T.; Bao, Y.K.; Hu, Z.Y. Multiple-output support vector regression with a firefly algorithm for interval-valued stock price index forecasting. *Knowl. Based Syst.* **2014**, *55*, 87–100. [CrossRef]
13. Luo, Q.W.; Shi, Y.B.; Wang, Z.G.; Zhang, W.; Zhang, Y. Approach for removing ghost-images in remote field eddy current testing of ferromagnetic pipes. *Rev. Sci. Instrum.* **2016**, *87*, 104707. [CrossRef] [PubMed]

14. Luo, Q.W.; Shi, Y.B.; Wang, Z.G.; Zhang, W.; Ma, D. Method for removing secondary peaks in remote field eddy current testing of pipes. *J. Nondestruct. Eval.* **2017**, *36*, 1. [CrossRef]

15. Luo, Q.W.; Shi, Y.B.; Wang, Z.G.; Zhang, W.; Ma, D. Location and inspection method for large area pipe defect based on RFEC testing. *Chin. J. Sci. Instrum.* **2015**, *36*, 2790–2797.

sensors

MDPI

Article

Defect Detection of Adhesive Layer of Thermal Insulation Materials Based on Improved Particle Swarm Optimization of ECT

Yintang Wen [1], Yao Jia [2], Yuyan Zhang [2], Xiaoyuan Luo [2,*] and Hongrui Wang [2]

[1] School of Science and Technology, Yanshan University, Qinhuangdao 066004, China; ytwen@ysu.edu.cn
[2] School of Electrical Engineering, Yanshan University, Qinhuangdao 066004, China; jiayao89@sina.cn (Y.J.);
     yyzhang@ysu.edu.cn (Y.Z.); hb_wang@ysu.edu.cn (H.W.)
* Correspondence: xyluo@ysu.edu.cn; Tel.: +86-136-4336-1077

Received: 1 September 2017; Accepted: 20 October 2017; Published: 25 October 2017

Abstract: This paper studies the defect detection problem of adhesive layer of thermal insulation materials. A novel detection method based on an improved particle swarm optimization (PSO) algorithm of Electrical Capacitance Tomography (ECT) is presented. Firstly, a least squares support vector machine is applied for data processing of measured capacitance values. Then, the improved PSO algorithm is proposed and applied for image reconstruction. Finally, some experiments are provided to verify the effectiveness of the proposed method in defect detection for adhesive layer of thermal insulation materials. The performance comparisons demonstrate that the proposed method has higher precision by comparing with traditional ECT algorithms.

Keywords: thermal insulation material; electrical capacitance tomography; defect detection; image reconstruction; PSO

1. Introduction

Thermal insulation materials are widely used in aeronautics and astronautics for their characteristics, such as light weight and heat insulation, etc. [1]. At present, the thermal insulation materials are usually glued to the surface of spacecrafts by adhesive. In the complex space environment, the adhesive layer defects of thermal insulation materials, such as cracks and bubbles in the rubber may cause the thermal insulation materials broken off during flying, and hence, it is important to detect the adhesive layer defects of thermal insulation materials for spacecraft safety. Along with the development of new adhesive processes, traditional defect detection technologies no longer satisfy the high accuracy requirements [1–3]. Therefore, developing new techniques and methods to detect defects of adhesive layer for thermal insulation materials is an urgent work.

Recently, some interesting defect detection methods for thermal insulation materials and composite materials with perfect physical properties, e.g., material uniformity, electrical conductivity, etc. have been reported. For example, Park and Kyu [2], Sun and Zhou [3] presented a method based on laser ultrasonic technology for defect detection of carbon fiber resin matrix composites pore fastening holes and composite materials of high temperature resistant layers. Guo and Jing proposed a method based on infrared thermal wave NDT for debonding flaws in some helicopter blades [4], which was analyzed by the thermogram and the peak amplitude image of the second derivative thermogram. A method for coating thickness testing and internal defects detection based on infrared thermal wave was developed in [5], where the interior defects can be detected through measuring the thickness of samples with infrared thermal wave. A method to find counterfeit drugs quickly and reliably based on transmission spectroscopic terahertz (THz) measurement technique was developed in [6]. Palka and Krimi [7], Li and Ding [8] proposed a method based on terahertz time-domain spectroscopy for thickness detection of composite materials, which can obtain the thicknesses of all of the layers of the composite materials based on a time-domain fitting procedure. A method based on laser ultrasonic detection of drilling-induced delamination was presented for the test of composite laminates in [9], where the laser ultrasonic C scan was used to test composite laminates, and the morphologies, dimensions, and positions of drilling-induced delamination can be obtained. Zhang and Gao proposed a method that applied wavelet transform and fuzzy pattern recognition to ultrasonic detection [10]. In this way, they can detect the bonding quality for thin composite plate.

The effectiveness of the aforementioned detection methods have been verified in some application fields. However, some limitations are obvious in the field of defect detection for adhesive layer of thermal insulation materials. Firstly, ultrasonic test is less satisfactory in performance for the insulation materials that have loose structures or uneven characteristics. Secondly, infrared thermal wave test cannot penetrate the adhesive layer defect for the materials which have strong heat-resistant. In other words, infrared thermal wave was unable to effectively obtain the information of adhesive layer of thermal insulation materials. Thirdly, terahertz test is greatly influenced by the characteristics of thermal insulation materials. In addition, terahertz test is poor in obtaining the information of the layer of medium distribution.

It is necessary to develop a new method for defect detection of adhesive layer of thermal insulation materials. Towards this end, a defect detection method based on electrical capacitance tomography is proposed in the paper. A detection system of planar electrode capacitance is adopted, and an Improved Particle Swarm Optimization (IPSO) algorithm is proposed for the defect detection of adhesive layer of thermal insulation materials. Then, an experiment of defect detection of the bonding layer of thermal insulating materials is provided to verify the effectiveness of the proposed defect detection algorithm. The obtained results demonstrate that the proposed defect detection method has higher performance than the traditional Electrical Capacitance Tomography (ECT) methods.

2. Defect Detection Principle of ECT

ECT is a new nondestructive testing technology developed in recent years based on the mechanism of capacitance sensitive, and it has been widely applied in the fields of industrial fluidized bed monitoring, multidirectional flow detection, and medical science [8,9]. The basic principle of ECT imaging technology is that the multiphase medium often has different dielectric constants, such that the medium distribution images can be obtained through capacitance sensors. In this paper, according to the dielectric properties of rubber insulation materials and the characteristics of material surface structure, planar capacitive electrode substrate is used. As shown in Figure 1, an ECT system contains three modules: capacitance sensor module, measurement and data acquisition module, and image reconstruction module. The working process of the ECT system is as follows: acquire the capacitance values via capacitance sensors firstly, and then transmit the values to the computer, and finally reconstruct the field distribution image in computer.

Planar electrode plates of 12 electrodes are used in the capacitance sensor unit. In order to guarantee the credibility and accuracy of measurement data, shielding processing is adopted between the electrodes, and the interface of detection electrode, as well as in the data acquisition unit. The different active electrodes are selected by a multiplexer. The capacitance measurement system of the Intertek Testing Services (ITS) company is used in the measurement and data acquisition unit for experimental data processing of capacitance plate collection. The sensitivity field of material distribution is generated by using the software of ANSYS.

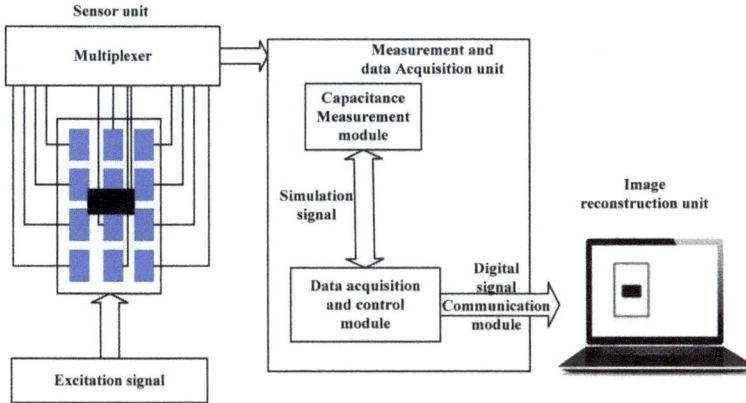

Figure 1. Electrical Capacitance Tomography (ECT) system structure diagram.

3. Reconstruction Algorithm of Image

Reconstruction algorithm of image is to use the collected data from the measurement and data acquisition module of ECT system to build the image projection, and then one can obtain the field distribution diagram, which is used to defect detection and defect analysis for thermal insulation materials. Two key computational problems are required to be solved in ECT: the forward problem and the inverse problem. For the forward problem, inter-electrode capacitances are to be determined by the permittivity distribution. In this paper, a planar capacitive sensor array containing $n = 12$ electrodes is used, and then one has $M = n(n - 1)/2 = 66$ independent capacitance measurements.

Without loss of generality, the effect of shielding layer to dielectric capacitor is neglected, and then based on the electrical principle, the capacitance can be computed as follows [10]

$$C_i = \iint_D \varepsilon(x,y) \cdot S_i(x,y,\varepsilon(x,y))dxdy, \ i = 1,2,\cdots,66 \tag{1}$$

where D, $\varepsilon(x,y)$ and $S_i(x,y,\varepsilon(x,y))$ are the electrode surface, the permittivity distribution of sensing field, and the sensitivity matrix of the sensor's imaging field, respectively. The capacitance differences, which are produced by different material properties on tiny pixel areas, can be distinguished by sensitivity matrix. If we segment the material small enough, then the function of sensitivity distribution is affected slightly by medium distribution [10], and thus Equation (1) can be simplified as follows

$$C_i = \iint_D \varepsilon(x,y) \cdot S_i(x,y)dxdy \tag{2}$$

where $S_i(x,y)$ is the sensitivity function of material capacitance C_i. Then, one can linearize and discretize Equation (2), as follows

$$\mathbf{C} = \mathbf{SG} \tag{3}$$

where $\mathbf{C}$ denotes the normalized capacitance vector, $\mathbf{G}$ is the normalized permittivity vector, and $\mathbf{S}$ represents the normalized sensitivity matrix. Thus, the forward problem is modeled by Equation (3).

For the inverse problem, one needs to acquire the permittivity distribution based on capacitance measurements. Usually, the result of this problem is shown by a visual image, and thus this process is also called image reconstruction.

If there exists the inverse of matrix $\mathbf{S}$, we can solve Equation (3) directly by

$$\mathbf{S}^{-1}\mathbf{C} = \mathbf{G} \tag{4}$$

Unfortunately, the matrix **S** cannot be obtained accurately, because there are three major difficulties for the inverse problem. The first one is the "soft field characteristics [11]", i.e., the measurement sensitive field of ECT sensor is affected greatly by medium distribution. The second one is that Equation (4) is an indeterminate equation, since the number of unknown variables N (i.e., the number of pixels) is usually much larger than the number of equations M (i.e., the number of capacitance measurements), and thus the solution is not unique. The third one is that Equation (4) is an ill conditioned equation [12].

In the past few years, a number of image reconstruction algorithms have been developed to address the ill posed and ill conditioned problems. In general, they can be categorized into two groups: non-iterative (or single step) algorithms (e.g., Tikhonov Algorithm [9], Linear Back Projection (LBP) Algorithm [10], etc.), and iterative algorithms (e.g., SIRT Algorithm [12], Landweber Algorithm [13], etc.). Here, we introduce two typical algorithms.

3.1. LBP Algorithm

Linear Back Projection (LBP) is a non-iterative algorithm and it is the earliest algorithm for ECT imaging technology [10], where if **S** is considered to be a linear mapping from the permittivity vector space to the capacitance vector space, $\mathbf{S^T}$ can be considered as a related mapping from the capacitance vector space to the permittivity vector space. Then the approximated solution can be given as follows.

$$\mathbf{G = S^T C} \tag{5}$$

LBP algorithm is still widely used for on-line image reconstruction because of its simplicity. However, it produces poor-quality image and can only provide qualitative information. LBP algorithm is commonly used in qualitative analysis. However, for complex media distribution error detection, its resolution accuracy for image reconstruction is relatively low.

3.2. Landweber Algorithm

The Landweber algorithm [13] is an iterative algorithm and is developed based on the foundation of steepest descent method. Up to now, the Landweber algorithm has been widely used in the field of ECT. The main principle is to correct the solutions of the equation in the minus gradient direction of data residuals. The data residual gradient is shown as follows

$$\nabla \cdot \frac{1}{2}\|\mathbf{SG - C}\|^2 = \mathbf{S^T(SG - C)} \tag{6}$$

and the iterative equation is

$$\mathbf{G}^{(k+1)} = \mathbf{G}^{(k)} + \alpha\mathbf{S^T}\left(\mathbf{C - SG}^{(k)}\right) \tag{7}$$

where α is the positive scalar, which plays an important role in the process of iteration. However, **G** ($\mathbf{G = S^T C}$) is regarded as the initial guess in the process of iterative calculation, which will produce a large error between the initial guess and real value [12]. Traditional ECT image reconstruction (either the iteration or non-iteration) is flat of sensitivity field. However, in practical applications, phase distribution of different types may cause the differences of sensitivity field. If the differences are ignored, the accuracy of defect detection will be influenced seriously.

3.3. Defect Detect Algorithm Based on Improved PSO

3.3.1. Data Processing Based on LS-SVM

In ECT system, the measuring capacitance **C** will have relatively subtle change for the small defects of adhesive layer of thermal insulation material; besides, the sensitivity matrix **S** is generally considered to be constant, which will result in certain errors in the rubber with different defects. Thus, the experimental value of the capacitance C_m has a certain deviation in the calculative value of the formula. The deviation is given as follows

$$\Delta C = C_m - S \cdot G \tag{8}$$

where C_m is the normalized measurement value of the capacitance [13], and G is the matrix of dielectric constant distribution.

By using multiphase medium with different permittivities, the image of medium distribution can be obtained through measuring the obtained permittivities by capacitance sensors.

The errors between capacitance measurements and theoretical simulation capacitance values can be obtained through training samples, which are trained by least squares support vector machine (LS-SVM).

The vector norm of Equation (8) is as follows

$$y = \|\Delta C\| = \|C_m - S \cdot G\| = f(x) : R^n \to R^1 \tag{9}$$

where R^n is the n dimensional real vector set and R^1 is the real set.

In Equation (9), the measurement capacitance can be viewed as the input vector while the norm of the difference vector between the measurement capacitance vector and the computed capacitance vector is viewed as the output vector.

As training samples used by LS-SVM, the measurement capacitance vectors of defects are used as the input samples, and the norm of capacitance deviation vectors of the same defects are used as the output samples. According to the theory of SVM, the more the training samples are used, the stronger the generalization ability is [14]. In this paper, three kinds of common defects of composite material bonding structure (i.e., fracture defect, bubbles, lack of glue) are considered as a sample set, where each kind of defects have 32 samples with each sample has only one type defect at a particular position, and totally 96 training samples are used.

3.3.2. Image Reconstruction Algorithm Based on Improved PSO

Particle swarm optimization (PSO) algorithm was first proposed in 1995 by the American social psychologist James Kennedy and electrical engineer Russell Eberhart [15]. After that, some other similar algorithms were further proposed. In these algorithms, the evolution of PSO algorithm is also used by the concept of "community" and "evolution". It is also based on the fitness of individuals (particles) size in these algorithms. The difference is that the particle swarm algorithm to each operator as in n dimensional search space does not have a weight and volume of small profit, and in the search space at a certain speed, it changes the speed by the individual's flight experience and group of flight dynamic adjustment [16]. Particle swarm optimization algorithm is a kind of self-adaptive random algorithm based on group hunting strategy, which is an algorithm of simple implementation and fast convergence with few parameters. At present, although the PSO algorithm has some limitations, it can be used after some appropriate improvements [15,17].

In this case, we set the search space in D dimensions with a total of N particles. The ith particle position is represented as $X_i = (x_{i1}, x_{i2}, \ldots, x_{iD})$ and the ith particle's position varying rate is represented as $V_i = (v_{i1}, v_{i2}, \ldots, v_{iD})$. The position of each individual particle changes as follows

$$v_{id}(t+1) = \omega \cdot v_{id}(t) + c_1 \cdot r_1 [p_{id}(t) - x_{id}(t)] + c_2 \cdot r_2 \cdot [p_{gd}(t) - x_{id}(t)] \tag{10}$$

$$x_{id}(t+1) = x_{id}(t) + v_{id}(t+1) \tag{11}$$

where c_1, c_2 are positive constants which are called as the acceleration factors, r_1, r_2 are random numbers between [0, 1], ω is called the inertial factor, i is the ith particle $(1 \leq i \leq N)$, and d is the dimension of each particle $(1 \leq d \leq D)$. The initial position and speed of particle swarm are randomly generated, and are iterated according to Equations (10) and (11). The Improved Particle

Swarm Optimization (IPSO) algorithm is presented based on the Basic Particle Swarm Optimization (BPSO), where the main improvements are shown as follows

1. According to the analysis of ECT imaging principle, capacitance testing equipment, and image reconstruction algorithm, a modified fitness function is presented as $\mathbf{F} = \min(\|\mathbf{C_M} - \mathbf{SG_k}\| - \|\Delta\mathbf{C}\|)$, where $\|\Delta\mathbf{C}\|$ is the output trained by LS-SVM. Then, with the LS-SVM training results, one can optimize the fitness function, and compensate or eliminate the errors induced by the sensitivity matrix $\mathbf{S}$ and measuring device.

2. Based on the principle of PSO, the initial value is randomly generated, and Equation (11) is expressed as $\mathbf{G}_{(t+1)} = \mathbf{G}_{(t)} + \mathbf{v_{id}}(\mathbf{t})$, where $v_{id}(t)$ is obtained by Equation (10). A nonlinear and dynamic adjustment method is then presented to adjust the inertia weight ω in Equation (10) as follows

$$\omega = \begin{cases} \omega_{\text{int}} - (\omega_{\text{int}} - \omega_{end}) \cdot \frac{F - F_{\min}}{F_{avg} - F_{\min}} & F < F_{avg} \\ \omega_{\text{int}} & F \geq F_{avg} \end{cases} \tag{12}$$

where F is the fitness value of particle at present, F_{avg} is the average fitness, and $F_{\min}$ is the minimum fitness, i.e., the fitness of the best particle. Firstly, the value of ω_{int} is kept unchanged, and the value of ω_{end} is the minimum value for cumulative value of ω in each iteration (a given initial value: $\omega_{\text{int}} = 1.2$, $\omega_{end} = 0.8$) [17]. The inertial factor is changed with the adaptation degree of each generation.

3. Population will search the extreme value that is decided by P_{igbest} and P_{gbest} after several iterations. If no better position of population than P_{gbest} is found in the iteration process, the algorithm will stagnate. Since the change of P_{gbest} can reflect the change of P_{igbest}, the change of the best personal position of each particle can be used as the only judgment foundation for variation. In this paper, the minimal fitness value is used as the benchmark, and the fitness value at the tth iteration is obtain as follows

$$F_{t,avg} = \frac{1}{N}\sum_{i=1}^{N} F_{t,p_{best},i} \tag{13}$$

where $F_{t,p_{best},i}$ is the best personal position of several particles at the tth iteration. When the condition $B : (F_{t+1} < F_{t,avg})$ is satisfied at the $t+1$th iteration, the optimal process is regarded as good (either too large or too small values of K will influence the result of genetic algorithm [18] (in this paper, $K = 3$). One can reduce the particle complexity and increase the speed of the algorithm by decreasing the particle number at this moment. When the adaptive condition does not satisfy condition B (i.e., $F_{t+1} = F_{t,avg}$, lasts up to three generations), the diversity of the population will be lost, and then the particle variability can be kept by increasing the number of particles.

4. One can keep the diversiform direction of movement for each particle while decreasing the computation complexity of the proposed algorithm by improving the number of particles. Firstly, we give two boundary values pop_min and pop_max. If population size has reached pop_max and it still needs to increase the particles, the population size is reduced by a particle. If population size has reached pop_min and it still needs to decrease the particles, the population size remains. The continuous generation (Consecutive Generations, CG) strategy is used to reduce the particles [18], i.e., we delete particles randomly which are not at the best locations in the current particle swarm as well as not at the optimal positions for the particle swarm. Then, the uniform mutation (Uniform Mutation, UM) [19] strategy is used to increase the particles.

v_{np}, x_{np} and $pBest_{np}$ are used as the new particle's speed, current location, and optimal position separately in history, respectively. The updates of variation are given as follows:

$$\begin{cases} \text{pBest}_{np}^d = X_{\min}^d + \lambda \times (X_{\max}^d - X_{\min}^d) & d = U[1, D] \\ \text{pBest}_{np}^d = \text{pBest}_{gb}^d & otherwise \end{cases} \quad (14)$$

where d is the dimension randomly selected to mutate, and $X_{\max}^d$ and $X_{\min}^d$ are upper and lower bounds of the search space, respectively.

The flat electrode substrate is chosen according to the characteristics of the ceramic porous thermal insulation material. Therefore, based on the previous analysis of basic principle of electrical capacitance tomography and the image reconstruction algorithms (such as LBP algorithm and Landweber algorithm), an Improved Particle Swarm Optimization (IPSO) algorithm is proposed.

The proposed IPSO algorithm considers the change of sensitivity matrix **S** caused by different defect fields. The change of **S** matrix is optimized by the training of LS-SVM, which is different from the traditional ECT imaging algorithm. On this basis, the results trained by the LS-SVM are added to the fitness function of IPSO algorithm. In order to find the optimal particle, a calculating strategy of nonlinear inertial parameter ω is adopted.

At the same time the evaluation index of the particle falls into local optimum and the method of the particle to overstep the local extremum are put forward. On this basis, the IPSO algorithm for ECT image reconstruction has been proposed, which fundamentally solves the issue of iteration caused by the initial information error. Moreover, the proposed algorithm can enhance the overall searching capability and local optimum jumping capability. Furthermore, based on the IPSO algorithm, we can reconstruct the ECT image for the defect detection of adhesive layer of thermal insulation materials.

4. Experiment Study

A nondestructive defect detection technology for adhesive layer of thermal insulation materials is presented based on the IPSO method for ECT image reconstruction in Section 3. Three types of glue line defects (Air bubbles, Irregular defect samples, Wide glue line), which always occur in aerospace applications, are studied in this section.

The experimental cases are as follows: 18×18 cm^2 materials of porous ceramic are used for the experiments. The glue line is 16×16 cm^2 epoxy resin, and the adhesive thickness is 3 mm. The full yard (full adhesive) and the empty yard (full air) are shown in Figures 2 and 3.

Figure 2. The sample without adhesion.

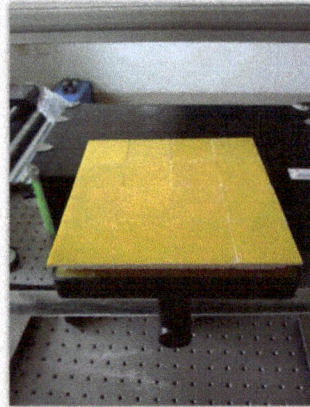

Figure 3. The complete sample with adhesion.

4.1. Experiment Results

4.1.1. Experiment for Imitation of Small Air Bubble Defects

We use a 16×16 cm^2 epoxy resin in this experiment. To simulate the traditional defects of cementing structure such as the bubbles and lacks of plastic induced by pressing, here we consider the Sample 1 with six pieces of 2×2 cm^2 square holes as the defect, as shown in Figure 4.

Four algorithms of LBP, Landweber, BPSO and improved PSO algorithms are applied for defect detection of Sample 1, respectively. The simulation results are shown in Figure 5. One can find from Figure 5a,b that LBP algorithm can only roughly detect location, size, and contour information of the defect. From Figure 5c–f, the Landweber and BPSO algorithms can reflect defect, which certainly show a little better detection performance than the LBP algorithm. However, the defect marginals of reconstructed images by the three algorithms are not very clear, which makes defect edge segmentation from the reconstructed images difficult.

It is worth pointing out that, seen from Figure 5g,h, the proposed IPSO algorithm in the paper shows the significant advantages to the defect location, size, in addition, contour information of the defect can be more clearly reconstructed by the proposed IPSO algorithm. It is obvious that the defect location, size, and contour information of the defect can be more clearly reconstructed by the proposed IPSO algorithm than the other three algorithms.

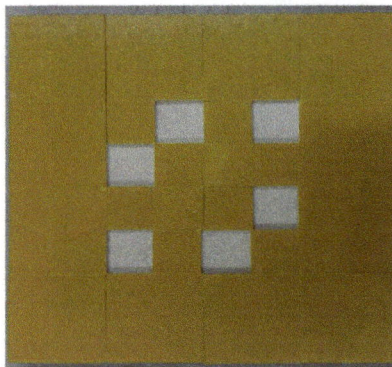

Figure 4. Sample 1 of defect.

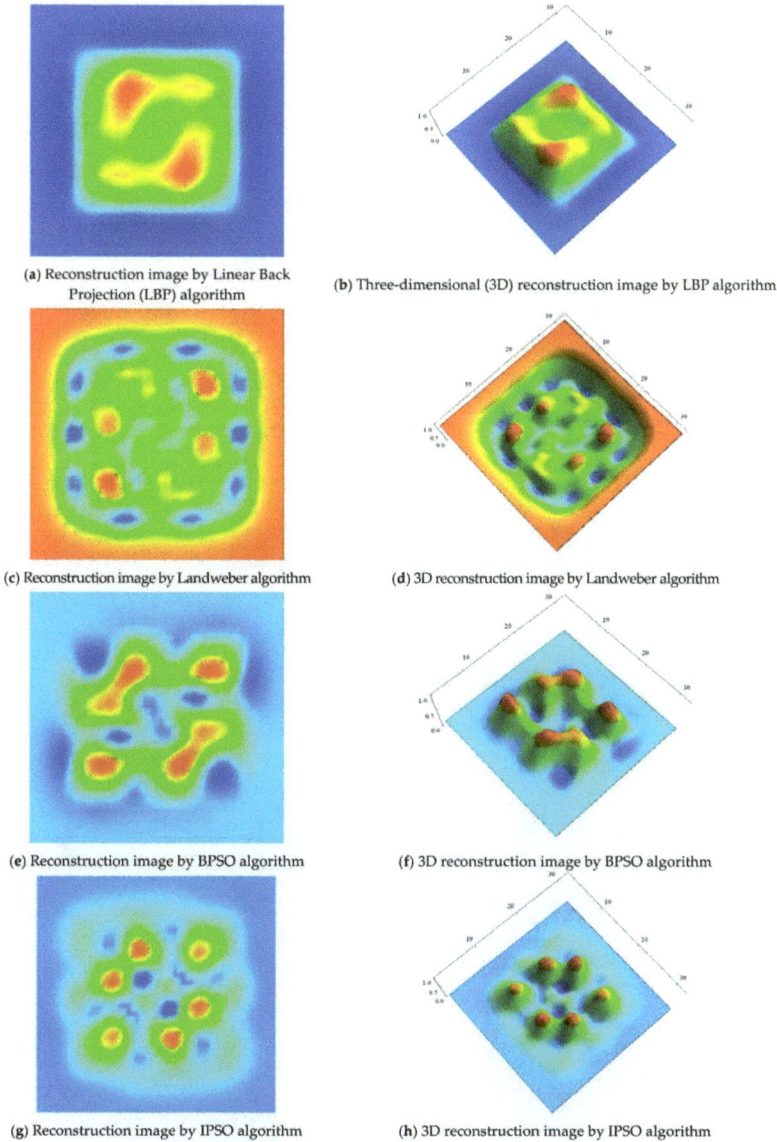

(a) Reconstruction image by Linear Back Projection (LBP) algorithm

(b) Three-dimensional (3D) reconstruction image by LBP algorithm

(c) Reconstruction image by Landweber algorithm

(d) 3D reconstruction image by Landweber algorithm

(e) Reconstruction image by BPSO algorithm

(f) 3D reconstruction image by BPSO algorithm

(g) Reconstruction image by IPSO algorithm

(h) 3D reconstruction image by IPSO algorithm

Figure 5. Images reconstructed by different algorithms for Sample 1.

4.1.2. Experiment for Rubber Fracture Defects

Another type defect, which has two bubbles in glue line with one large and another small, as shown in Figure 6, is considered in this subsection. We call the defect as Sample 2. In order to assess the quality of image easily, two defects are separately replaced by two small circles with the areas of 7 cm^2 and 1 cm^2, respectively. The rubber block is surrounded by foam rubber with 2 cm wide and 16 cm length foam (the permittivity of foam rubber is similar to the permittivity of air).

LBP, Landweber, BPSO, and the proposed IPSO algorithms are applied for the defect detection of Sample 2. The experimental results are shown in Figure 7. It can be seen from Figure 7 that when compared with Figure 7a–f using LBP, Landweber and BPSO algorithms, the performance of reconstructed images Figure 7g,h, using the proposed IPSO are significantly improved and more detailed information of defects, such as the size, shape, edge, etc. are displayed clearly, which demonstrates the effectiveness and superiority of the proposed algorithm.

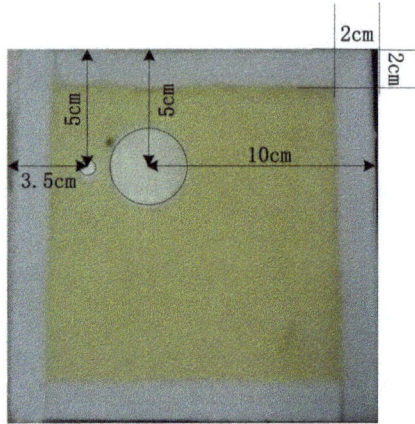

Figure 6. Sample 2 of defect.

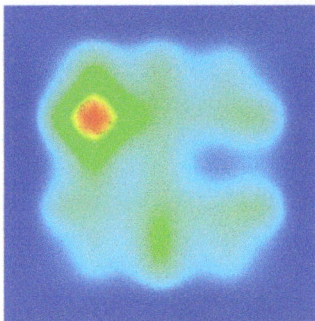

(**a**) Reconstruction image by LBP algorithm   (**b**) 3D reconstruction image by LBP algorithm

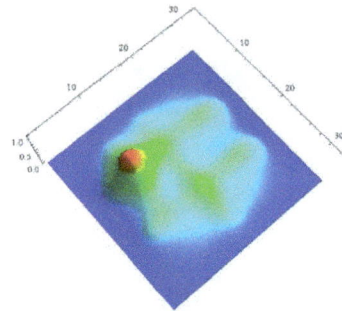

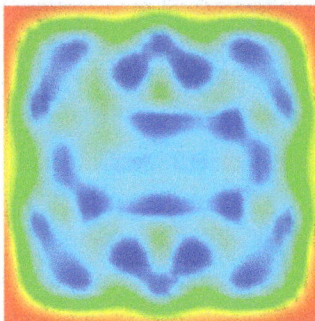

(**c**) Reconstruction image by Landweber algorithm   (**d**) 3D reconstruction image by Landweber algorithm

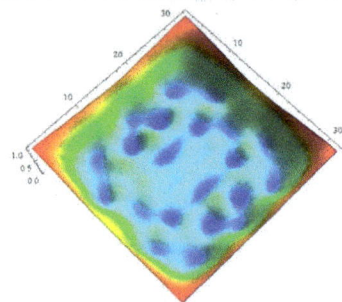

Figure 7. *Cont.*

(e) Reconstruction image by BPSO algorithm   (f) 3D reconstruction image by BPSO algorithm

(g) Reconstruction image by IPSO algorithm   (h) 3D reconstruction image by IPSO algorithm

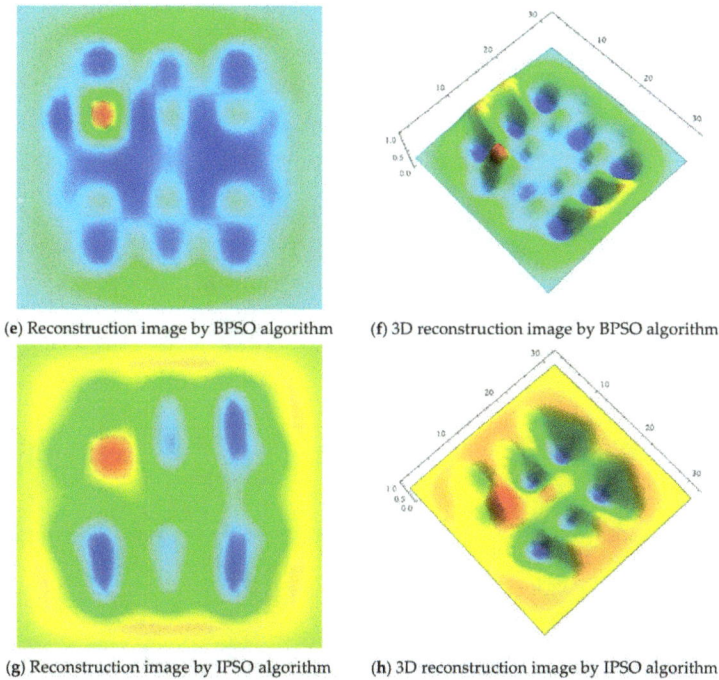

Figure 7. Images reconstructed by different algorithms for Sample 2.

In addition, it is worth pointing out that, from all of the subfigures in Figure 7, all the 4 algorithms cannot be able to detect the smaller defect, which makes the smaller one missed in the reconstructed images. Our analysis of the results may be that the reconstructed images of the smaller defect are submerged by the ones of the large defect. Besides, to the best of our knowledge, at present, small defect below 1 cm^2 cannot be detected by using current detection algorithms including our proposed algorithm.

4.1.3. Experiment for Rubber Fracture Defects

To simulate the traditional defects of cemented structure, e.g., the fracture of adhesive layers, u slot defect with the entire length being 15 cm and the width 5 mm, as shown in Figure 8, is considered. We call the defect as Sample 3.

Similarly, LBP, Landweber, BPSO and the proposed IPSO algorithms are applied to detect the defects of Sample 3. The experimental results are shown in Figure 9. It can be seen from Figure 9 that, when compared with Figure 9a–f using LBP, Landweber, and BPSO algorithms, the effect of reconstructed images Figure 9g,h using the proposed IPSO are significantly improved, and more detailed information of defects, such as the size, shape, edge, etc. are displayed more clearly, which also demonstrates the effectiveness and the superiority of the proposed IPSO algorithm.

It is worth noting that because the capacitance distribution from 12-electrode capacitance sensors is used, the quality of reconstructed images is not effective to some defects with sizes being below the cm-level. The accuracy of reconstructed images is influenced by the less original data of capacitance to a certain extent. However, in the real applications, the destructive effect of below the cm-level's defect of adhesive layer of thermal insulation materials is far less than that of the above cm-level's defect. Obviously, for the defect detection of defect area greater than 1 cm^2, the IPSO algorithm can achieve significantly better performance than the classical algorithms.

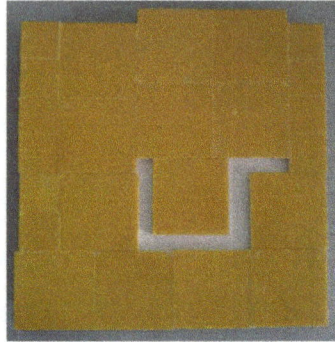

Figure 8. Sample 3 of defect.

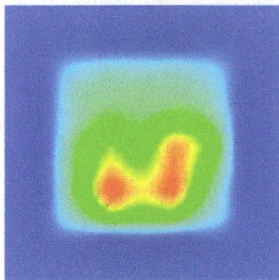

(a) Reconstruction image by LBP algorithm

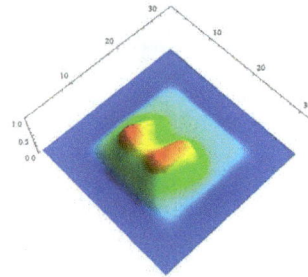

(b) 3D reconstruction image by LBP algorithm

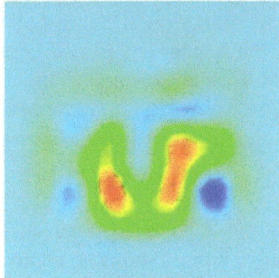

(c) Reconstruction image by Landweber algorithm

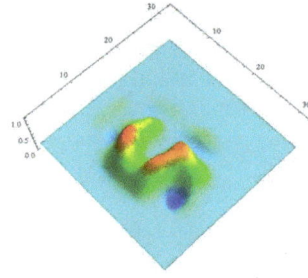

(d) 3D reconstruction image by Landweber algorithm

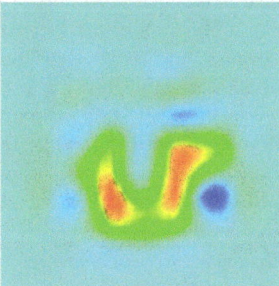

(e) Reconstruction image by BPSO algorithm

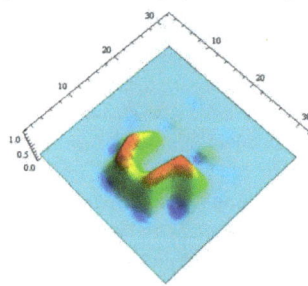

(f) 3D reconstruction image by BPSO algorithm

Figure 9. *Cont.*

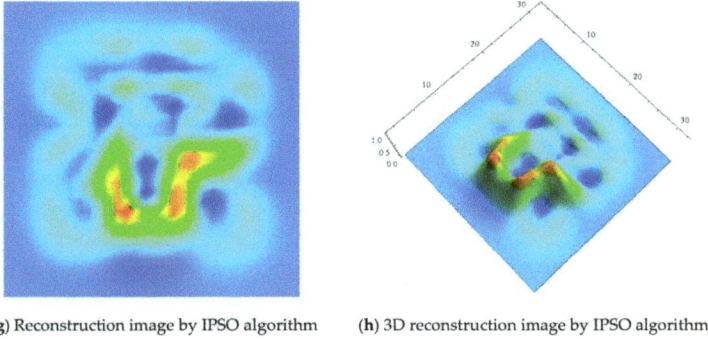

(**g**) Reconstruction image by IPSO algorithm   (**h**) 3D reconstruction image by IPSO algorithm

Figure 9. Images reconstructed by different algorithms for Sample 2.

4.2. Evaluation of Algorithms by Experiment

In this subsection, we will further evaluate the performance of the four algorithms. Two evaluation criteria will be used in this paper: the image correlation I_c, which is the similarity degree between the reconstruction image and the test object image, and the relative image error I_e. They are shown as follows

$$I_c = \frac{\sum_{i=1}^{m} [(e_i^{(k)} - \overline{E}^{(k)})(e_i^0 - \overline{E}^0)]}{\sum_{i=1}^{m} \sqrt{\left(e_i^{(k)} - \overline{E}^{(k)}\right)^2} \sum_{i=1}^{m} \sqrt{\left(e_i^0 - \overline{E}^0\right)^2}} \tag{15}$$

$$I_e = \frac{\sqrt{\sum_{i=1}^{m} \left(e_i^{(k)} - e_i^0\right)^2}}{\sqrt{\left(\overline{E}^0\right)^2}} \tag{16}$$

where $e_i^{(k)}$ is the ith element in the final reconstructed image $E^{(k)}$, $\overline{E}^{(k)}$ is the average of pixels in the $E^{(k)}$ space, and e_i^0 is the ith element in the original simulated image E^0. The clearer the reconstruction image is, if the lower I_e is and the higher I_c is [19].

The computing results under the four algorithms are shown in Tables 1 and 2. Table 1 shows the image correlation coefficients of the reconstructed images for Sample 1, Sample 2, and Sample 3 using LBP, Landweber, BPSO and IPSO, respectively. Table 2 shows the errors of the reconstructed images of Sample 1, Sample 2 and Sample 3 based on LBP, Landweber, BPSO and IPSO, respectively. Obviously, the simulation results in Tables 1 and 2 demonstrate that the proposed IPSO scheme has higher significant effect for defect image reconstruction than the other three algorithms.

Table 1. Image correlation I_c (%) by different algorithms.

	Sample 1	Sample 2	Sample 3
LBP	0.5680	0.6087	0.4787
Landweber	0.4680	0.6829	0.6853
BPSO	0.7648	0.6946	0.6992
IPSO	0.8484	0.7737	0.7712

Sensors **2017**, *17*, 2440

Table 2. Relative image error I_e by different algorithms.

	Sample 1	Sample 2	Sample 3
LBP	2.1927	1.2625	3.6358
Landweber	2.3209	0.8923	4.7879
BPSO	2.1230	0.9321	4.5827
IPSO	1.0999	0.8317	3.5059

5. Conclusions and Future Work

This paper develops a novel non-destructive method for defect detection of adhesive layer of thermal insulation materials based on the proposed IPSO algorithm. In the method, the least squares support vector machine is utilized at first for data processing of measured capacitance values, and then the improved PSO algorithm is proposed for the optimization of image reconstruction. Simulation and experiment results demonstrate that when compared with the traditional algorithms such as LBP algorithm and Landweber algorithm, the proposed IPSO algorithm can display the defect information including size, shape, and edge of the defects more clearly. Future work will be implemented to further identify the feature of defects.

Acknowledgments: This work is supported in part by the National Natural Science Foundation of China (61403333).

Author Contributions: Yintang Wen, Yao Jia and Xiaoyuan Luo conceived of the original idea of the paper. Yintang Wen, Yuyan Zhang and Yao Jia performed the experiments. Yuyan Zhang and Hongrui Wang collaborated to the development of the idea, critical revision and improvement of the paper. Yintang Wen and Xiaoyuan Luo wrote the paper.

Conflicts of Interest: The authors declare no conflict of interest.

References

1. Wu, R.; Pan, W.; Ren, X.; Wan, C.; Qu, Z.; Du, A. An extremely low thermal conduction ceramic: $RE_{9.33}$ $(SiO_4)_6O_2$ silicate oxyapatite. *Acta Mater.* **2012**, *60*, 5536–5544. [CrossRef]
2. Park, B.; An, Y.K.; Sohn, H. Visualization of hidden delamination and debonding in composites through noncontact laser ultrasonic scanning. *Compos. Sci. Technol.* **2014**, *100*, 10–18. [CrossRef]
3. Sun, G.; Zhou, Z. Application of laser ultrasonic technique for non-contact detection of drilling-induced delamination in aeronautical composite components. *Optik—Int. J. Light Electron Opt.* **2014**, *125*, 3608–3611. [CrossRef]
4. Guo, Q.; Jing, P.; Qi, G. Infrared thermal wave NDT for helicopter blades debonding. In *Proceedings of the First Symposium on Aviation Maintenance and Management—Volume I*; Springer: Berlin/Heidelberg, Germany, 2014; Volume 296, pp. 331–336.
5. Huo, Y.; Zhang, C. Quantitative infrared prediction method for defect depth in carbon fiber reinforced plastics composite. *Acta Phys. Sin.* **2012**, *61*, 144204. (In Chinese)
6. Kissi, E.O.; Bawuah, P.; Silfsten, P.; Peiponen, K. A tape method for fast characterization and identification of active pharmaceutical ingredients in the 2–18 THz spectral range. *J. Infrared Millim. Terahertz Waves* **2014**, *36*, 278–290. [CrossRef]
7. Palka, N.; Krimi, S.; Ospald, F.; Beigang, R. Precise determination of thicknesses of multilayer polyethylene composite materials by terahertz time-domain spectroscopy. *J. Infrared Millim. Terahertz Waves* **2015**, *36*, 578–596. [CrossRef]
8. Li, Q.; Ding, S.; Li, Y.; Xue, K.; Wang, Q. Research on reconstruction algorithms in 2.52 THz off-axis digital holography. *J. Infrared Millim. Terahertz Waves* **2012**, *33*, 1039–1051. [CrossRef]
9. Wei, D.; Zhou, Z. Application of non-linear frequency-modulation based pulse compression in air-coupled ultrasonic testing. *J. Mech. Eng.* **2012**, *48*, 8–13. [CrossRef]
10. Zhang, Z.; Gao, Z. Application of wavelet transform and fuzzy pattern recognition in ultrasonic detection. *Lect. Notes Electr. Eng.* **2012**, *136*, 521–529.

11. Polydorides, N. Image Reconstruction Algorithms for Soft-Field Tomography. Ph.D. Thesis, University of Manchester Institute of Science and Technology, Manchester, UK, 2002; pp. 4–8.

12. Hansen, P.C. Rank-deficient and discrete ill posed problems: Numerical aspects of linear inversion. *Am. Math. Mon.* **1997**, *4*, 491.

13. Lei, J.; Liu, W.; Liu, Q.; Wang, X.; Liu, S. Robust dynamic inversion algorithm for the visualization in electrical capacitance tomography. *Measurement.* **2014**, *50*, 305–318. [CrossRef]

14. Suykens, J.A.K.; Vandewalle, J. Least squares support vector machine classifiers. *Neural Process. Lett.* **1999**, *9*, 293–300. [CrossRef]

15. Li, M.; Kang, H.; Zhou, P.; Hong, W. Hybrid optimization algorithm based on chaos, cloud and particle swarm optimization algorithm. *J. Syst. Eng. Electron.* **2013**, *24*, 324–334. [CrossRef]

16. Luo, D.; Chu, Z.; Luo, L.; Liu, Q. Applications study of particle swarm optimization neural network in CFRD dam deformation monitoring. In Proceedings of the 2010 Second International Asia Symposium on Intelligent Interaction and Affective Computing and Second International Conference on Innovation Management (ASIA-ICIM 2010), Hong Kong, China, 15–17 September 2010; Volume 8, pp. 6–12.

17. Lu, Y.; Yan, D.; Zhang, J.; Levy, D. A variant with a time varying PID controller of particle swarm optimizers. *Inf. Sci.* **2015**, *297*, 21–49. [CrossRef]

18. Li, C.; Yang, X.; Wang, Y. Image reconstruction algorithm based on fixed-point iteration for electrical capacitance tomography. In *Communication Systems and Information Technology*; Springer: Berlin/Heidelberg, Germany, 2011; pp. 325–332.

19. Yang, W.; Peng, L. Image reconstruction algorithms for electrical capacitance tomography. *Meas. Sci. Technol.* **2002**, *14*, R1–R13. [CrossRef]

Article

Towards Intelligent Interpretation of Low Strain Pile Integrity Testing Results Using Machine Learning Techniques

De-Mi Cui [1], Weizhong Yan [2,*] , Xiao-Quan Wang [1] and Lie-Min Lu [1]

[1] Anhui and Huaihe River Institute of Hydraulic Research, No. 771 Zhihuai Road, Bengbu 233000, China; cdm@ahwrri.org.cn (D.-M.C.); wxq@ahwrri.org.cn (X.-Q.W.); llm@ahwrri.org.cn (L.-M.L.)

[2] GE Global Research Center, Niskayuna, New York, NY 12309, USA

* Correspondence: yan@ge.com; Tel.: +1-518-387-5704

Received: 4 August 2017; Accepted: 13 October 2017; Published: 25 October 2017

Abstract: Low strain pile integrity testing (LSPIT), due to its simplicity and low cost, is one of the most popular NDE methods used in pile foundation construction. While performing LSPIT in the field is generally quite simple and quick, determining the integrity of the test piles by analyzing and interpreting the test signals (reflectograms) is still a manual process performed by experienced experts only. For foundation construction sites where the number of piles to be tested is large, it may take days before the expert can complete interpreting all of the piles and delivering the integrity assessment report. Techniques that can automate test signal interpretation, thus shortening the LSPIT's turnaround time, are of great business value and are in great need. Motivated by this need, in this paper, we develop a computer-aided reflectogram interpretation (CARI) methodology that can interpret a large number of LSPIT signals quickly and consistently. The methodology, built on advanced signal processing and machine learning technologies, can be used to assist the experts in performing both qualitative and quantitative interpretation of LSPIT signals. Specifically, the methodology can ease experts' interpretation burden by screening all test piles quickly and identifying a small number of suspected piles for experts to perform manual, in-depth interpretation. We demonstrate the methodology's effectiveness using the LSPIT signals collected from a number of real-world pile construction sites. The proposed methodology can potentially enhance LSPIT and make it even more efficient and effective in quality control of deep foundation construction.

Keywords: deep foundation; defect detection; extreme learning machine; neural network; non-destructive evaluation; pile integrity testing; wavelet decomposition

1. Introduction

Assessing the structural integrity of deep foundation elements such as drilled or driven piles has always been a critical quality control task in the construction industry. Over the years, many nondestructive evaluation (NDE) methods have been developed for reliably assessing the integrity of piles, for example, low strain pile integrity testing (LSPIT), high strain pile integrity testing (HSPIT), cross-hole sonic logging (CSL), single hole sonic logging (SSL), and gamma-gamma density logging (GDL) [1]. Among these different integrity testing methods, LSPIT, also called the sonic echo test, is probably the most popular one widely used in various parts of the world. The popularity of LSPIT comes from the fact that it is effective in detecting major discontinuities or defects, such as cavities, cracks, necking, bulging, and soil inclusions, and relatively simple to perform in the field [2].

LSPIT works by following one-dimensional wave propagation theory [3]. A stress wave introduced by the blow of a hand-held hammer on the pile top propagates axially along the pile, and reflections are generated whenever the stress wave encounters impedance changes (discontinuities). Theses reflections are measured with the acceleration transducer installed on pile top. These reflections are later analyzed based on one dimensional stress wave analysis.

An entire LSPIT involves two parts: (1) field testing—signal acquisition and (2) signal interpretation—qualitatively and quantitatively assessing pile integrity by interpreting the signals (velocity reflectograms) collected from the field test. While field testing is relatively simple and can be performed fairly quickly by qualified personnel, assessing the pile integrity by interpreting the test signals is still quite challenging and involving. That is because many factors affect the wave propagation and thus the reflection signals. In particular, interpreting the signals for detecting pile defects to determine defect types and locations requires experienced personnel with good knowledge of wave propagation theory, soil mechanics, and piling construction techniques, in addition to a good understanding of LSPIT itself. As a result, interpreting field-obtained signals is currently performed manually by experienced experts only. When the number of piles tested is large, the experts may be overwhelmed by the large amount of manual interpretation work, which leads to one of the two potential consequences: (1) the inability to complete the pile integrity assessment on time, resulting in delays of the foundation construction schedule, and (2) an increased possibility of errors (e.g., mis-detection of defect piles) due to lack of time for the expert to perform a thorough interpretation of test signals and in-depth analysis of soil conditions. Therefore, techniques that enable speedy and reliable interpretation of LSPIT signals, while minimizing experts' effort at the same time, are of great business value.

There have been a few studies on intelligent interpretation of LSPIT signals in recent years. For example, in [4], a neural network classifier was used for detecting and identifying defects of concrete piles. However, their method worked based on both the known ideal reflectograms and the field test PIT signals, where the ideal reflectograms were numerically obtained through finite element method (FEM) and scaled boundary FEM (SBFEM). The requirement of ideal reflectograms limits its broad applications. Other works, e.g., [5], also used neural network and worked on numerically simulated reflectograms.

In this paper we proposed a computer-aided reflectogram interpretation (CARI) methodology that interprets field-generated LSPIT signals/reflectograms directly without a numerical model of the piles. Our proposed CARI methodology is based on advanced signal processing and machine learning techniques to analyze and interpret PIT signals more effectively and efficiently. Specifically, wavelet analysis is used to extract important features from the raw LSPIT signals, and such extracted features are then used as input to extreme learning machines (ELM) [6], an advanced artificial intelligence technique, for pile defect detection. Since many factors affect the reflectograms, fully automatic interpretation without human intervention is practically infeasible. That is exactly why ASTM D5882 specifically requires "Engineers with specialized experience in this field are to make final integrity evaluation" [7]. Realizing this, we deliberately develop our CARI methodology such that it is not completely free of human experts' involvement in interpretation, but rather it is designed to greatly reducing human experts' effort in interpretation. Specifically, the proposed methodology eases experts' interpretation burden by quickly screening all tested piles and identifying a small number of suspected piles for experts to do manual interpretation. Note that, in a typical real-world pile foundation construction, the number of defect piles is normally small compared to the number of normal piles.

ELM, a new family of neural networks, has been actively studied in the past a few years [6]. The applications of ELM cover diverse domains, including image analysis [8,9], medical science [10], and text analysis [11]. Recently, ELM has also been applied to fault detection and diagnosis of mechanical systems [12,13]. Using ELM as a means for intelligent interpretation of PIT signals has never been done, to the best of our knowledge. Thus, our contribution in this paper is primarily to introduce ELM-based CARI methodology for intelligent and fast interpretation of LSPIT signals. We also demonstrate the effectiveness of the proposed methodology using a large number of LSPIT signals collected from various real-world engineering construction sites. The proposed CARI methodology can potentially enhance LSPIT as a NDE tool and make it even more efficient and effective in quality control of deep foundation construction.

The remainder of our paper is organized as follows. In Section 2, we provide background information about LSPIT, including the principle of LSPIT and the prior work related to intelligent interpretation of LSPIT signals. Details of the proposed methodology are given in Section 3. Section 4 provides experimental results and discussion, while conclusions are given in Section 5.

2. Background

2.1. Principle of LSPIT

The low strain pile integrity test is an echo method for qualitative evaluation of the physical dimensions, continuity of a pile, and consistency of the pile material. The low strain integrity test has been used since the 1970s and has been standardized by ASTM D5882—Standard Test Method for Low Strain Impact Integrity Testing of Deep Foundations [7]. As specified in ASTM D5882, there are two testing methods for the low strain integrity test. One is the Pulse Echo Method (PEM), and another is the Transient Response Method (TRM). This study is concerned with the PEM where only pile head motion is measured and analyzed for pile integrity evaluation. As illustrated in Figure 1, to perform LSPIT with PEM, the pile head is taped with a hammer, which generates the stress wave (sound wave) that travels through the pile length and reflects back to the pile head. The acceleration transducer placed on top of pile head measures the response of the stress wave. The measured acceleration is integrated into a velocity signal, popularly called "velocity reflectogram" or simply "reflectogram," which offers a great amount of information for both the qualitative and quantitative assessment of pile integrity.

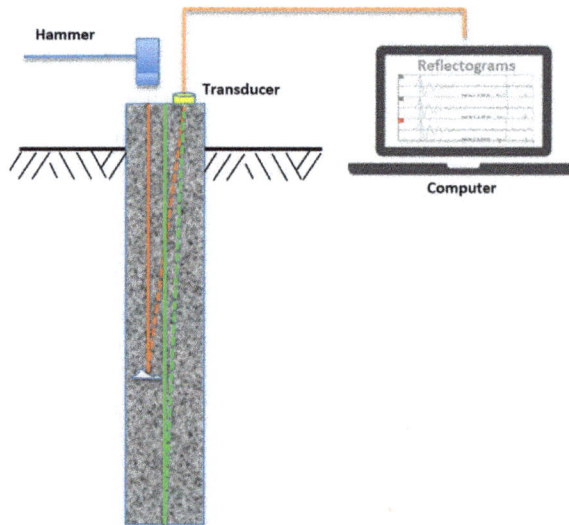

Figure 1. Schematic view of low strain pile integrity testing.

The well-known wave equation in a one-dimensional elastic rod (ignoring the soil resistance) is given as (assuming pile material is homogeneous).

$$\frac{\partial^2 u(x,t)}{\partial t^2} - C^2 \frac{\partial^2 u(x,t)}{\partial x^2} \tag{1}$$

$C = \sqrt{\frac{E}{\rho}}$ is the wave propagation velocity, where E and ρ are the dynamic Young's modulus and the mass density of the pile material, respectively, and $u(x,t)$ is the axial displacement of a mass point at section x and time t.

The impedance, defined as $Z = \frac{EA}{C} = A\sqrt{E\rho}$, where A is the cross-sectional area of the pile, is a metric for measuring pile resistance change with respect to velocity. Any change in A, E, ρ, or a combination of them will result in an impedance change or discontinuity. When a wave traveling in a rod meets such a discontinuity, one part of it will be reflected back while another part will go on beyond the discontinuity. Assume at the discontinuity section the impedance changes from Z_1 to Z_2. Then, the amplitude of reflected wave, V_R, is related to the amplitude of the incoming wave, V_I, as follows:

$$V_R = \frac{Z_1 - Z_2}{Z_1 + Z_2}V_I = \frac{\rho_1 C_1 A_1 - \rho_2 C_2 A_2}{\rho_1 C_1 A_1 + \rho_2 C_2 A_2}V_I \tag{2}$$

2.2. Related Work

As one of the most popular NDE methods in the pile foundation construction industry, LSPIT has been widely used for qualitative evaluation of the physical dimensions, continuity of a pile, and consistency of the pile material since 1970s. Over the years, research efforts on improving the LSPIT method have been focused on two separate directions: (1) theoretical analysis and understanding of LSPIT mechanism and (2) intelligent and automatic interpretation of test results. While the first direction is important and has attracted much research interests (e.g., [14]), our paper is concerned with the second direction, i.e., automatic interpretation of LSPIT signals. Toward intelligent interpretation of LSPIT signals, a limited number of studies have been on using advanced signal processing and artificial intelligence (pattern recognition) techniques for LSPIT signal analysis.

In his PhD thesis [15], Watson applied three different types of neural networks to LSPIT for aiding test signal interpretation. However, the data he used for training the neural network models was generated from the FEM models, rather than from field test signals. Others have also applied neural networks to PIT signals. For example, Zhang & Zhang [16] proposed using two ANN models for processing the PIT signals for diagnosing pile integrity. They used the first ANN model for identifying defect patterns and the second ANN model for assessing severity of the defects. The inputs to the ANN models include time-domain signals, the pile length, the cross-sectional area, and the wave velocity. Tam et al. [5] investigated using PNN (probabilistic neural networks) for diagnosing prestressed concrete pile defects.

More recently, Protopapadakis et al. [4] applied a genetically optimized neural classifier to identify neck and bulk defects of concrete piles. Instead of directly using test signals as inputs to the neural classifier, they used the difference between the test signals and the ideal waveform where the ideal waveform was generated numerically using FEM and scaled boundary finite element method (SBFEM). They also used the island generic algorithm (GA) to optimize the neural network structure.

In [17], Garcia et al. proposed using recurrence plots, a technique from Chaos theory for analyzing and interpreting pile test signals. They converted 1D reflectograms into 2D RP images and then found different characteristics of the 2D RP images associated with different conditions (normal and different defects).

Wavelet analysis is an advanced signal processing technique [18]. Being able to "look" at the signals through both a temporal and scale lens simultaneously, wavelet analysis can handle noisy and non-stationary signals much better than traditional Fourier transformation. Zhang et al. [19] proposed using wavelet analysis and neural networks for pile defect diagnosis.

Machine learning techniques have been widely used for condition-based maintenance (CBM), structural health monitoring (SHM), and non-destructive evaluation (NDE). References [20–23] provide an in-depth overview of applications of machine learning techniques. Extreme learning machines (ELM), an advanced artificial intelligence technique, has been successfully used in various applications in recent years [8–11,24–27]. Using ELM as a means for intelligent interpretation of PIT signals, however, has never been done before, to the best of our knowledge.

3. Proposed Methodology

Aiming for automatic and intelligent interpretation of a large number of field test signals, we proposed the CARI methodology. The proposed methodology, built on the advanced signal processing and artificial intelligence techniques, consists of three components: (1) signal preprocessing, (2) Wavelet-based feature extraction, and (3) ELM-based defect detection, as shown in Figure 2. Detailed descriptions of these three components are given as follows.

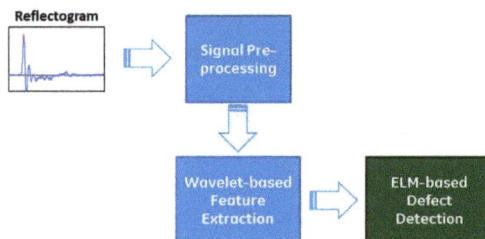

Figure 2. Overall structure of the proposed CARI methodology.

3.1. Signal Preprocessing

To strengthen the reflectogram signal, thus for better interpretation of test results, several signal enhancement strategies are often required. First, the signals measured at the pile head need to be amplified exponentially for compensating stress wave signal weakening due to the shaft friction influence during wave traveling through the pile. Then, high frequency reflections exist in reflectograms due to the shear wave influence at the pile top and steel reinforcement inside the pile. Thus, low-pass filtering is also needed most of time to remove those high frequency reflections. We use wavelet decomposition to perform the filtering during the feature extraction process (See Section 3.2).

As will be discussed in Section 3.2, our feature extraction is wavelet decomposition–based, and feature calculations will be performed over the portion of reflectograms between pile top and toe reflections. Thus identification of reflections from reflectograms is needed for our proposed methodology. Reflections in reflectograms appear as waveform peaks. Reflection identification thus becomes a peak detection problem. In the literature, there are many peak detection algorithms. In this study, we use a simple peak detection algorithm that is based on the sign of first-order difference of the signal [28]. A peak occurs when the signal changes directions, that is, a peak is defined as the sign of signal differences changes from a streak of positives and zeros to negative. The pseudo code of our peak detection is shown in Algorithm 1.

Algorithm 1. Pseudo-code of our peak detection algorithm.
Inputs: $x = [x_1, x_2, \ldots, x_n]$ // a n-point waveform
    T   // A threshold value
Outputs: *pK, pIdx* // peaks and corresponding indices

1. Calculate the 1st-order difference of x, $dx[i] = x[i] - x[i-1]$, $i \in 2, 3, \ldots, n$
2. Determine the sign of $dx[i]$, $i \in 2, 3, \ldots, n$
    $sdx[i] = +1$ for $dx[i] > 0$; $sdx[i] = -1$ for $dx[i] < 0$; $sdx[i] = 0$ for $dx[i] = 0$
3. Search for the sign changes from +1 to −1
    $pIdx = \varnothing$; $pK = \varnothing$
    for $j = 2$ to n
    if $(sdx[j] - sdx[j-1]) = -2$) *then* $pIdx = pIdx \cup j$; $pK = pK \cup x[j]$; *end*
    end
4. Remove those peaks with values being less than the threshold, T

3.2. Wavelet-Based Feature Extraction

Extracting a set of good features from the raw signals is almost always required in order to achieve better predictive models. In the domain of machine learning, feature extraction is regarded as the critical and labor-intensive task [29,30]. We have seen a wide range of feature extraction methods in literature, including the traditional statistical-based methods as well as the modern deep learning methods [31,32]. Given the fact that LSPIT signals (reflectograms) are highly non-stationary and noisy, in this paper, we propose using wavelet analysis to extract features from the reflectograms. Wavelet analysis is an advanced signal processing technique [18] and has been popularly used in various domains and applications [33–40].

Multi-resolution wavelet decomposition, a type of wavelet analysis, is to decomposes a signal into a bunch of orthonormal bases with different time and frequency resolutions [18]. As illustrated in Figure 3, for 3-level wavelet decomposition, the signal is represented by an approximation that contains the high-scale, low-frequency components of the signal and three details that represent the low-scale, high-frequency components of the signals.

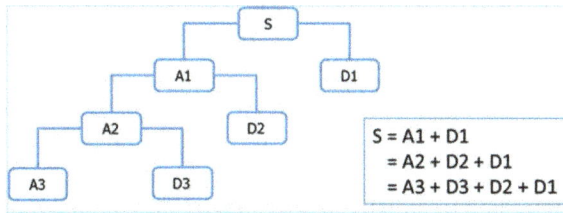

Figure 3. An illustration of 3-level wavelet decomposition.

For the LSPIT reflectograms concerned in this paper, we adopt 4-level wavelet decomposition to achieve a reasonable balance between the time and frequency resolutions. Based on visual analysis of reflectograms, we choose the seventh order "symlet" (see Figure 4 above) as the mother wavelet, which gives the wavelet shape that best matches the reflections of the reflectograms. Figure 5 shows an example of the reflectograms and the approximation and details resulted from the 4-level wavelet decompositions. We perform feature extraction on the fourth approximation (A4) and the fourth and thirrd details (D4 & D3) and ignore other higher frequency components. For each of the three selected bases, we extract seven features (defined below) from its wavelet coefficients, its reconstructed waveforms, and the spectrums of the reconstructed waveforms, respectively, which gives us a total of 21 features.

Let x_n, $n = 1, 2, \ldots, N$ be the time domain signals and $[p_i, f_i]$, $i = 1, 2, \ldots, M$ be its corresponding spectrum, where p_i and f_i are the amplitude and the frequency at ith frequency bin, respectively. The seven features are defined as follows:

(1) Energy: $E = \sum_{i=1}^{N} x_i^2$; (2) Total power: $TP = \sum_{i=1}^{M} p_i$; (3) Mean power: $MP = \frac{TP}{M}$; (4) first spectral moment (centroid): $M_1 = \sum_{i=1}^{M} p_i f_i / TP$; (5) 2nd spectral moment (standard deviation):

$$M_2 = \sqrt{\sum_{i=1}^{M} (f_i - M_1)^2 \cdot p_i / TP}. \tag{3}$$

(6) third spectral moment (skewness):

$$M_3 = \sum_{i=1}^{M} (f_i - M_1)^3 \cdot p_i \left/ M_2^3 \cdot TP \right. \tag{4}$$

and (7) fourth spectral moment (kurtosis):

$$M_4 = \sum_{i=1}^{M} (f_i - M_1)^4 \cdot p_i \Big/ M_2{}^4 \cdot TP \tag{5}$$

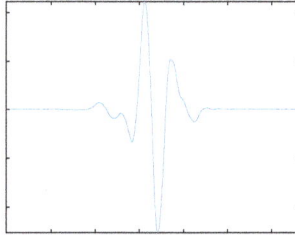

Figure 4. The seventh "symlet" wavelet function.

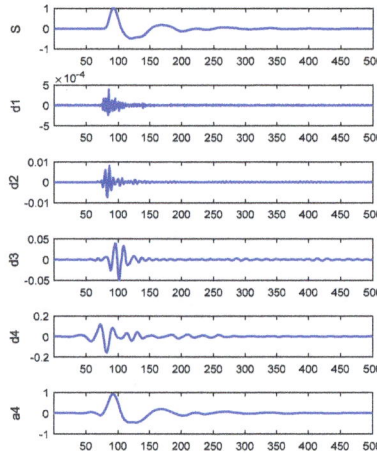

Figure 5. An illustration of 4-level wavelet decomposition of a reflectogram: s = original signal; d1~d4 are the first through fourth details and a4 is the fourth approximation.

3.3. Defect Detection Using ELM

The CARI methodology proposed in this paper is for quickly assessing the pile integrity status based on the LSPIT signals collected from the field tests, that is, to determine whether or not the test pile has any defect, which is often called defect detection. Treating defect detection as a binary classification problem, we can then apply a classification method to solve it.

There are numerous classification methods available, including decision trees, various neural networks, different types of support vector machines, and ensemble learning based method, e.g., random forests and adBoost. In this study, we choose ELM as our defect detection (classification) model, simply because ELM has several unique properties that are well suited for our CARI methodology. The unique ELM properties are summarized as follows.

ELM, a new family of neural networks [6], involves a different way to determine the network parameters—the connection weights and biases between layers. More specifically, instead of learning all parameters as in conventional feed-forward neural networks, ELM's connection weights and biases between the input and hidden layers are randomly generated and are kept fixed. ELM learning is then simply to determine the connection weights between the hidden and the output layers through solving a linear least squares problem [6]. Because of the special design of ELM, there is no need to go through an iterative optimization process for finding the optimal network parameters, as required in conventional neural networks. As the result, ELM is much faster than conventional neural networks with regard to learning. Also ELM seems to be more effective in handling problems with sparse data, i.e., small number of training samples. More importantly, ELM can achieve better performance than other machine learning methods (including SVM), based on several of recent studies (both empirical and analytical) [41,42].

For completeness in the paper, we briefly describe the essence of ELMs as follows. Please refer to [6,41,42] for more thorough discussion of ELMs.

Consider a set of M training samples, (x_i, y_i), $x_i \in \mathbb{R}^d$, $y_i \in \mathbb{R}^k$, where d and k are the dimensions of input and output spaces, respectively. The output of a single layer ELM network with L hidden neurons for an input vector, x, can be expressed as

$$f(x) = \sum_{i=1}^{L} \beta_i h_i(x) = h(x)\beta \tag{6}$$

In the above equation, $h_i(x) = G(w_i, b_i, x)$ (w_i and b_i are the randomly generated weights and bias) is the ith hidden neuron output of the input x; $G(w, b, x)$ is the activation function, which can be any nonlinear piecewise continuous function that satisfies the ELM universal approximation capability theorems [41]; β is the weight matrix that contains the weights connecting the hidden and output layers. $h(x) = [h_1(x), \ldots, h_L(x)]$ is often called the L-dimension random feature space, or the ELM feature space.

ELM learning is simply to find the optimal β through optimizing the objective function as defined [42]:

$$\text{Minimize} : L_p = \frac{1}{2}\| \beta \|^2 + \frac{1}{2}C\sum_{i=1}^{N} \| \xi_i \|^2 \tag{7}$$

$$\text{Subject to} : h(x_i)\beta = y_i^T - \xi_i^T, i = 1, \ldots, N$$

where $\xi_i = [\xi_{i,1}, \ldots, \xi_{i,k}]^T$ is the error vector of the training sample x_i, and C is a constant regularizing the ELM network complexity and its prediction performance.

The equivalent dual optimization objective function is

$$L_d = \frac{1}{2}\| \beta \|^2 + \frac{1}{2}C\sum_{i=1}^{N} \| \xi_i \|^2 - \sum_{i=1}^{N} \sum_{j=1}^{k} \alpha_{i,j}(h(x_i)\beta_j - y_{i,j} + \xi_{i,j}). \tag{8}$$

Solving the above optimization utilizing the Karush–Kuhn–Tucker (KKT) condition, we obtain the ELM's input-output relationship function $f(x)$ as follows [42]:

$$f(x) = h(x)\beta = h(x)H^T \left(\frac{I}{C} + HH^T\right)^{-1} Y, \text{ when } N \text{ is not too big} \tag{9}$$

$$\text{and } f(x) = h(x)\beta = h(x)\left(\frac{I}{C} + H^T H\right)^{-1} H^T Y, \text{ when } N \gg L \tag{10}$$

where H is the hidden layer output matrix.

$$H = \begin{bmatrix} h(x_1) \\ \vdots \\ h(x_N) \end{bmatrix} = \begin{bmatrix} h_1(x_1) & \ldots & h_L(x_1) \\ \vdots & \vdots & \vdots \\ h_1(x_N) & \ldots & h_L(x_N) \end{bmatrix} \tag{11}$$

To handle the situation where data has imbalanced class distribution, weighted ELM (WELM) has been proposed [43]. Let W be a $N \times N$ diagonal matrix and $W_{ii} = 1/\#(C_i)$, where $\#(C_i)$ is the number of samples in the class that ith sample belonging to. With the W defined, the ELM output function (Equation (8)) becomes $f(x) = h(x)\beta = h(x)\left(\frac{I}{C} + H^T W H\right)^{-1} H^T W Y$.

4. Experimental Results and Discussion

4.1. The Reflectogram Data

To demonstrate the capability of the proposed CARI methodology for identifying defect piles based on LSPIT signals, in this section we apply the proposed CARI to a large number of reflectograms collected from various real-world foundation construction sites. Table 1 summarizes the piles considered for our experiments. A total of 923 piles from 27 construction sites are considered. All the piles are friction piles and have four different types: Type I—pre-cast RC pipe pile, Type II—prestressed high strength concrete driven pipe pile, Type III—RC cast-in-place (bored) pile, and Type IV—RC cast-in-place (dogged) pile. Pile lengths vary from 5 m to 18.8 m (see Table 1). For all of the piles considered, the LSPIT tests were performed by experienced experts from our institute—Anhui and Huaihe River Institute of Hydraulic Research (AHRIHR). Manual interpretation of the test signals (reflectograms) were also performed by the experts. The detailed test reports for each of the construction sites and associated test signals have been well documented and archived in the institute's database.

Table 1. Pile and construction site summary.

No	Site Name	# of Piles	# of Defect Piles	Pile Length (m)	Pile Type
1	Jing-Ao Bldg.#7	21	0	11	I
2	Jing-Ao Bldg.#9	19	0	14	I
3	Jing-Ao Bldg.#10	20	0	12	II
4	Jing-Ao Bldg.#26	15	0	11	II
5	Jing-Ao Bldg.#28	20	0	12	II
6	Jing-Ao Bldg.#29	38	0	13	II
7	Jing-Ao Bldg.#30	19	0	13	II
8	Jing-Ao Bldg.#31	20	0	13	II
9	Jing-Ao Bldg.#32	37	0	14	II
10	Jing-Ao Bldg.#35	36	0	12	II
11	Jing-Ao Bldg.#36	36	0	12	II
12	Jing-Ao Bldg.#37	20	0	13	II
13	Jing-Ao Bldg.#38	35	0	12	II
14	Ye-Ji 35kvRoad	66	8	6.8–10.5	III
15	Fong-Fang Railroad Bldg.#4	46	3	11, 12, 14	III
16	Fong-Fang Railroad Bldg.#5	47	2	11, 12, 14	III
17	Fong-Fang Railroad Bldg.#7	50	6	11, 12, 14	III
18	Fong-Fang Railroad Pump Station	34	3	16, 18	III
19	Shang-Shui-Guang	37	6	10.5	III
20	Yi-Shi-Jia Bldg.#3	27	4	16, 17	II
21	Yi-Shi-Jia Package Bldg.	33	3	15, 16	II
22	Yi-Shi-Jia Bldg.#2	56	3	17	II
23	Ying-Chao-Yang	6	6	9.8–18.8	II
24	Yi-Shi-Jia Bldg.#1	65	3	16, 17	III
25	Lu-An FongHuanBldg.# 5	87	7	5–9.47	IV
26	Long-Hua 35 KV Engr. Site	19	8	7.5–13.5	IV
27	Bing-He Shuandung Power Station	14	1	9–12	IV
	Total	923	63	N/A	N/A

Field LSPIT testing was performed using the RSM-PRT Low Strain Pile Integrity Tester manufactured by Wuhan Sinorock Technology Co., Ltd., Hubei, China. Figure 6 shows the test equipment.

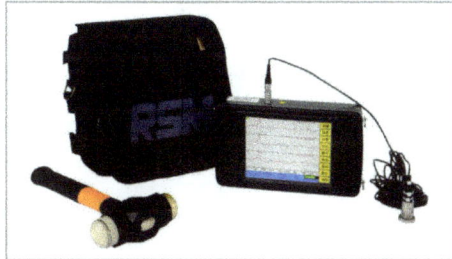

Figure 6. Our LSPIT equipment.

The sampling rate is 50 kHz and the total number of samples per pile is 1024. During field testing, for each pile three reflectograms were obtained by tapping three different spots of the pile top with different hammers with different weights. Only one reflectogram containing clear pile features was chosen for processing and storing in the database.

Since all defective piles have been verified by the experts and/or in the field, we can assume they are truly defective, thus our ground truth for defect detection model evaluation. We will evaluate our proposed CARI by comparing the classification results of CARI against the ground truths (normal and defective piles) described in the test report.

As shown in Table 1, in total we have 860 normal piles and 63 defective piles. Our goal here is to design a binary classifier to distinguish defective piles from normal piles. Since the number of defective piles is much smaller than the number of normal piles, we have so-called "imbalanced data distribution" issue, an important machine learning issue that requires special attention in classifier building. In this work we address the data imbalance issue by using the weighted ELM described in Section 3.

4.2. Detection Performance Evaluation and Methods

To assess the performance of anomaly detection models, we use the Receiver Operating Characteristic (ROC) curves and the related area-under-curve (AUC), the well-known performance metrics, as the classification performance measures for performance comparison. ROC curves represent the tradeoffs between true positive rate (TPR) and positive rate (FPR) [44,45]. ROC curves are good for visual comparison of classifier performance. To quantitatively assess classifier performance, the area under the curve (AUC) calculated from the ROC curve is often used. The AUC is not sensitive to the class sample distribution and represents the classification performance at various decision thresholds.

In terms of actual model evaluation method, we use five-fold cross-validation. In fact, k-fold cross-validation is a well-known model evaluation method that has been popularly used in many predictive modeling applications [46]. To ensure a robust comparison we run the five-fold cross-validation 10 times, each time with different randomly splitting of the five folds of the data.

4.3. Results and Discussion

For ELM classifier design, the number of hidden neurons is fixed to 500, as suggested in [6]. The activation function for the hidden neurons is the sigmoid function, $G(w, b, x) = 1/(1 + \exp(-(w^T x + b)))$. The model parameter, C, is empirically determined via cross-validation by trying 20 different values, i.e., $C = [2^{-9}, 2^{-8}, \ldots 2^{10}]$. For comparison purpose, we also implement a conventional feed-forward neural network (FFNN) as the classification model, using the same data, the same extracted features and the same evaluation method. For the FFNN design we also use the sigmoid function as the activation function, and the number of hidden neurons varies from 5 to 50 with an increment of 5.

Figure 7 shows the ROCs of the 10 5-fold cross-validation runs for both ELM and FFNN models, respectively, where y-axis (sensitivity) is the TPR and x-axis (1—specificity) is the FPR. From Figure 7, one can visually see that the ELM model not only has better classification performance (higher ROC curves), but also is more robust (smaller variation for different runs) than the FFNN does. To perform a quantitative comparison of the ROCs, area-under-curve (AUC) for each of the ROCs is calculated. Table 2 shows the means and the standard deviations of AUCs of the 10 random runs for the two models compared, which confirms our visual observation, that is, ELM outperforms FFNN in terms of classification accuracy (larger mean value) and robustness (smaller variance).

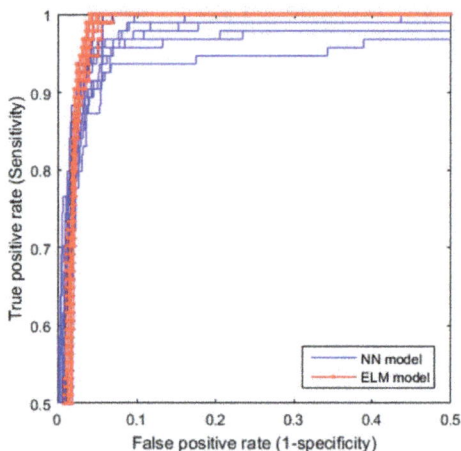

Figure 7. Receiver Operating Characteristic (ROC) curves for extreme learning machines (ELM) and feed-forward neural network (FFNN) models.

Table 2. Areas-under-curve (AUCs).

	AUCs
ELM	0.9841 $\pm$ 0.0022
FFNN	0.9780 $\pm$ 0.0112

For reflectogram interpretation concerned in this paper, since our goal is to identify a small number of suspected piles from a large number of the tested piles in order to reduce the expert's effort on analysis and interpretation, we would like our CARI methodology to have the highest sensitivity (TPR) as possible to minimize misdetection, and our false positive rate, equivalent to how many piles the human expert needs to interpret, to be as small as possible. If we set our true positive rate to be 100%, i.e., no misdetection at all, the average false positive rate obtained from the ROC curves is 5.55% as shown in the confusion matrix (Table 3).

Table 3. Confusion matrix averaged over the 10 random runs.

		Predicted	
		Normal	**Defective**
True	**Normal**	94.45%	5.55%
	Defective	0.00%	100.00%

To help better understand the value of the proposed CARI methodology, let use a hypothetic example. Assume a construction site has a total of 101 piles (1 defective and 100 normal piles). Without using the proposed CARI, the human expert has to examine LSPIT signals of all 101 piles in order to identify the one defective piles. Now with the proposed CARI, he or she only needs to interpret signals of seven piles: one defective pile and the 6 ($100 \times 5.55\% \cong 6$) false identified piles.

That is, the proposed CARI methodology reduces human experts' interpretation effort from 101 piles to seven piles, while ensuring the one defective pile is correctly identified.

The results shown above are the cross-validation outcomes without discerning different pile types, which represent an overall classification performance. These results also give a good indication on how well the proposed CARI generalizes across different construction sites. To assess how well our proposed methodology generalizes cross different pile types, we conduct the pile type-wise cross-validation. That is, for the dataset concerned in this paper, which has for pile types (see Table 1), we perform 4-time type-wise cross-validation. Specifically, each time we leave out all samples associated with one pile type and train the ELM detection model with all remaining samples; and then test the model on the samples of the leave-out type. Table 4 shows the type-wise cross-validation results.

Table 4. Model performance summary based on the pile type-wise cross validation.

Pile Type	# of Piles	# of Defect Piles	TPR (%)	FPR (%)
I	40	0	-	0.20
II	418	16	93.75	4.78
III	345	31	96.77	5.51
IV	120	16	87.50	5.83
Total	923	63		

Comparing Tables 3 and 4, one can see that our model classification performance (true detection rate and false positive rate) degrades under the type-wise cross-validation, indicating that reflectogram characteristics among different pile types are significantly different. Essentially, the type-wise cross-validation assesses how well our detection model performs on new, unseen pile types that have different reflectogram characteristics. We argue that making our detection model generalize well to a new pile type with different characteristics is practically unnecessary in real-world applications. Given the fact that different pile types have different reflectogram characteristics, in order to apply our methodology to a new pile type, we just need to update our detection model whenever samples (i.e., reflectograms of LSPIT tests) of the new pile type are available. An alternative solution would be to simply build a specific detection model for each of the pile types. Hence, the less effectiveness of our methodology in generalizing to new pile types should not hinder the applicability of our methodology in real-world applications.

5. Conclusions

Low strain pile integrity test (LSPIT) is a mature method and has been popularly used worldwide for assessing pile integrity. Test results interpretation, an important task of LSPIT, is currently still a manual process performed by experienced experts. Such a manual process is labor-demanding and also becomes a great burden in situations when field testing and the pile integrity assessment results need to be completed on a tight schedule. Technologies that can enhance LSPIT by speeding up the interpretation process, thus being able to have a quick turnaround in completing the test, are of great business value and are in great need. Our study in this work is an effort toward addressing this need. Realizing that fully automated interpretation is practically infeasible since many factors affect the reflectograms, in this study, we propose the CARI methodology that does not completely free human experts' involvement in interpretation, but rather greatly reduces human experts' effort in interpretation. Since human experts only need to look at a few suspected piles identified by our CARI methodology, they can afford to perform more thorough analysis by considering soil condition and pile construction information, thus obtaining more reliable assessment results. As the result, we would expect the integrity assessment obtained would be more reliable and more accurate, enabling LSPIT to be more efficient and more effective.

Using a reasonable number of reflectograms collected from real-world piles in various foundation construction sites, we have demonstrated that the proposed CARI methodology is effective in detecting defective piles while maintaining the false positive rate reasonably low. We also noted that different pile types have different reflectogram characteristics, and thus cautious measures are required when applying the methodology to a new, unseen pile type.

In future, we will continue to validate the proposed methodology using more real-world piles with more diverse types and soil conditions. We will also explore other different modeling techniques, for example, ensemble of ELM models, to further improve the classification performance. It is also our interest to expand the capabilities of the proposed methodology to cover defect identification, i.e., identifying defect types of piles.

Acknowledgments: This research work is partially supported by the Civil Engineering Quality Insurance and Assessment Grants of the Scientific Research Special Interests of Anhui Province. Those research grants also cover the costs to publish in open access journals.

Author Contributions: D.-M.C. initiated the idea and provided supervision and guidance in experiments and analysis; W.Y. performed machine learning modeling and was the primary writer of the paper; X.-Q.W. performed the PIT field tests and manual reflectogram interpretation; and L.-M.L. was in charge of the project.

Conflicts of Interest: The authors declare no conflict of interest.

References

1. Rausche, F. Non-Destructive Evaluation of Deep Foundations. In Proceedings of the 5th International Conference on Case Histories in Geotechnical Engineering, New York, NY, USA, 13–17 April 2004.

2. Likins, G.E.; Rausche, F.; Miner, R.; Hussein, M.H. Verification of Deep Foundations by NDT Methods. In Proceedings of the ASCE Annual Meeting, Irvine, CA, USA, 19–21 April 1993.

3. Massoudi, N.; Teferra, W. Non-Destructive Testing of Piles Using the Low Strain Integrity Method. In Proceedings of the Fifth International Conference on Case Histories in Geotechnical Engineering, New York, NY, USA, 13–17 April 2004.

4. Protopapadakis, E.; Schauer, M.; Pierri, E.; Doulamis, A.D.; Stavroulakis, G.E.; Böhrnsen, J.; Langer, S. A genetically optimized neural classifier applied to numerical pile integrity tests considering concrete piles. *Comput. Struct.* **2016**, *162*, 68–79. [CrossRef]

5. Tam, C.M.; Tong, T.K.L.; Lau, T.C.T.; Chan, K.K. Diagnosis of prestressed concrete pile defects using probabilistic neural networks. *Eng. Struct.* **2004**, *26*, 1155–1162. [CrossRef]

6. Huang, G.B.; Zhu, Q.Y.; Siew, C.K. Extreme learning machine: Theory and applications. *Neurocomputing* **2006**, *70*, 489–501. [CrossRef]

7. *ASTM D5882-07: Standard Test Method for Low Strain Impact Integrity Testing of Deep Foundations*; ASTM International: West Conshohocken, PA, USA, 2007.

8. He, B.; Xu, D.; Nian, R.; van Heeswijk, M.; Yu, Q.; Miche, Y.; Lendasse, A. Fast face recognition via sparse coding and extreme learning machine. *Cognit. Comput.* **2014**, *6*, 264–277. [CrossRef]

9. Bazi, Y.; Alajlan, N.; Melgani, F.; AlHichri, H.; Malek, S.; Yager, R.R. Differential evolution extreme learning machine for the classification of hyperspectral images. *IEEE Geosci. Remote Sens. Lett.* **2014**, *11*, 1066–1070. [CrossRef]

10. Kaya, Y.; Uyar, M. A hybrid decision support system based on rough set and extreme learning machine for diagnosis of hepatitis disease. *Appl. Soft Comput.* **2013**, *13*, 3429–3438. [CrossRef]

11. Yang, X.; Mao, K. Reduced ELMs for causal relation extraction from unstructured text. *IEEE Intell. Syst.* **2013**, *28*, 48–52.

12. Yan, W.Z.; Yu, L.J. On Accurate and Reliable Anomaly Detection for Gas Turbine Combustors: A Deep Learning Approach. In Proceedings of the Annual Conference of the Prognostics and Health Management Society, San Diego, CA, USA, 18–24 October 2015.

13. Wong, P.K.; Yang, Z.X.; Vong, C.M.; Zhong, J.H. Real-time fault diagnosis for gas turbine generator systems using extreme learning machine. *Neurocomputing* **2014**, *128*, 249–257. [CrossRef]

14. Ertel, J.P.; Niederleithinger, E.; Grohmann, M. Advanced in pile integrity testing. *Near Surface Geophysics.* **2016**, *14*, 503–512. [CrossRef]

15. Watson, J.N. The Application of Neural Networks to Non-destructive Testing Techniques. Ph.D. Thesis, School of the Build Environment, Napier University, Edinburgh, UK, 2001.

16. Zhang, C.; Zhang, J. Application of Artificial Neural Network for Diagnosing Pile Integrity Based on Low Strain Dynamic Testing. In *Computational Structural Engineering*; Yuan, Y., Cui, J., Mang, H.A., Eds.; Springer: Dordrecht, The Netherlands, 2009; pp. 857–862.

17. Garcia, S.; Romero, J.; Lopez-Molina, J. An intelligent pattern recognition model to automate the categorization of pile damage. In Proceedings of the 19th International Conference on Soil Mechanics and Geotechnical Engineering, Seoul, Korea, 17–22 September 2017.

18. Daubechies, I. *Ten Lectures on Wavelets*; Society for Industrial and Applied Mathematics: Philadelphia, PA, USA, 1992.

19. Zhang, G.; Jiang, X.L.; Liu, Z.J.; Chen, C.C. Pile Defect Intelligent Identification Based on Wavelet Analysis and Neural Networks. *Appl. Mech. Mater.* **2014**, *608–609*, 899–902. [CrossRef]

20. Ying, Y.; Garrett, J., Jr.; Oppenheim, I.; Soibelman, L.; Harley, J.; Shi, J.; Jin, Y. Toward Data-Driven Structural Health Monitoring: Application of Machine Learning and Signal Processing to Damage Detection. *J. Comput. Civ. Eng.* **2013**, *27*, 667–680. [CrossRef]

21. Farrar, C.R.; Worden, K. *Structural Health Monitoring: A Machine Learning Perspective*; John Wiley and Sons: Hoboken, NJ, USA, 2012.

22. Srivastava, A.N.; Han, J. *Machine Learning and Knowledge Discovery for Engineering Systems Health Management*; CRC Press: Boca Raton, FL, USA, 2011.

23. Dackermann, U.; Skinner, B.; Li, J. Guided-wave-based condition assessment of in-situ timber utility poles using machine learning algorithms. *Struct. Health Monit.* **2014**, *13*, 374–388. [CrossRef]

24. Peng, C.; Yan, J.; Duan, S.; Wang, L.; Jia, P.; Zhang, S. Enhancing Electronic Nose Performance Based on a Novel QPSO-KELM Model. *Sensors* **2016**, *16*, 520. [CrossRef] [PubMed]

25. Jian, Y.; Huang, D.; Yan, J.; Lu, K.; Huang, Y.; Wen, T.; Zeng, T.; Zhong, S.; Xie, Q. A Novel Extreme Learning Machine Classification Model for e-Nose Application Based on the Multiple Kernel Approach. *Sensors* **2017**, *17*, 1434. [CrossRef] [PubMed]

26. Zhang, J.; Xiao, W.; Zhang, S.; Huang, S. Device-Free Localization via an Extreme Learning Machine with Parameterized Geometrical Feature Extraction. *Sensors* **2017**, *17*, 879. [CrossRef] [PubMed]

27. Zhang, J.K.; Yan, W.; Cui, D.M. Concrete Condition Assessment Using Impact-Echo Method and Extreme Learning Machines. *Sensors* **2016**, *16*, 447. [CrossRef] [PubMed]

28. Nijm, G.M.; Sahakian, A.V.; Swiryn, S.; Larson, A.C. Comparison of Signal Peak Detection Algorithms for Self-Gated Cardiac Cine MRI. In Proceedings of the Computers in Cardiology, Durham, NC, USA, 30 September–3 October 2007.

29. Brownlee, J. 2014. Discover Feature Engineering, How To Engineer Features and How to Get Good at it. Available online: Machinelearningmastery.com/discover-feature-engineering-how-to-engineer-features-and-how-to-get-good-at-it/ (accessed on 5 August 2015).

30. Domingos, P. A few useful things to know about machine learning. *Commun. ACM* **2012**, *55*, 78. [CrossRef]

31. NIPS (2014). Deep Learning and Representation Learning Workshop: NIPS 2014. Available online: http://www.dlworkshop.org/ (accessed on 7 August 2015).

32. ICLR (2015), International conference on learning representations. Available online: http://www.iclr.cc/doku.php (accessed on 7 August 2015).

33. Samuel, P.D.; Pines, D.J. Classifying helicopter gearbox faults using a normalized energy metric. *Smart Mater. Struct.* **2001**, *10*, 145–153. [CrossRef]

34. Lotfollahi, Y.M.A.; Hesari, M.A. Using Wavelet Analysis in Crack Detection at the Arch Concrete Dam Under Frequency Analysis with FEM. *Res. India Publ. J. Wavelet Theory Appl.* **2008**, *2*, 61–81.

35. Noh, H.Y.; Nair, K.; Lignos, D.G.; Kiremidjian, A. Application of Wavelet Coefficient Energies of Stationary and Non-Stationary Response Signals for Structural Damage Diagnosis. In Proceedings of the 7th International Workshop on Structural Health Monitoring, Stanford, CA, USA, 9–11 September 2009.

36. Luk, B.L.; Jiang, Z.D.; Liu, L.K.P.; Tong, F. Impact Acoustic Non-Destructive Evaluation in Noisy Environment Based on Wavelet Packet. In Proceedings of the International Multi Conference of Engineers and Computer Scientists (IMECS 2008); Volume II, Hong Kong, China, 19–21 March 2008.

37. Yeh, P.L.; Liu, P.L. Application of the Wavelet Transform and the Enhanced Fourier Spectrum in the Impact Echo Test. *NDT E Int.* **2008**, *41*, 382–394. [CrossRef]

38. Navarro, P.; Fernández-Isla, C.; Alcover, P.M.; Suardíaz, J. Defect Detection in Textures through the Use of Entropy as a Means for Automatically Selecting the Wavelet Decomposition Level. *Sensors* **2016**, *16*, 1178. [CrossRef] [PubMed]

39. Boukabache, H.; Escriba, C.; Fourniols, J.Y. Toward Smart Aerospace Structures: Design of a Piezoelectric Sensor and Its Analog Interface for Flaw Detection. *Sensors* **2014**, *14*, 20543–20561. [CrossRef] [PubMed]

40. Yu, B.; Liu, D.; Zhang, T. Fault Diagnosis for Micro-Gas Turbine Engine Sensors via Wavelet Entropy. *Sensors* **2011**, *11*, 9928–9941. [CrossRef] [PubMed]

41. Huang, G.B.; Zhou, H.M.; Ding, X.J.; Zhang, R. Extreme learning machine for regression and multiclass classification. *IEEE Trans. Syst. Man Cybern. Part B* **2012**, *42*, 513–529. [CrossRef] [PubMed]

42. Huang, G.B. An Insight into Extreme Learning Machines: Random Neurons, Random Features and Kernels. *Cognit. Comput.* **2014**, *6*, 376–390. [CrossRef]

43. Zong, W.; Huang, G.-B.; Chen, Y. Weighted extreme learning machine for imbalance learning. *Neurocomputing* **2013**, *101*, 229–242. [CrossRef]

44. Bradley, A.P. The use of the area under the ROC curve in the evaluation of machine learning algorithms. *Pattern Recognit.* **1997**, *30*, 1145–1159. [CrossRef]

45. Provost, F.; Fawcett, T. Analysis and visualization of classifier performance: Comparison under imprecise class and cost distributions. In Proceedings of the 3rd International Conference on Knowledge Discovery and Data Mining (KDD-97), Newport Beach, CA, USA, 14–17 August 1997; pp. 43–48.

46. Yu, C.H. Resampling Methods: Concepts, Applications, and Justification. *Pract. Assess Res. Eval.* **2003**, *8*, 1–23. Available online: http://PAREonline.net/getvn.asp?v=8&n=19 (accessed on 17 October 2017).

sensors

MDPI

Article

Infrared Thermography Sensor for Temperature and Speed Measurement of Moving Material

Rubén Usamentiaga * and Daniel Fernando García

Department of Computer Science and Engineering, University of Oviedo, Campus de Viesques,
33204 Gijón, Spain; dfgarcia@uniovi.es
* Correspondence: rusamentiaga@uniovi.es; Tel.: +34-985-182626; Fax: +34-985-181986

Academic Editor: Vittorio M. N. Passaro
Received: 7 April 2017; Accepted: 17 May 2017; Published: 18 May 2017

Abstract: Infrared thermography offers significant advantages in monitoring the temperature of objects over time, but crucial aspects need to be addressed. Movements between the infrared camera and the inspected material seriously affect the accuracy of the calculated temperature. These movements can be the consequence of solid objects that are moved, molten metal poured, material on a conveyor belt, or just vibrations. This work proposes a solution for monitoring the temperature of material in these scenarios. In this work both real movements and vibrations are treated equally, proposing a unified solution for both problems. The three key steps of the proposed procedure are image rectification, motion estimation and motion compensation. Image rectification calculates a front-parallel projection of the image that simplifies the estimation and compensation of the movement. Motion estimation describes the movement using a mathematical model, and estimates the coefficients using robust methods adapted to infrared images. Motion is finally compensated for in order to produce the correct temperature time history of the monitored material regardless of the movement. The result is a robust sensor for temperature of moving material that can also be used to measure the speed of the material. Different experiments are carried out to validate the proposed method in laboratory and real environments. Results show excellent performance.

Keywords: temperature monitoring sensor; motion compensation; infrared thermography

1. Introduction

Temperature is one of the most measured physical properties. It describes the average kinetic energy of the molecules and atoms that make up a substance. Temperature provides information about the internal energy of an object, thus measurement, monitoring and control are crucial in most industrial processes.

Many different types of temperature sensors have been developed [1]. However, the most common are based on four different technologies: mechanical, electrical, ultrasonic and infrared. Most mechanical sensors are based on the volume of a fluid that changes with temperature. Mercury and alcohol are commonly used, although mercury based sensors are not sold any more due the toxicity and potentially harmful effects from broken thermometers. Electrical sensors are mostly thermocouples or thermoresistors. Thermocouples contain a junction of two dissimilar metal wires where voltage varies with temperature. Thermoresistors are made of semiconductors where the resistance varies rapidly and predictably with temperature. Ultrasonic sensors generate an ultrasonic wave and measure temperature based on the variation in the speed of propagation. Infrared sensors are based on the infrared radiation emitted by objects, which is mainly a function of their temperature.

Temperature sensors based on infrared thermography have many advantages over other types of sensors [2]. Infrared sensors are non-contact, thus they do not intrude upon the measurement. Moreover, they can measure the temperature of extremely hot objects. These sensors are also very fast, producing temperature readings in microseconds. They can also be grouped in an array of sensors called a focal plane array, in which each sensor provides information about the radiation at a single point, combining to produce a 2D thermal image. These advantages make sensors based on infrared thermography extremely useful in many different applications, such as electrical inspection [3], mechanical inspection [4], non-destructive testing [5], building inspection [6], industrial processes monitoring [7], medicine [8], cultural heritage diagnostics [9] or even pest detection [10].

Monitoring temperature using 2D thermal images equipped with infrared cameras means measurements can be taken at different areas of the scene simultaneously [2]. Moreover, these devices are able to acquire images at very high frame rates. Consequently, the temperature time history in these areas can be recorded and analyzed. This approach is commonly used in many different applications. For example, in [11] the temperature time history of pig iron is monitored while it is being poured. Analyzing the temperature time history is especially important for non-destructive testing applications. In these applications, objects are thermally stimulated to induce contrast between regions of interest [5]. The temperature time history at each point describes the thermal propagation of the external stimulation. Subsurface anomalies produce thermal variations during heating or cooling which sound areas do not. Hence, the analysis of the temperature time history can be used to detect thermal contrast, i.e., defects.

Under controlled conditions where the position of the camera does not change relative to the monitored object, the analysis of the temperature time history at different positions consists of the selection of different pixels, or an area of pixels, in the image. The intensity value of these pixels, or the average intensity if an area is selected, is calculated from the sequence of images, providing the required temperature time history. However, in many different environments measurements are affected by vibrations [12], which can be described as periodic or random motion from an equilibrium position. Vibrations can affect the camera or the monitored object. In either case, the selected position in the image will not correspond to the same area of the monitored object in consecutive images, which invalidates the calculation of the temperature time history. A different scenario, but with similar consequences is when the monitored object or material is moving. Again, a static selection of points in the image cannot be used to calculate the temperature time history of the regions of interest.

The compensation for unwanted camera motion, generally called image stabilization has been widely studied with visible images [13–15]. However, in the case of infrared images research works is scarce, and generally focused on particular applications. In [16], the authors propose an image stabilization algorithm for infrared images based on the 2D Fourier Transform. In this case, the problem is focused on the analysis of the temperature distribution in biomedical applications, where motion appears because patients move due to breathing, pulse and other voluntary and involuntary reactions. In [17] a similar approach is applied to compensate for vibrations in online welding monitoring. In this case, a combination of point tracking and direct phase substitution is used. Both works assume that vibrations only provoke slight movements of the camera relative to the inspected object. Moreover, they do not consider the problem of temperature monitoring when the object is really moving, i.e., when the position of the region of interest in the image also changes because the object is moving, not just affected by vibrations.

Vibration control and compensation is an active research field with numerous developments [18,19]. Mechanical sensors include different components to compensate for vibrations. Generally, a sensor measures the vibrations using an accelerometer. Then, the resulting signal is transformed so that different actuators can generate movements that compensate for the detected vibrations. Digital sensors use the acquired images to extract features that can be used to detect the vibrations. Then, they transform subsequent images to compensate for the movements.

In this work, a general solution is proposed for the problem of temperature monitoring of moving material. The proposed procedure includes tracking. Therefore, not only the image stabilization problem is solved, but also the calculation of the temperature time history of moving material. Both vibrations and movement are treated equally. The solution to these problems is unified in a single method composed of three steps: image rectification, motion estimation and motion compensation. Image rectification calculates a transformed image with a front-parallel projection where measurements in real-world units (mm), rather than pixels, can be carried out. Assuming all points of interest lie on the same plane, then motion can be described accurately using a 2D rigid body transformation. This approach greatly simplifies the mathematical model used to describe motion and also the compensation. Motion is estimated using a combination of feature detection applied to the rectified images and robust model estimation. The proposed method is evaluated in different applications in laboratory experiments and also in real industrial environments.

The main contribution of this paper is the proposal of a new sensor for temperature of moving objects. The sensor is based on infrared thermography, and keeps track of the movements between the infrared camera and the material to calculate the temperature time history accurately. Processing infrared images is a challenging task because standard algorithms do not provide good results. Therefore, specific procedures are proposed. The proposed solution can also be applied to measurement scenarios where the camera or the inspected object are affected by vibrations. This work includes camera calibration, therefore, it produces the correct temperature time history using a simple yet accurate linear mathematical model. Moreover, the real speed of the material can be calculated at any point in time. The intelligent sensor proposed in this work can provide accurate readings regardless of the movement. Excellent performance is obtained in terms of accuracy and robustness.

Including the compensation for the movement of the monitored material using rectified images present a novel approach to solve the considered problem. It provides not only a very accurate method to model motion but also a robust method to measure the speed of the material. This provides a major advantage when designing a sensor that needs to provide accurate information about a signal, temperature in this case, of a material that is moving at variable speed. Moreover, it can also be used to detect when the material is moved or stopped, which can be key to detecting abnormal measurement patterns correctly.

The remainder of this paper is organized as follows. Section 2 presents the proposed approach for temperature monitoring; Section 3 discusses the results obtained with real data; and finally, Section 4 reports conclusions.

2. Monitoring Procedure

The temperature monitoring procedure proposed in this work first acquires the images using an infrared camera. Images are rectified to calculate a front-parallel projection, removing perspective distortion. This step requires the estimation of the camera projection parameters. The proposed method is based on the extraction of the contour of the inspected object and an iterative approximation to the reference shape. Next, motion estimation and compensation is applied to the rectified images. This requires a preprocessing procedure to enhance the contrast in the images. Features are extracted from these enhanced images and used to estimate the movement model robustly. Finally, the temperature time history of the inspected object is calculated. The following sections describe the details of these steps. Figure 1 shows a summary of the steps.

2.1. Image Acquisition

The first required step in order to monitor the temperature is the acquisition of the infrared images. The images are acquired using an infrared camera. These devices measure infrared radiation in a particular wavelength, typically from 8 to 12 µm, or from 2 to 5 µm. The measured infrared radiation is converted into temperature based on the properties of the inspected material, including the emissivity, and the conditions in which the image was acquired, including reflected temperature, ambient temperature, distance, or relative humidity. The accuracy of the calculated temperature values is greatly affected by errors in the estimation of these parameters.

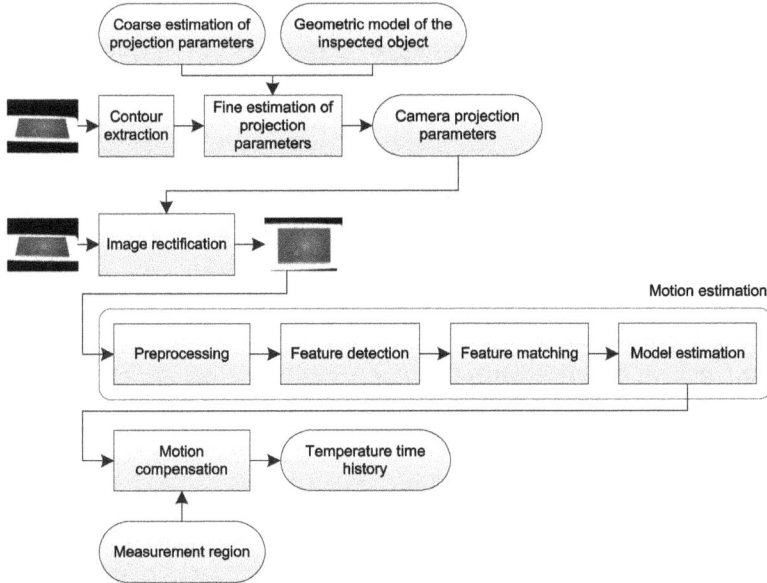

Figure 1. Summary of the proposed approach.

The most common cameras used to acquire infrared images in industrial applications are long-wavelength infrared cameras based on uncooled microbolometers that operate in the range from 8 to 12 µm. They do not require cooling. However, the acquisition rate is low compared with high-end mid-wavelength infrared cameras that operate in the range from 2 to 5 µm. These cameras are usually based on cooled semiconductor detectors that provide much better temperature resolution and higher speed, but they are also more expensive and require more maintenance. Thus, both camera types have their advantages and disadvantages. In the case of fast moving objects, the proposed procedure would require high-end cameras to operate correctly and avoid blurred images. However, the proposed monitoring procedure can be applied using any type of camera.

Vibrations or camera motion cause not only a shifting of the monitored object in the image, they can also cause blurring of the image. Cooled cameras require short integration times (around 1 to 1.5 ms at room temperature). However, a microbolometer detector usually requires an integration time ten times higher. Therefore, depending on the speed of the movement between the camera and monitored object, motion blurring could appear in the acquired images. This work does not deal with this issue. It is assumed that moving objects are exposed sharply and edges can be detected accurately, either because objects move slowly or because a high-end camera based on cooled semiconductor detectors is used. In case motion blurring cannot be avoided, motion deblurring should be applied to the images before applying the proposed procedure. This issue is studied with detail in [20] for visual images. Reference [21] proposes a procedure for motion deblurring of infrared images from a microbolometer camera.

2.2. Image Rectification

In this work image rectification is used to calculate a front-parallel projection of the images that can be used to estimate the motion between images accurately. Image rectification requires an estimation of the parameters that control camera projection. These parameters can be classified as extrinsic or intrinsic camera parameters.

Extrinsic parameters control the transformation of points in world coordinates to points in camera coordinates, and include three rotations and three translations. This transformation can be expressed as (1).

$$T_{ext} = \begin{pmatrix} r_{11} & r_{12} & r_{13} & t_x \\ r_{21} & r_{22} & r_{23} & t_y \\ r_{31} & r_{32} & r_{33} & t_z \\ 0 & 0 & 0 & 1 \end{pmatrix} \tag{1}$$

Intrinsic parameters determine the projection of points from camera coordinates to pixels in the image, and include the focal length (f), the size of the pixels (width and height: S_x and S_y), and the position of the central pixel (C_x and C_y). This projection can be described as the combination of a perspective projection from 3D to 2D, expressed as (2), and a 2D affine transformation, expressed as (3).

$$T_{proj} = \begin{pmatrix} f & 0 & 0 & 0 \\ 0 & f & 0 & 0 \\ 0 & 0 & 1 & 0 \end{pmatrix} \tag{2}$$

$$T_{aff} = \begin{pmatrix} \frac{1}{S_x} & 0 & C_x \\ 0 & \frac{1}{S_y} & C_y \\ 0 & 0 & 1 \end{pmatrix} \tag{3}$$

The transformation of a 3D point $P = (x^w, y^w, z^w)^T$ in world coordinates into a 2D point $p = (r, c)^T$ in pixel coordinates can be expressed as (4), where r and c stand for pixel row and column in the image.

$$p = T_{aff} T_{proj} T_{ext} P$$

$$\begin{pmatrix} r \\ c \end{pmatrix} = \begin{pmatrix} \frac{1}{S_x} & 0 & C_x \\ 0 & \frac{1}{S_y} & C_y \\ 0 & 0 & 1 \end{pmatrix} \begin{pmatrix} f & 0 & 0 & 0 \\ 0 & f & 0 & 0 \\ 0 & 0 & 1 & 0 \end{pmatrix} \begin{pmatrix} r_{11} & r_{12} & r_{13} & t_x \\ r_{21} & r_{22} & r_{23} & t_y \\ r_{31} & r_{32} & r_{33} & t_z \\ 0 & 0 & 0 & 1 \end{pmatrix} \begin{pmatrix} x^w \\ y^w \\ z^w \\ 1 \end{pmatrix} \tag{4}$$

In order to estimate the optimal values for the projection camera parameters, observations of a known target are required. Feature extraction from the images provides the position of known reference points in the calibration target. The parameters of the camera projection model are then estimated by using direct or iterative methods based on this set of reference points. This approach for the estimation of the camera parameters requires calibration targets with features of known dimensions. In visible cameras, accurate calibration targets can be accurately printed using off-the-shelf printers. However, infrared cameras require calibration targets with distinguishable features in terms of infrared radiation.

A recent work on infrared camera calibration estimates the projection parameters in (4) without using specific calibration targets [22]. In this work the projection parameters are estimated with iterative approximations based on the position of the edges in the image, which represent a great advantage for infrared image because objects of interest in infrared images can be easily distinguished from the environment due to temperature differences. This procedure does not consider distortions, but provides accuracy acceptable for common infrared applications. This method is easy to apply, and it does not require specific calibration targets because it is based on information that can be extracted from objects in the image. Moreover, it can be applied from only one image of a known object. Therefore, this is the method used to estimate camera projection parameters in this work.

The considered rectification procedure assumes that the area where the measurement is performed is flat. Therefore, the extracted points from the images lie on the same plane. This plane, the measurement plane, appears in many different applications where infrared thermography is used, such as building inspection or non-destructive testing where the inspected specimens are usually flat. Considering this plane $Z = 0$ then all world points have a z^w equal to zero. Thus, (4) can be expressed as (5).

$$
\begin{pmatrix} c \\ r \\ 1 \end{pmatrix} = \begin{pmatrix} \frac{f}{S_x} & 0 & C_x \\ 0 & \frac{f}{S_y} & C_y \\ 0 & 0 & 0 \end{pmatrix} \begin{pmatrix} r_{11} & r_{12} & t_x \\ r_{21} & r_{22} & t_y \\ r_{31} & r_{32} & t_z \end{pmatrix} \begin{pmatrix} x^w \\ y^w \\ 1 \end{pmatrix}
$$
$$
\begin{pmatrix} c \\ r \\ 1 \end{pmatrix} = H \begin{pmatrix} x^w \\ y^w \\ 1 \end{pmatrix}
$$

(5)

The iterative method proposed to estimate the coefficients of H requires a coarse estimation of the coefficients. The initial coarse estimation of the intrinsic parameters is provided by the manufacturer of the particular camera used (focal length, detector pitch and IR resolution). The initial estimation of the extrinsic parameters consists of the estimation of the displacement (vertical, horizontal and distance) and rotation (pan, tilt and roll) of the measurement plane relative to the camera.

The estimation of the projection parameters continues from the coarse estimation of H. A contour of the inspected object is extracted from the image and transformed into world coordinates using H^{-1}. The proposed method to extract the contour of the inspected object is the Canny edge detector [23]. Then, correspondences are estimated by computing the closest points from the model to the object after applying the transformation to world coordinates. Incorrect correspondences bias the procedure, thus they must be filtered using robust statistics. The final step is the estimation of a homography using the correspondences. The procedure is repeated until convergence is reached. In each iteration the distance from the extracted contour to the real shape of the object is reduced. The result is a homography that describes the projection parameters accurately. This transformation can be directly applied to the original infrared image in order to obtain the rectified image in world coordinates.

In order to illustrate this procedure, a solid object is manually moved while temperature monitoring is performed. The goal is to measure the temperature in a particular location regardless of the movement. This experiment simulates the temperature monitoring of material that is moved, for example hot metal stones on a conveyor belt, or affected by vibrations. Next section will extend the tests with images acquired in real environments. In this example, a test piece made up of metal is used. The dimensions of the test piece are 300 mm $\times$ 199 mm $\times$ 5 mm. A visible image of the test piece can be seen in Figure 2. The test piece is placed on a hot plate (electric griddle), which is at 150 °C approximately. The experiment is performed when the test piece is around 100 °C.

Figure 2. Visible spectrum image of the test piece.

An infrared image of the test piece placed on the hot plate can be seen in Figure 3a. This image is the first of a sequence of images acquired while the test piece is moved within the measurement plane to simulate movements of the material or vibrations. A piece of electrical tape is stuck on the surface for later tests. The temperature of the electrical tape is nearly identical to the underlying test piece and it is used for temperature monitoring. The electrical tape can be clearly distinguished in the images because the emissivity of the tape is higher than the emissivity of the surface of the metal test piece. Therefore, at the same temperature it emits more infrared radiation.

In order to extract the contour of the test piece, an edge detector is applied to the image. The result can be seen in Figure 3b. The extracted contour in Figure 3c is used for the estimation of the projection parameters.

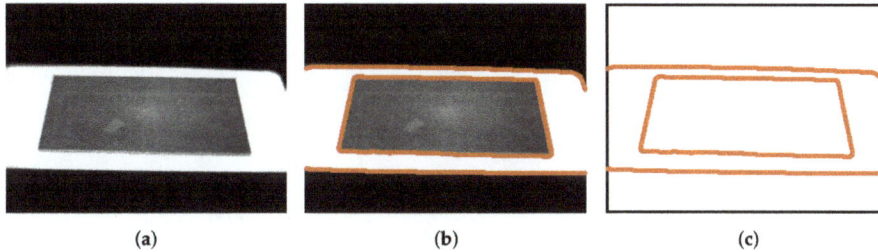

| (a) | (b) | (c) |

Figure 3. Test piece used to illustrate the rectification procedure. (**a**) Infrared image of the heated test piece; (**b**) Edges in the image; (**c**) Extracted contour of the objects in the image.

The infrared camera used in this experiment is a FLIR T450sc (FLIR Systems, Wilsonville, OR, USA). The manufacturer provides information that can be used to obtain a coarse estimation of the projection parameters: 18 mm focal length, 25 μm detector pitch and 320 × 240 image resolution. For the initial values of the extrinsic parameters the following values are roughly estimated: 3° pan, 54° tilt, 180 mm and 180 mm horizontal and vertical displacements, and 1200 mm distance. This camera can acquire raw infrared images, and lossless videos. Therefore, the images used in the tests are not corrupted by noise, for example due to JPEG compression. The technical specifications of this camera are given in Table 1.

Table 1. Technical specifications of the infrared camera FLIR T450sc used in the experiments.

Camera	FLIR T450sc
Temperature range	−20 to +120 °C
Thermal sensitivity/NETD	30 mK at 30 °C
Detector	320 × 240 Uncooled Focal Plane Array (UFPA)
Spectral range	7.5 − 13 μm
Image frequency	60 Hz
Spatial resolution	1.36 mrad
Field of view (FOV)	25° × 19°
Detector pitch (μm)	25

The initial values of the projection parameters are used to transform the extracted contour to world coordinates. The result can be seen in Figure 4a. The shape of the inspected object (test piece is 300 mm × 199 mm) is included in the figure. As can be seen, the transformation of the extracted contour does not match the real shape of the object, as it is only an approximation.

The fine estimation of the projection parameters is carried out by minimizing the distance from the extracted contour in world units to the real shape of the object. The procedure runs iteratively until convergence. In each iteration the approximation improves, that is, the estimation of the projection parameters is more accurate. Figure 4b shows the results after 5 iterations. Figure 4c shows the results when convergence is reached, where an accurate estimation of the projection parameters is obtained.

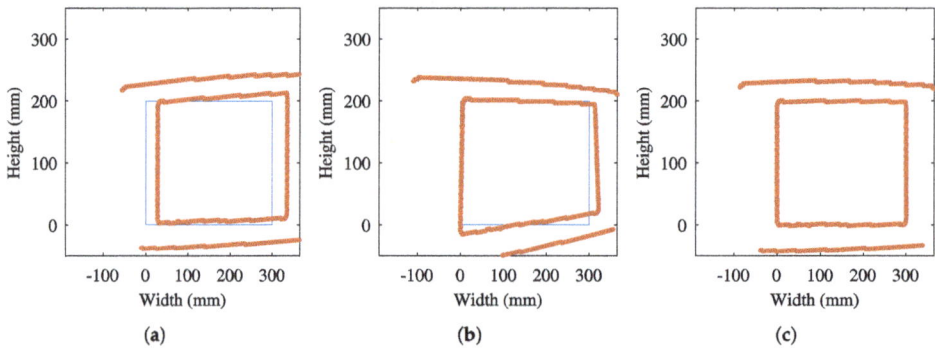

Figure 4. Iterative estimation of the projection parameters. (**a**) Initial estimation; (**b**) Iteration 5; (**c**) Final iteration.

Not all the points in the extracted contour produce valid correspondences. As can be seen in Figure 4, the edges of the base plate are not part of the shape of the test piece. Therefore, these points are discarded [24].

The final valid correspondences are used to accurately estimate the projection parameters, that is, the homography that describes the projection of points from world coordinates to image coordinates, and vice versa. The estimated homography is used to rectify the infrared image. In this procedure the image is interpolated according to a rectangular grid in order to calculate a front-parallel projection of the image using the projection parameters described in the homography. This procedure is generally available in most image processing packages as a projective transformation of an image [25]. The result can be seen in Figure 5, where an image with pixel coordinates is transformed into an image with a front-parallel projection in real-world units. The coordinates of the image in Figure 5b are real-world units. Thus, useful geometric information can be easily extracted from the image.

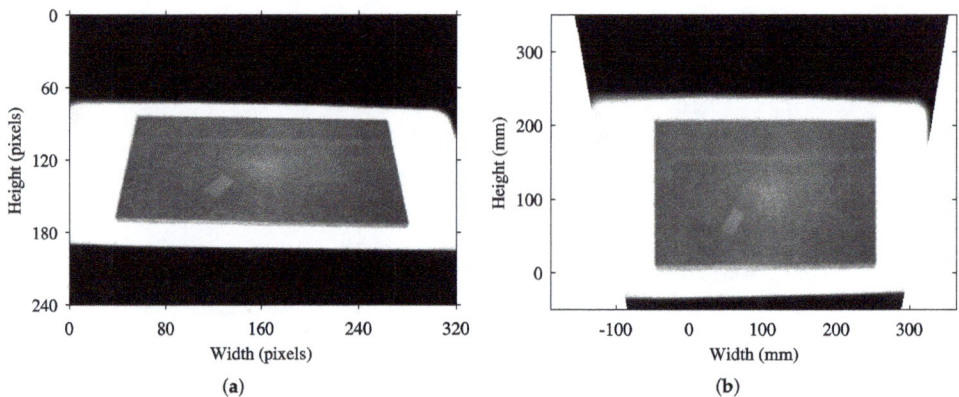

Figure 5. Result of the image rectification procedure. (**a**) Original image; (**b**) Rectified image.

2.3. Motion Estimation

Motion estimation is required to compensate for the movement of the monitored material in the sequence of images. This movement must be described with a mathematical model. Therefore, motion estimation requires the estimation of the values of the mathematical model. Once the model is estimated, it can be used to compensate for the movement of the object, by applying the inverse transformation to the objects that have moved in the image.

2.3.1. Mathematical Model

Modeling the movement between two images is complex, as an image provides a 2D representation of a 3D scene. However, rectified images provide a major advantage in this aspect as pixels represent world coordinates in the measurement plane. This way, the mathematical model required to describe movement is greatly simplified, yet accurate and complete. In a rectified image, the movements of the objects is 2D. Thus, it can be modeled using a 2D rigid transformation. This transformation has three coefficients: the rotation angle θ; and the horizontal and vertical translations: tx and ty. This transformation can be expressed as (6).

$$M = \begin{pmatrix} \cos(\theta) & -\sin(\theta) & tx \\ \sin(\theta) & \cos(\theta) & ty \\ 0 & 0 & 1 \end{pmatrix} \tag{6}$$

Working with rectified images has many advantages. One of them is that the model of the movement is very simple.

2.3.2. Feature Detection

This step detects salient and distinctive features from the images. Features must be distributed over the image. Also, the same features must be efficiently detectable in consecutive images. The goal is to find matching features between consecutive images, that is, a feature that identifies the same point in the scene in the two images. Generally, these features are detected from distinctive locations in the images, such as region corners or line intersections.

Feature detection includes two parts: the detection of the points of interest, and the description of these points. Points of interest in the image are stable and repeatable positions in the image. In visible images, these points can correspond to corners. A vector of features is calculated then for each of these points. These features include derivatives, or moment invariants. One of the most used method for feature detection is SURF (Speeded Up Robust Features) [26]. This method is based on Hessian detectors and use the Haar wavelet to calculate the features of the detected points.

SURF does not provide good results using raw infrared images because in most cases the contrast in the region of interest is not enough to detect the features required to estimate movement. Moreover, when the image contains information about the moving material but also about non-moving objects, such as the background, features can also be detected in non-moving areas. This mixture of features cannot be used to estimate movement. Therefore, a preprocessing stage is proposed.

The first step of the preprocessing stage is to extract the region of interest from the image, that is, the part where the moving material is located. This step is application dependent, but can be carried out in most cases using thresholding techniques [27]. The moving material inspected using infrared thermography usually has a different temperature from the rest of the image. Thus, thresholding the image based on the temperature level is an effective solution that works for most applications. The example presented in Figure 3 is slightly different because in the image three parts can be distinguished based on temperature: the background, the hot plate and the test piece. In this case an effective approach is to apply thresholding twice: a first thresholding to distinguish the plate and the test piece from the background, and then a second thresholding applied only to the extracted region in the first thresholding to distinguish the plate from the test piece.

The second step of the preprocessing stage is the enhancement of the contrast in the image. This step enables SURF to extract meaningful features from the region of interest in the image. Applying SURF to the raw image can result in a low number of features focused on the corners of the material that do not provide the required information to estimate movement correctly. One of the most common methods to enhance contrast in images is a method known as CLAHE (Contrast Limited Adaptive Histogram Equalization) [28]. This is the proposed method for contrast enhancement in this work.

Figure 6 shows the results of the feature detection procedure for the test piece used in the previous example. Figure 6b shows the results of the first thresholding, where a region that includes the hot plate and the test piece is obtained. This first thresholding is applied to distinguish the hot plate and the test piece from the background. The result is a binary image, where the white part represents the foreground and the black part the background that is ignored in next steps. The results of the second thresholding are shown in Figure 6c. In this case, the obtained regions distinguish the test piece from the plate. The white area in this image represent the region of interest for the considered problem: the region in the image where the test piece is located. Using this region in the original image in Figure 6a produces the result shown in Figure 6d. This image is obtained by multiplying the images in Figure 6a,c (in some references this is described as the application of the and logical operator to the images). Figure 6e shows the result of the next step in the preprocessing: contrast enhancement using CLAHE. The resulting image can now be used to detect the features required to estimate the movement. The location of the features for the example can be seen in Figure 6f.

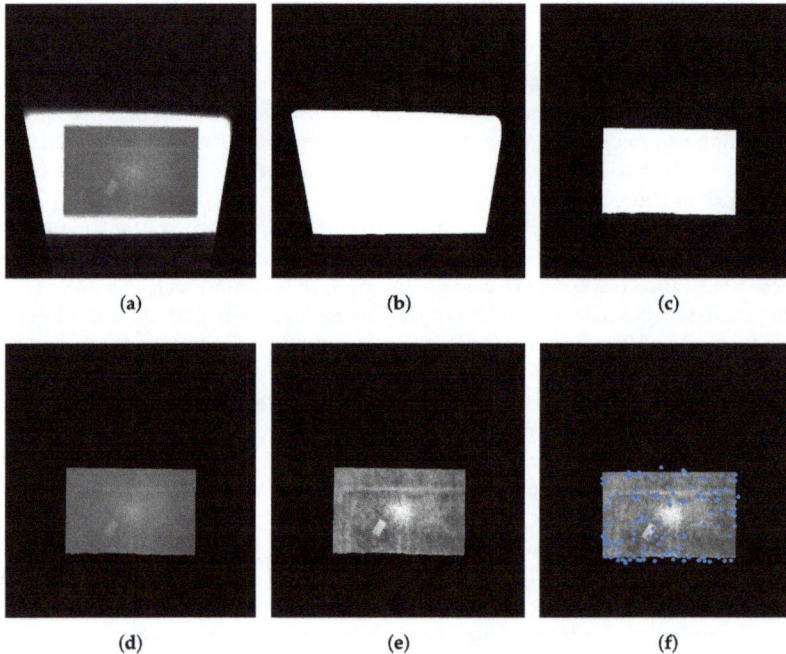

Figure 6. Result of the feature detection procedure. (**a**) Rectified image; (**b**) Region extracted after the first thresholding; (**c**) Region extracted after the second thresholding; (**d**) Extracted object of interest; (**e**) Contrast enhancement of the image for the object of interest; (**f**) Location of the detected features.

As can be seen in Figure 6f, some features are located outside the boundary of the test piece. This is because features are calculated based on derivatives that use windows of pixels around the pixel in which the derivative is calculated. Cropping the image around the test piece would solve this problem, but some interesting features in the corners could be missed.

2.3.3. Feature Matching

Feature matching looks for correspondences between two set of features. Features from the two considered images are compared and linked by minimizing the sum of squared differences. The result is a set of possible correspondences, in most cases containing outliers.

Figure 7 shows an example of feature matching. In this example a second image of the same test piece acquired later is shown. When the second image is acquired the test piece is slightly moved to the left and upwards. The movement of the test piece is performed within the measurement plane, thus, the same projection parameters are used to rectify the second image. The feature detection procedure is applied to the two images, including preprocessing and enhancement. The results are shown in Figure 7a,b. The goal of the feature matching procedure is to find the corresponding features between the two images. The initial result of the feature matching procedure can be seen in Figure 7c. A line connects the matched features between the two images. They also indicate the estimated movement, from the crosses to the circles. The initial result includes many outliers that do not provide the correct information about the movement of the test piece. Ideally, all the matched features should identify the same movement. Part of these outliers can be removed using heuristics. For example, distances between feature vectors can be sorted. Then, only a percentage of the closest distances can be selected as valid in order to reject ambiguous matches. Multiple features in the first image matching the same feature in the second image can also be removed to reduce the number of outliers. Using these two heuristics, the result of the feature matching procedure reduces the number of outliers, as can be seen in Figure 7d. However, no heuristic can guarantee there will not be outliers in the result of the matching. In the example, there are still clearly visible outliers.

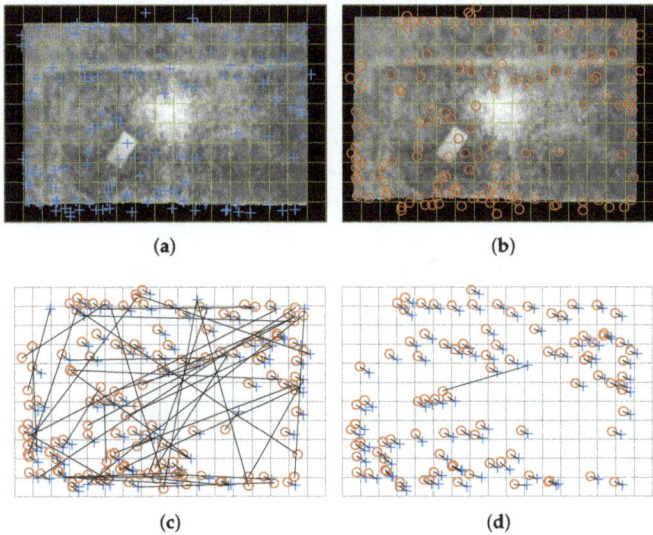

Figure 7. Result of the feature matching procedure. (**a**) Features from the first image; (**b**) Features from the second image; (**c**) Results of the feature matching; (**d**) Results of the feature matching using heuristics.

When using infrared images, features can also change with time due to temperature differences that can diminish due to heat diffusion. Therefore it is not possible to find matching features between images acquired at distant time periods. In this work feature matching is applied between images acquired consecutively, where the features are expected to remain constant. However, temperature differences could generate some outliers that need to be considered for the model estimation.

2.3.4. Model Estimation

The movement model is described using a 2D rigid transformation. The coefficients of this model must be estimated using the result of the feature matching procedure: a set of point correspondences. These correspondences provide information about the movement of the material in the image. In this work, the method used to estimate a rigid transformation is a fast 2D method [29].

Considering a set of n points $\mathcal{P} = \{p_1, p_2, \ldots, p_n\}$, and $\mathcal{Q} = \{q_1, q_2, \ldots, q_n\}$ in $\mathbb{R}^2$, where $p_i = (p_{ix}, p_{iy})^T$ and $q_i = (q_{ix}, q_{iy})^T$ represents the 2D coordinates of the i-th point in $\mathcal{P}$ and $\mathcal{Q}$, the rigid transformation that maps $\mathcal{P}$ into $\mathcal{Q}$ can be described as (7), where R is the rotation and t the translation.

$$\mathcal{Q} = \mathcal{P}R + t \tag{7}$$

Solving (7) requires minimizing E, which is obtained using the least squares error criterion and can be defined as (8).

$$E = \sum_{i=1}^{n} |\mathcal{Q} - \mathcal{P}R - t|^2 \tag{8}$$

The value of t that minimizes E must satisfy (9).

$$0 = \frac{\partial E}{\partial t} = -2 \sum_{i=1}^{n} |\mathcal{Q} - \mathcal{P}R - t| \tag{9}$$

Therefore, t can be calculated using (10), where $\bar{p}$ and $\bar{q}$ are the centroids of $\mathcal{P}$ and $\mathcal{Q}$

$$t = \bar{q} - R\bar{p} \tag{10}$$

Substituting the centered points $\mathcal{P}^z = \{p_1^z = p_1 - \bar{p}, p_2^z = p_2 - \bar{p}, \ldots, p_n^z = p_n - \bar{p}\}$, and $\mathcal{Q}^z = \{q_1^z = q_1 - \bar{q}, q_2^z = q_2 - \bar{q}, \ldots, q_n^z = q_n - \bar{q}\}$ in (8) yields (11).

$$E = \sum_{i=1}^{n} |\mathcal{Q}^z - \mathcal{P}^z R|^2 \tag{11}$$

The angle of rotation θ defines the rotation matrix. The rotation of point p_i^z using this angle is (12).

$$Rp_i^z = \begin{pmatrix} \cos(\theta) p_{ix}^z - \sin(\theta) p_{iy}^z \\ \sin(\theta) p_{ix}^z + \cos(\theta) p_{iy}^z \end{pmatrix} \tag{12}$$

Substituting (12) in (11) gives an equation where E only depends on θ. Solving for θ results in (13).

$$\theta = \tan^{-1} \left(\frac{\sum_{i=1}^{n} (p_{ix}^z q_{iy}^z - p_{iy}^z q_{ix}^z)}{\sum_{i=1}^{n} (p_{ix}^z q_{ix}^z + p_{iy}^z q_{iy}^z)} \right) \tag{13}$$

In order to calculate the translation t, the value of R must be substituted in (10).

The method used to estimate the rigid transformation between correspondences should only be applied when there are no outliers in the data. Correspondence outliers would lead to major errors in the resulting estimated transformation. Therefore, the method to estimate the rigid transformation cannot be applied to the matched features directly.

The proposed solution for the estimation of the rigid transformation using noisy correspondences is MLESAC [30]. This robust estimator is an enhanced version of the Random Sample Consensus

(RANSAC) algorithm [31], widely applied to estimate mathematical models robustly. The algorithm randomly samples the available correspondences and estimates rigid transformations using the previously described method. Not all point correspondences are used, just the strictly required number to estimate the rigid transformation. Among all the putative solutions, the solution that maximizes the likelihood is chosen.

Figure 8 shows the results of the motion estimation procedure for the considered example. As can be seen in Figure 8a, only some of the correspondences in Figure 7d are truly considered for the robust estimation of the movement model. The final result represented in Figure 8b is an accurate estimation of the movement in the test piece between the two images. The result of the robust estimation of the movement is a 2D rigid transformation that perfectly describes the movement of the material in the measurement plane.

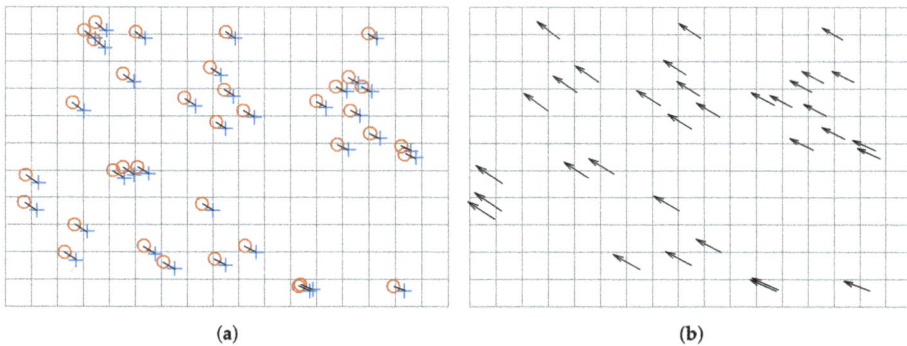

(a)   (b)

Figure 8. Result of the motion estimation procedure. (**a**) Results of the feature matching after the robust model estimation; (**b**) Estimation of the movement (arrows are not to scale).

2.4. Motion Compensation

Motion compensation can be applied in two equivalent ways. The first possible approach is to move the pixels in the second image according to the inverse of the estimated movement. This approach requires image reinterpolation.

Figure 9 shows an illustration of the image reinterpolation approach. The first row of images shows the raw infrared images acquired while the test piece is moved. In the first image (Figure 9a), a circular measurement region is established on the electrical tape stuck on the test piece. As expected, while the test piece is moved the position of the measurement region misses the location of the center of the tape. The second row in the figure shows the images after motion compensation. In this case, the circular measurement region always stays at the same position relative to the electrical tape, regardless of the movement.

Monitoring the temperature in the circular measurement region of the previous example provides the results shown in Figure 10. When the raw images are used, the position of the circular measurement region fails to identify the position of the center of the tape, as can be seen in Figure 9. Therefore, the resulting signal does not provide the correct temperature of the tape over time. However, when the movement is compensated using the proposed approach, the position of the measurement region is always correct, resulting is an accurate signal representing the temperature time history of the inspected material.

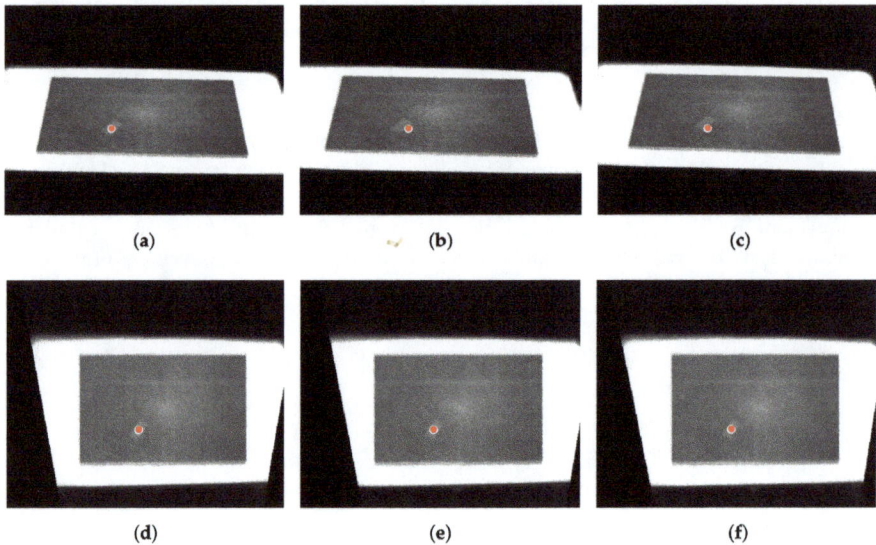

Figure 9. Result of the motion compensation procedure. (**a**) Raw image at t_0; (**b**) Raw image at t_1; (**c**) Raw image at t_2; (**d**) Motion compensated image at t_0; (**e**) Motion compensated image at t_1; (**f**) Motion compensated image at t_2.

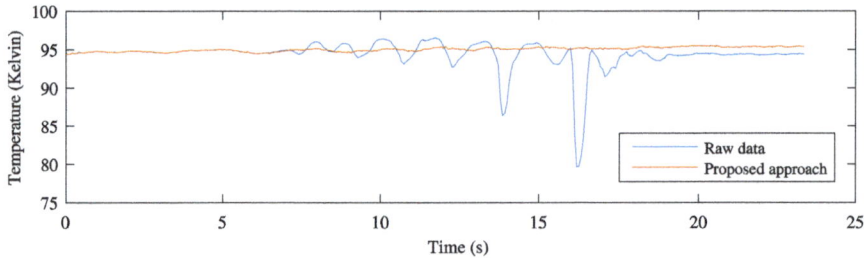

Figure 10. Comparing temperature monitoring using raw images and the proposed approach.

The second approach to image reinterpolation is to move the measurement regions according to the estimated movement of the monitored material. This approach does not require image reinterpolation, thus, it is faster and produces the same results.

The motion estimation procedure produces a 2D rigid transformation H_i between every two consecutively acquired images, I_{i-1} and I_i. The obtained transformations can be composed to obtain the transformation from the first image to the current image i using (14).

$$H_{T,i} = \prod_{j=0}^{j=i} H_j \tag{14}$$

Using (14) any single point in one image can be transformed back and forth between any other image. Therefore, it can be used to compensate for the movement of the material.

3. Results and Discussion

In order to test the proposed procedure, a first experiment is performed with the same test piece with a different orientation and movement. The test piece is placed on a hot plate. The goal is to monitor the temperature of a piece of electrical tape stuck on the test piece while it is moved, simulating vibrations.

The results of the experiment can be seen in Figure 11. The first step is the image rectification using the estimated projection parameters. The results of this procedure are similar to those described for the test piece in the original orientation. Using the rectified images, motion is estimated and compensated. Images are reinterpolated in order to compensate for the movement. In this experiment, the movement of the test piece is increased, as can be seen in the figure. However, this does not affect the estimation and compensation of the movement, producing accurate results. Therefore, robust temperature monitoring can be performed regardless of the movement. As can be seen, the measurement region is always in the same position relative to the test piece.

(a) (b) (c)

(d) (e) (f)

Figure 11. Result of the motion compensation procedure for the test piece with different orientation and movement. (a) Raw image at t_0; (b) Raw image at t_1; (c) Raw image at t_2; (d) Motion compensated image at t_0; (e) Motion compensated image at t_1; (f) Motion compensated image at t_2.

The monitored temperature can be seen in Figure 12. The resulting temperature signal when using the proposed approach provides the correct information about the temperature in the region of interest. This result can be compared with the temperature signal obtained from the raw images. In this case, the temperature time history is incorrect because it is affected by the movement of the material.

In order to calculate the temperature of the test piece in the experiments, the infrared camera was configured using the emissivity of the electrical tape: 0.96 (Scotch[TM] Premium Vinyl Electrical Tape 88, 3M, Maplewood, MN, USA); the reflected temperature estimated using the reflector method [32]: 22.4 °C; and the distance, ambient temperature and relative humidity.

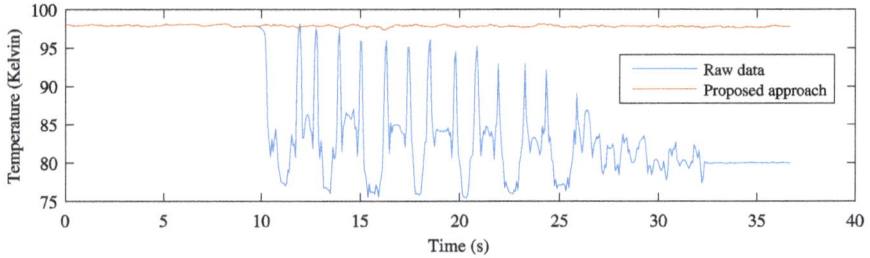

Figure 12. Comparing temperature monitoring using raw images and the proposed approach for the test piece with different orientation and movement.

Experiments have also been performed in a real environment: a sinter cooler. Sinter is a solidified porous material used in the steel industry. It is created by applying heat and pressure to a mixture of different raw materials including fine particles of iron, and other materials such as limestone and coke [33]. The material is later moved to a rotatory cooler where the temperature must be monitored before the final transportation using a conveyor belt to the next step of the industrial process in the blast furnace, where pig iron is produced.

The sinter cooler is a 3.2 m wide circular rotatory ring where air is blown from fans. The cooler moves slowly while the sinter material cools. Temperature monitoring in the cooler is critical to ensure that the cooling pattern is correct. Moreover, temperature monitoring is also used to prevent fires due to excessive temperature in the transportation by conveyor belt. Figure 13 shows an image of the rotatory cooler and the camera used for monitoring.

Figure 13. Infrared camera for sinter monitoring.

Figure 14 shows an infrared image of the cooler and the contour extraction procedure. Due to the optics of the camera and the distance from the camera to the object, only a partial view of the cooler is available. However, this visible part is enough to apply the proposed rectification procedure. The first step is to extract the contour of the cooler. This can be performed by applying an edge detector to the image. The sinter material is hotter than the rest of the image. Thus, it can be clearly distinguished. The extracted contour is the boundary of the visible part of the cooler in the image.

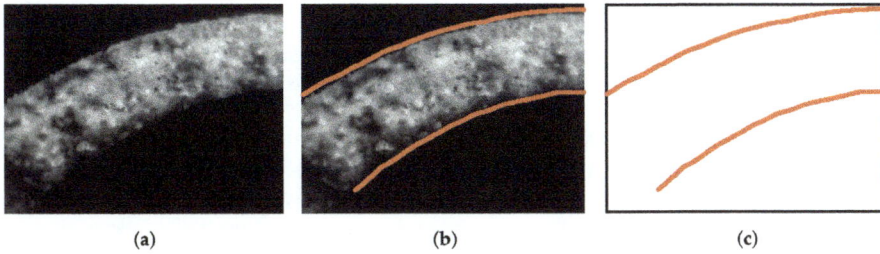

Figure 14. Result of the contour extraction for the sinter material. (**a**) Infrared image of the sinter in the cooler; (**b**) Edges in the image; (**c**) Extracted contour of the cooler in the image.

In this example a FLIR A315 infrared camera (FLIR Systems, Wilsonville, OR, USA) is used. The information provided by the manufacturer and an estimation of the pan, tilt and distance to the cooler is used for the coarse estimation of the projection parameters. The technical specifications of this camera are given in Table 2.

Table 2. Technical specifications of the infrared camera FLIR A315 used in the experiments.

Camera	FLIR A315
Temperature range	0 to + 500 °C
Thermal sensitivity/NETD	50 mK at 30 °C
Detector	320 × 240 Uncooled Focal Plane Array (UFPA)
Spectral range	7.5 −14 μm
Image frequency	60 Hz
Spatial resolution	1.36 mrad
Field of view (FOV)	25° × 18.8°
Detector pitch (μm)	25

Figure 15a shows the transformation of the extracted contour to world coordinates. In this same figure, a model of the cooler ring is represented. The information about the shape of the cooler is obtained from the plans of the factory in which it is installed.

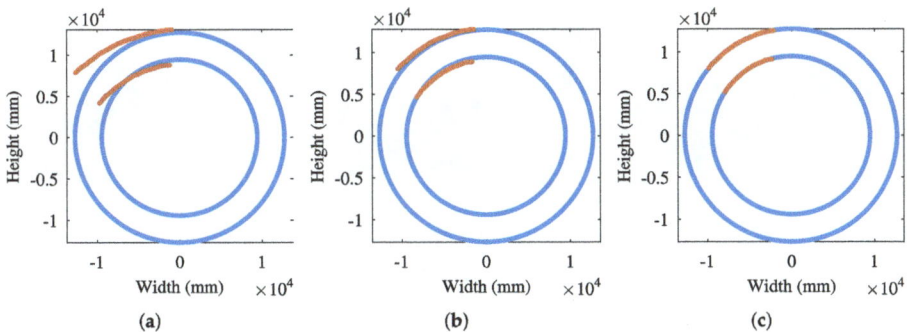

Figure 15. Iterative estimation of the projection parameters for the sinter material. (**a**) Initial estimation; (**b**) Iteration 10; (**c**) Final iteration.

As can be seen in Figure 15, the initial transformation of the extracted contour to world coordinates is just a rough approximation. The fine estimation of the projection parameters is applied next.

The iterative procedure minimizes the distance between the extracted contour and the model until convergence. Finally, a nearly perfect match between the extracted contour in world coordinates and the visible part of the cooler ring is obtained. The error obtained after the iterative procedure converges is 13.36 RMS, with a mean error of 19.89 mm. This error is very low compared with the size of the rotatory cooler, with a diameter of 13 m. The result of this iterative procedure is an accurate estimation of the projection parameters that can be used to rectify the infrared images. The estimation of the projection parameters is valid while the position of the camera is not changed.

The next step is the estimation and compensation of the movement of the material in the cooler. In this case the approach used to monitor temperature is to update the position of the measurement region according to the estimated movement. Therefore, in this case movement compensation is applied as an equal movement to the position of the measurement region.

Figure 16 shows the result of the estimation and compensation of the movement of the material in the cooler. The goal is to monitor the temperature of the hot spot in the images. As expected, in the raw images monitoring cannot be performed because as soon as the material moves, the position of the measurement region is incorrect, as seen in Figure 16b,c. However, applying the proposed procedure monitoring is possible since motion is estimated and compensated accurately. As seen in Figure 16d–f the position of the circular measurement region is updated correctly according to the motion of the material. Therefore, this makes the temperature monitoring of the selected region possible.

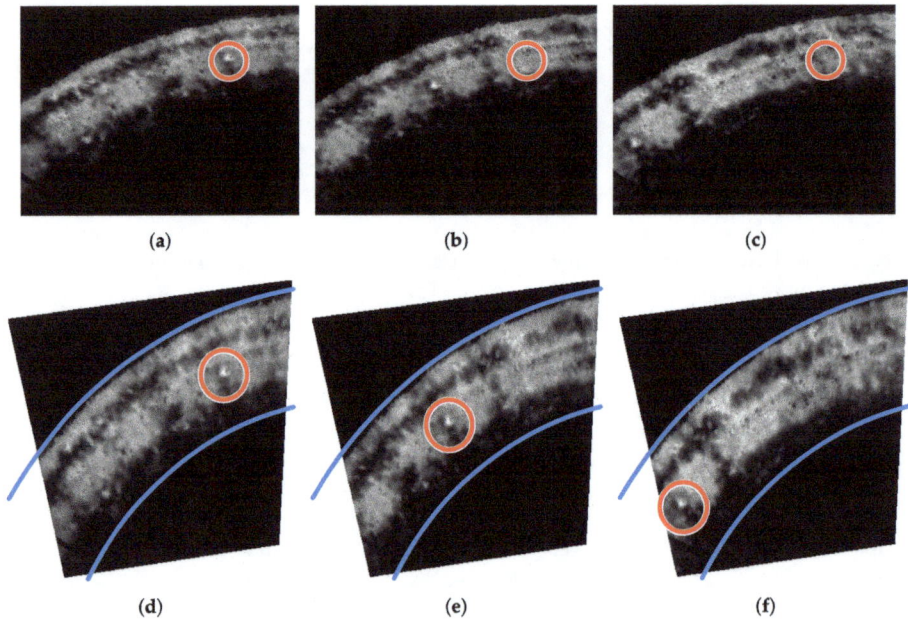

Figure 16. Result of the motion compensation procedure for the sinter material. (**a**) Raw image at t_0; (**b**) Raw image at t_1; (**c**) Raw image at t_2; (**d**) Rectified image at t_0; (**e**) Rectified image at t_1 with the position of the measurement region updated; (**f**) Rectified image at t_2 with the position of the measurement region updated.

Figure 17 shows the result of the temperature monitoring procedure when using the raw images and the proposed procedure. The temperature values shown in this figure represent real temperature readings using a calibrated infrared camera. The signal extracted from the raw images provides information about changes in the temperature at a fixed position of the cooler. On the other hand, the signal extracted when using the proposed procedure contains an accurate representation of the temperature time history of the material at a selected region. This signal provides information about the cooling behavior and cooling per time unit, thus, it can be used to control the variables of the industrial process, including speed or air flow of the fans. Temperature monitoring at a fixed position is useless because the temperature of the material changes, and the temperature decay curve cannot be calculated. The proposed procedure solves this problem by compensating for the movement and measuring the temperature in the same area of material while it moves.

The proposed procedure not only provides the opportunity to monitor the temperature of the material as it moves, it also produces images that can be used to extract useful geometric information. For example, it is possible the measure the size of a region of interest in the image in real-world units. Also, measurement regions in the material can be established with specific sizes.

Using rectified images to estimate motion greatly simplifies the definition of the mathematical model of the movement and its estimation and compensation. However, there is another great advantage: the estimated movement is in real-world units. Therefore, it provides information that can be used to control the industrial process. Not only does the proposed procedure produce the correct temperature time history, but also the real speed of the material at any point in time.

The proposed procedure is not without limitations. It assumes that the acquisition speed of the infrared camera is much higher than the movement speed of the monitored object. This way image blurring does not affect negatively the motion estimation. Moreover, it assumes that the acquisition rate is also much higher than the speed at which the temperature of the monitored object changes.

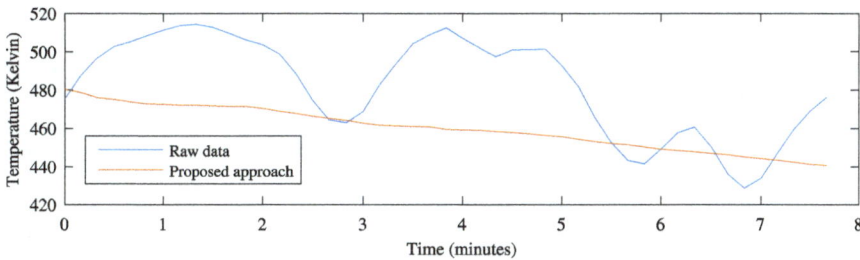

Figure 17. Comparing temperature monitoring using raw images and the proposed approach for the sinter material.

4. Conclusions

This work proposes a solution for temperature monitoring when the inspected object is moved or it is affected by vibrations, very common in many scenarios. The first step is an image rectification procedure that calculates a transformed image in real-world units. This transformation produces a front-parallel projection of the image which greatly simplifies the estimation and compensation of the movements. Using this approach, motion is perfectly described using a simple 2D rigid transformation. The procedure to estimate motion proposes a robust method adapted to infrared images, but also based on well-known techniques successfully proven in visible images, such as feature detection using SURF and robust model estimation using MLESAC. The result is an accurate and robust method that provides the temperature time history of the inspected material without being affected by the movements of the material. The proposed approach assumes that the region of interest is flat, which is the case in many different types of applications.

The proposed method has been tested in laboratory and in real environments. Laboratory tests consist of a test piece that is manually moved, simulating vibrations. The proposed method robustly estimates and compensates for the simulated vibrations, providing movement free images that can be used to monitor temperature easily. The method has also been tested in a real environment: a sinter cooler. In this scenario, the material moves inside a circular ring and temperature monitoring is critical to calculate the cooling pattern and to avoid fires. The result demonstrated that the proposed work can be used to calculate the required temperature time history of the material as it moves. Moreover, additional geometric information can be extracted from result, such as the real speed of the cooler. The results of these tests validate the performance of the proposed work as a robust and accurate method to monitor the temperature of moving material. Tests are performed using long-wavelength infrared cameras, but the proposed approach could also be used with high-end mid-wavelength infrared cameras to monitor the temperature of fast moving objects.

Acknowledgments: This work has been partially funded by the project TIN2014-56047-P of the Spanish National Plan for Research, Development and Innovation.

Author Contributions: Rubén Usamentiaga took the leadership in this work. Daniel F. García contributed to some of the tests.

Conflicts of Interest: The authors declare no conflict of interest.

References

1. Michalski, L.; Eckersdorf, K.; Kucharski, J.; McGhee, J. *Temperature Measurement*; Wiley: New York, NY, USA, 2001.
2. Usamentiaga, R.; Venegas, P.; Guerediaga, J.; Vega, L.; Molleda, J.; Bulnes, F.G. Infrared thermography for temperature measurement and non-destructive testing. *Sensors* **2014**, *14*, 12305–12348.
3. Jadin, M.S.; Taib, S. Recent progress in diagnosing the reliability of electrical equipment by using infrared thermography. *Infrared Phys. Technol.* **2012**, *55*, 236–245.
4. Wang, H.; Jiang, L.; Liaw, P.; Brooks, C.; Klarstrom, D. Infrared temperature mapping of ULTIMET alloy during high-cycle fatigue tests. *Metall. Mater. Trans. A* **2000**, *31*, 1307–1310.
5. Maldague, X. *Theory and Practice of Infrared Technology for Nondestructive Testing*; Wiley: New York, NY, USA, 2001.
6. Hong, T.; Koo, C.; Kim, J.; Lee, M.; Jeong, K. A review on sustainable construction management strategies for monitoring, diagnosing, and retrofitting the building's dynamic energy performance: Focused on the operation and maintenance phase. *Appl. Energy* **2015**, *155*, 671–707.
7. Usamentiaga, R.; Molleda, J.; Garcia, D.F.; Bulnes, F.G. Monitoring sintering burn-through point using infrared thermography. *Sensors* **2013**, *13*, 10287–10305.
8. Lahiri, B.; Bagavathiappan, S.; Jayakumar, T.; Philip, J. Medical applications of infrared thermography: A review. *Infrared Phys. Technol.* **2012**, *55*, 221–235.
9. Arena, G.; Rippa, M.; Mormile, P.; Grilli, M.; Paturzo, M.; Fatigati, G.; Ferraro, P. Concurrent studies on artworks by digital speckle pattern interferometry and thermographic analysis. In Proceedings of the SPIE OPTO International Society for Optics and Photonics, San Francisco, CA, USA, 13 February 2016; p. 977107.
10. Al-doski, J.; Mansor, S.B.; Shafri, H.Z.B.M. Thermal Imaging for Pests Detecting—A Review. *Int. J. Agric. For. Plant.* **2016**, *2*, 10–30.
11. Usamentiaga, R.; Molleda, J.; García, D.F.; Granda, J.C.; Rendueles, J.L. Temperature measurement of molten pig iron with slag characterization and detection using infrared computer vision. *IEEE Trans. Instrum. Meas.* **2012**, *61*, 1149–1159.
12. Usamentiaga, R.; Molleda, J.; Garcia, D.F. Structured-Light Sensor Using Two Laser Stripes for 3D Reconstruction without Vibrations. *Sensors* **2014**, *14*, 20041–20063.
13. Morimoto, C.; Chellappa, R. Evaluation of image stabilization algorithms. In Proceedings of the 1998 IEEE International Conference on Acoustics, Speech and Signal Processing, Washington, DC, USA, 12–15 May 1998; Volume 5, pp. 2789–2792.
14. Ertürk, S. Real-time digital image stabilization using Kalman filters. *Real-Time Imaging* **2002**, *8*, 317–328.
15. Ertürk, S. Digital image stabilization with sub-image phase correlation based global motion estimation. *IEEE Trans. Consum. Electron.* **2003**, *49*, 1320–1325.
16. Moderhak, M. FFT spectra based matching algorithm for active dynamic thermography. *Quant. InfraRed Thermogr. J.* **2011**, *8*, 239–242.

17. Hidalgo-Gato, R.; Mingo, P.; López-Higuera, J.M.; Madruga, F.J. Pre-processing techniques of thermal sequences applied to online welding monitoring. *Quant. InfraRed Thermogr. J.* **2012**, *9*, 69–78.

18. Croft, D.; Devasia, S. Vibration compensation for high speed scanning tunneling microscopy. *Rev. Sci. Instrum.* **1999**, *70*, 4600–4605.

19. Usamentiaga, R.; Garcia, D.; Molleda, J.; Bulnes, F.; Bonet, G. Vibrations in steel strips: Effects on flatness measurement and filtering. *IEEE Trans. Ind. Appl.* **2014**, *50*, 3103–3112.

20. Rajagopalan, A.; Chellappa, R. *Motion Deblurring: Algorithms and Systems*; Cambridge University Press: Cambridge, UK, 2014.

21. Oswald-Tranta, B.; Sorger, M.; O'Leary, P. Motion deblurring of infrared images from a microbolometer camera. *Infrared Phys. Technol.* **2010**, *53*, 274–279.

22. Usamentiaga, R. Easy rectification for infrared images. *Infrared Phys. Technol.* **2016**, *76*, 328–337.

23. Canny, J. A computational approach to edge detection. *IEEE Trans. Pattern Anal. Mach. Intell.* **1986**, *8*, 679–698.

24. Phillips, J.M.; Liu, R.; Tomasi, C. Outlier robust ICP for minimizing fractional RMSD. In Proceedings of the IEEE Sixth International Conference on 3-D Digital Imaging and Modeling (3DIM'07), Montreal, QC, Canada, 21–23 August 2007; pp. 427–434.

25. Trucco, E.; Verri, A. *Introductory Techniques for 3-D Computer Vision*; Prentice Hall: Englewood Cliffs, NJ, USA, 1998; Volume 201.

26. Bay, H.; Ess, A.; Tuytelaars, T.; Van Gool, L. Speeded-up robust features (SURF). *Comput. Vision Image Underst.* **2008**, *110*, 346–359.

27. Sezgin, M.; Sankur, B. Survey over image thresholding techniques and quantitative performance evaluation. *J. Electron. Imaging* **2004**, *13*, 146–168.

28. Zuiderveld, K. Contrast limited adaptive histogram equalization. In *Graphics Gems IV*; Academic Press Professional, Inc.: San Diego, CA, USA, 1994; pp. 474–485.

29. Usamentiaga, R.; García, D.F.; Molleda, J. Efficient registration of 2D points to CAD models for real-time applications. *J. Real-Time Image Process.* **2015**, 1–19, doi:10.1007/s11554-015-0485-7.

30. Torr, P.H.; Zisserman, A. MLESAC: A new robust estimator with application to estimating image geometry. *Comput. Vision Image Underst.* **2000**, *78*, 138–156.

31. Fischler, M.A.; Bolles, R.C. Random sample consensus: A paradigm for model fitting with applications to image analysis and automated cartography. *Commun. ACM* **1981**, *24*, 381–395.

32. ISO 18434-1:2008. *Condition Monitoring and Diagnostics of Machines —Thermography—Part 1: General Procedures*; ISO: Geneva, Switzerland, 2011.

33. Kang, S.J.L. *Sintering: Densification, Grain Growth and Microstructure*; Butterworth-Heinemann: Boston, MA, USA, 2004.

sensors

MDPI

Article

Nondestructive Evaluation of Carbon Fiber Bicycle Frames Using Infrared Thermography

Rubén Usamentiaga [1,*] , Clemente Ibarra-Castanedo [2,3], Matthieu Klein [3], Xavier Maldague [2], Jeroen Peeters [4] and Alvaro Sanchez-Beato [5]

[1] Department of Computer Science and Engineering, University of Oviedo, 33204 Gijón, Asturias, Spain
[2] Computer Vision and Systems Laboratory, Laval University, Quebec City, QC G1V 0A6, Canada;
     clemente.ibarra-castanedo@gel.ulaval.ca (C.I.-C.); xavier.maldague@gel.ulaval.ca (X.M.)
[3] Visiooimage Inc., Infrared Thermography Testing Systems, 2604, Rue Lapointe,
     Quebec City, QC G1W 1A8, Canada; research@visiooimage.com
[4] Op3Mech Research Group, University of Antwerp, Groenenborgerlaan 171, B-2020 Antwerp, Belgium;
     Jeroen.peeters2@uantwerpen.be
[5] Actia Digital Ventures SRLU (THEBIKESPLACE.COM), 28413 Madrid, Spain; alvarosb@thebikesplace.com
*   Correspondence: rusamentiaga@uniovi.es; Tel.: +34-985-182626

Received: 29 September 2017; Accepted: 14 November 2017; Published: 20 November 2017

Abstract: Bicycle frames made of carbon fibre are extremely popular for high-performance cycling due to the stiffness-to-weight ratio, which enables greater power transfer. However, products manufactured using carbon fibre are sensitive to impact damage. Therefore, intelligent nondestructive evaluation is a required step to prevent failures and ensure a secure usage of the bicycle. This work proposes an inspection method based on active thermography, a proven technique successfully applied to other materials. Different configurations for the inspection are tested, including power and heating time. Moreover, experiments are applied to a real bicycle frame with generated impact damage of different energies. Tests show excellent results, detecting the generated damage during the inspection. When the results are combined with advanced image post-processing methods, the SNR is greatly increased, and the size and localization of the defects are clearly visible in the images.

Keywords: optical active infrared inspection; carbon fibre bicycle frame; nondestructive evaluation

1. Introduction

Nondestructive evaluation (NDE) is nowadays a fundamental technology to determine the quality and investigate the integrity of materials without damaging them. Nondestructive tests are used to detect defects, but also to prevent failures in order to ensure safe long-term operation [1]. Major innovations have been carried out in this field in recent decades, leading to important competitive advantages.

Many different techniques have been proposed for NDE. The most important can be classified as mechanical and optical, penetrating radiation, electromagnetic and electric, sonic and ultrasonic, thermal and infrared, chemical and analytical, image generation and signal and image analysis [2]. Infrared evaluation, in particular, has been proven to provide outstanding advantages compared to other techniques [3]. Infrared evaluation is fast, which enables high-speed scanning and major savings in time and cost. In addition, it is safe and suitable for prolonged and repeated use, with no harmful radiation effects such as in X-ray evaluation. Moreover, infrared evaluation is a non-invasive technique, where the inspected material is not affected or altered in any way [4].

NDE using infrared thermography is based on the acquisition and analysis of temperature and heat flow in the inspected material in order to detect subsurface anomalies. The most common approach for NDE is active infrared thermography. In this evaluation method, an external thermal stimulus is applied to the inspected material using optical flash lamps, heat lamps or other devices [5]. The thermal waves penetrate the surface of the material, producing a thermal contrast in areas with subsurface anomalies during the transient phase, which makes subsurface defect detection possible. The most common stimulation methods are pulse thermography, step heating thermography, lock-in thermography and ultrasound thermography. These methods can provide different results based on the material and the type of defect. NDE using active infrared thermography has been applied successfully to different materials, including carbon fibre-reinforced composites [6], steel [7], aluminium [8], walls [9], concrete [10] or cotton fibres [11], just to name a few.

One material with vast potential in many different industries is carbon fibre-reinforced polymer (CFRP), which is a composite material reinforced by carbon fibres. This material has excellent mechanical properties, including the modulus of elasticity and strength. Moreover, the material is not only strong, but also very lightweight. Additionally, it has far superior fatigue properties and corrosion resistance than metals, when combined with the proper resins [12]. Therefore, it is used in the manufacturing of numerous products where all these properties are of particular interest, such as the aerospace industry. The use of this material in other industries is still limited due to the cost, although in recent years, it has become more affordable. New high-performance products are taking advantage of CFRP's superior features, such as sporting goods or automobiles. One of these new products is CFRP bicycle frames, which have become the standard material in the performance cycling world during the last decade, overtaking steel, aluminium and titanium.

CFRP is the most popular material for high-performance cycling due to the stiffness-to-weight ratio. CFRP bicycle frames are not only light; they are also extremely stiff, which enables greater power transfer. However, carbon fibre-derived products are also comparatively brittle, susceptible to damage caused by low energy impact loading, during manufacturing and also during service [13]. Low energy impact loading can create cracks and delamination due to the propagation of the mechanical energy inside the material, causing extensive subsurface damage invisible on the surface [14]. Numerous scenarios can provoke impact damage in a CFRP bicycle frame: a bicycle may fall and hit the floor; a small stone on the road can be projected by a nearby car; or simply due to tight clamps used for transportation. This type of damage in CFRP bicycle frames can lead to severe consequences: the frame can break, exposing the rider to a probable injury or even death [15]. On the surface, the bicycle frame may seem flawless, but the internal and hidden damage can provoke a very sudden and catastrophic failure. This is why the evaluation and inspection of CFRP bicycle frames are extremely important.

Different techniques have been applied to the nondestructive evaluation of CFRP bicycle frames [16], including pulse thermography, radiography, ultrasonic, acoustic and tap testing. However, although techniques to predict, detect and quantify impact damage in composite materials have been already studied [17–20], defect detection sensors for CFRP bicycle frames have not been analysed with rigour. This work proposes a novel inspection system for impact damage detection based on active infrared thermography. The proposed inspection method is fast, and it can be used to inspect a bicycle frame in seconds. It only requires an infrared camera and a source of heat for the thermal stimulus of the material. The resulting evaluation procedure is an effective method to detect damage in the bicycle frame. The proposed method is applied to real CFRP bicycle frames with generated impact damage of different energies to test the detectability and characterization of defects. Advanced image post-processing techniques are applied to the acquired data, and the results are evaluated using the signal-to-noise (SNR) metric. This work demonstrates the feasibility of the inspection procedure using an inexpensive sub-20k Euros camera and provides clear guidelines for an adequate configuration for a thermal stimulus and post-processing algorithm that greatly enhances the SNR.

Compared with other techniques, X-ray inspection could provide images where subsurface anomalies could be appreciated much more clearly than thermography. However, X-ray inspection is much more expensive, and there are serious radiation hazards for the technicians [21]. Ultrasound inspection could also be applied. Although, due to the complexity of the shape of a bicycle frame, many spots would be difficult to reach with the ultrasound heads. Besides, most of the defects occur very close to the surface, where ultrasound inspection is less accurate. Tap testing can be used as a complementary method on CFRP bikes. A tap test consists of gently tapping the area under inspection with a small hammer, listening for significant changes in sound. Dullness can indicate delaminations, but it requires a highly skilled operator to be a cost-effective method. Infrared thermography definitely offers interesting compromises for any CFRP inspection: it is fast, contact-less and efficient for a thin material like CFRP. This work demonstrates the feasibility of the inspection based on infrared thermography, with a very easy-to-apply method and good results.

The remainder of this paper is organized as follows. Section 2 presents the experimental design to evaluate the detectability of impact damage in CFRP bicycle frames; Section 3 discusses the results obtained with real data and image post-processing techniques; and finally, Section 4 reports the conclusions.

2. Experimental Investigation

2.1. Description of the Specimen

The bicycle frame used in the experiments has been manufactured by Specialized™, one the most popular high-performance bicycle brands. Figure 1 shows an image of the inspected carbon fibre bicycle frame. The materials used in this bicycle are common for a carbon fibre bicycle frame. Thus, the results in this work can be easily extrapolated to most bicycles made of CFRP.

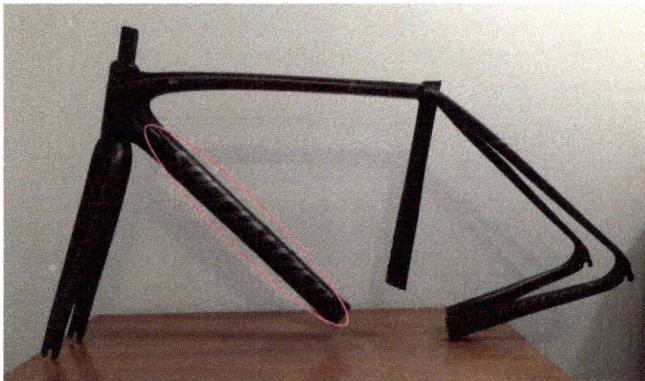

Figure 1. Inspected bicycle frame.

The bicycle frame is made of carbon fibre with reinforcements in some areas where different tubes are connected. The thickness of the tubes in the bicycle frame varies from 1.4 to 1.8 mm due to the bending of the carbon fibre composite during manufacturing. Experiments were performed in the down tube (highlighted in the image), which is the tube that connects the head tube and the bottom bracket. This tube can be clearly identified in the image because it contains the name of the brand. This tube is very close to the front wheel, where stones from the road can be projected, causing impact damage.

2.2. Infrared Camera

The infrared camera is a fundamental part of infrared evaluation. The camera is used to record the infrared radiation resulting from the thermal stimulus. The infrared camera used in the experiments is a Flir A655sc. The camera has a 24.6-mm lens and a sensor with a resolution of 640×480. The temperature range is configurable within several available ranges. In the experiments, the range [$-40, 140\,°C$] was selected. The manufacturer reports measurement accuracy of $\pm 2\,K$ and sensitivity lower than 30 mK at 30 °C. The long-wave infrared camera operates at 7.5 to 14 μm. The detector type used in the camera is an uncooled microbolometer. The complete technical specifications are given in Table 1.

Table 1. Technical specifications of the infrared camera FLIR A655sc used in the experiments.

Camera	FLIR A655sc
Temperature range	−40 to +140 °C
Thermal sensitivity/Noise Equivalent Temperature Difference (NETD)	30 mK at 30 °C
Detector	640 × 480 UFPA
Spectral range	7.5–14 μm
Image frequency	50 Hz
Spatial resolution	0.68 mrad
Field of view (FOV)	25° × 19°
Detector pitch	17 μm

Uncooled cameras are not as sensitive as cameras based on cooled detectors that usually work in the mid-wavelength infrared band. However, the price of uncooled cameras is much lower, and the maintenance is greatly reduced. Therefore, the inspection of the bicycle frame using an uncooled camera presents a more practical approach, as NDE sensors based on this type of camera could be manufactured much more easily.

2.3. Estimation of Emissivity

The emissivity of the surface of the bicycle frame has been estimated using the reference emissivity material method [22]. Based on this method, the bicycle frame is heated with a piece of electrical tape with known emissivity stuck on the surface. The reference temperature on the electrical tape is used to measure the emissivity of the surface of the bicycle frame. Figure 2 shows an infrared image acquired during the experiment to measure emissivity using a thermal colour palette [23].

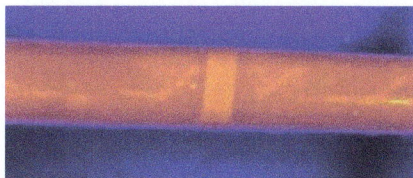

Figure 2. Infrared image of the experiment to measure emissivity using an electrical tape stuck on the surface.

The result of the emissivity measurement procedure indicates that the global emissivity of the surface of the bicycle frame perpendicular to the camera is 0.82.

2.4. Impact Damage

Impact damage is generated using a steel ball. The ball is designed and manufactured to perform the tests specified in different standards (IEC 60335, IEC 60065, IEC 60745, IEC 61029, IEC 60950). The ball is suspended from a pivot using a rod with negligible mass, creating a pendulum. The ball is displaced sideways from its equilibrium position with the rod fully extended. When released, the potential energy of the ball is transformed into kinetic energy, which is transferred to the bicycle frame during the impact. Figure 3 shows an illustration of the procedure to generate damage. The collision is considered elastic.

The energy of the impact depends on the mass of the ball, the height from the initial position to the impact position and the acceleration due to gravity. The ball used in the experiments is made of steel with a diameter of 50 mm and a mass of 0.5 kg. The acceleration due to gravity is also a constant.

Thus, by suitably adjusting the height of the ball, different impact energies can be generated. As can be seen in the figure, when the ball is at 0.2 m, the energy is 1 J, and when the ball is at 1.84 m, the energy is 9 J. These are the ranges of impacts generated in the bicycle frame: from 1 to 9 J, increasing by 1 J.

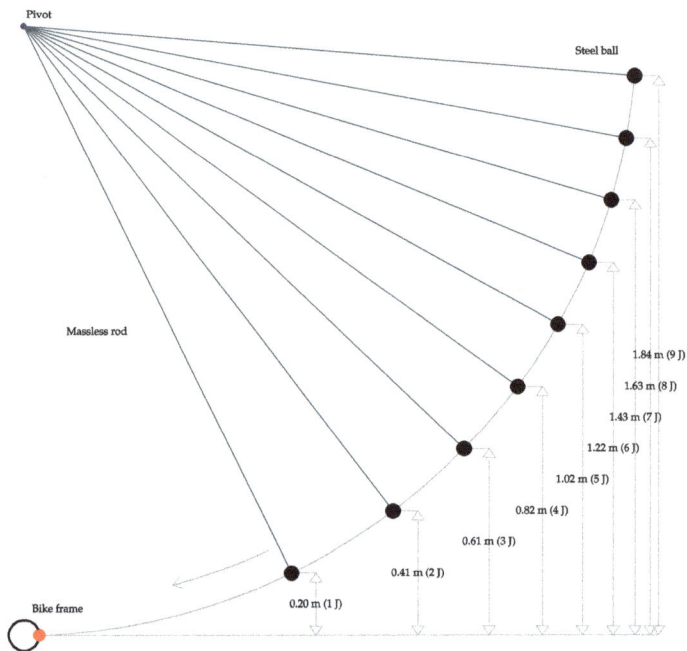

Figure 3. Procedure used to generate impact damage.

Figure 4 shows the calibrated steel ball used in the experiments. The position of the bicycle frame was suitably adjusted so the ball hit the frame close to the centre of the tube.

(a)   **(b)**

Figure 4. Calibrated steel ball used in the experiments. (**a**) Steel ball and rod. (**b**) Ball hitting a similar bicycle frame to produce impact damage.

The amount of damage generated in the bicycle frame is based on possible scenarios in which small stones in the road can be projected by a nearby car, the bicycle may fall and hit the floor or it simply has been generated during the manufacturing and assembly process. Accidents involving automobiles and bicycles can also be the source of impact damage. Damage can also result from abusive stress, overtightened devices and brackets over the carbon frame by users not respecting the manufacturer recommended torque or not using a torque wrench. In the case of mountain bikes, especially Enduro and Gravity (also known as downhill), there is a huge risk of impacts in the bottom bracket, downtube and chainstay. It is thus difficult to estimate what an average or typical defect on a carbon bike should be. Forces and areas of the damage can greatly vary, so can the resistance of the carbon frame for a given impact. CFRP thickness varies from 0.8 mm to 4 mm, with an average of 2 mm. Frame builders modify the thickness depending on the expected maximum load of each area. The idea proposed in this work is thus to cover a range of impacts from 1 to 9 joules, 1 joule being roughly the energy of impact of a standard full 33 cl aluminium can dropped from 30 cm height and 9 joules being the same can bottle dropped from about a 270-cm height.

The impact damage was generated on both sides of the down tube. On Side A, impacts of 1 to 6 J were generated. On the other side, Side B, impacts of 7 to 9 J were generated. Figure 5 illustrates the defect map on both sides of the down tube.

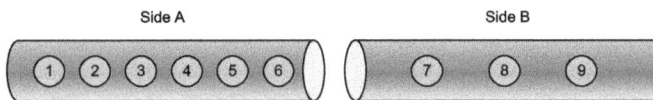

Figure 5. Map of defects on both side of the down tube and impact energy in joules.

2.5. Infrared Inspection

In the infrared inspection, optical stimulation is applied to thermally stimulate the bicycle frame. Two halogen lamps are used. These lamps provide 1000 W at maximum capacity each. Two different configurations are used to compare the results: 1000 W at maximum capacity and 500 W at medium capacity. Therefore, in the first configuration, the bicycle frame is stimulated with a total of 2000 W and in the second with 1000 W.

The inspection method used in the experiments is usually referred to as optical step heating, or long or square pulse. Optical step heating uses a much longer pulse than optical pulse heating. The steps can last from a few seconds to a minute, and both the heating and the cooling responses are of interest. This type of inspection generates more heating than flashes, and it can be used to detect deeper defects. During heating or cooling, deviations from the temperature evolution of a sound area indicate subsurface anomalies or defects. Three configurations are used: 5-, 10- and 30-s pulses. After the lamps are turned off, the temperature decay is recorded for another 60 s, making a total of 65, 70 and 90 s. All experiments are performed with the camera operating at 50 Hz.

Considering the heating power with the halogen lamps (1000 and 2000 W), and the time they are turned on (5, 10 and 30 s), a total of 6 different configurations is applied to the inspected of each side of the frame.

The experiments are performed in reflection, i.e., the halogen lamps and the infrared camera are on the same side. This is the most appropriate configuration when the defects are close to the surface. Figure 6 shows an illustration of the configuration used. As can be seen, halogen lamps are positioned at both sides of the camera, and the inspected tube of the bicycle frame is in the middle, receiving the stimulus from the two halogen lamps.

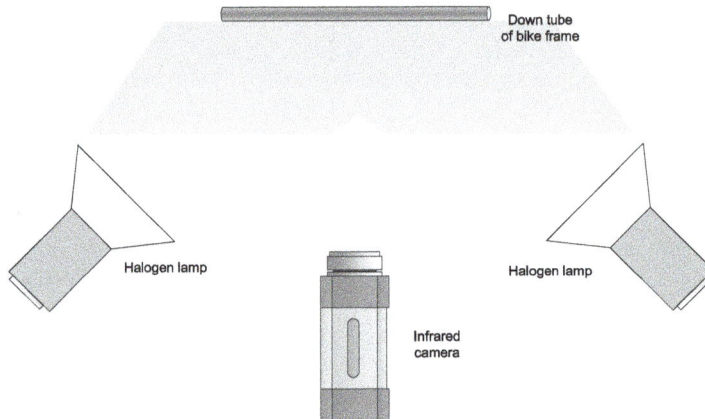

Figure 6. Infrared inspection in reflection mode.

Figure 7 shows two images documenting the experimental setup. The bicycle frame was slightly rotated backward to reduce the reflections, as they can degrade the acquired infrared radiation by the camera. The camera and the halogen lamps are slightly tilted downwards as well to reduce the reflections of the halogen lamps on the bicycle frame.

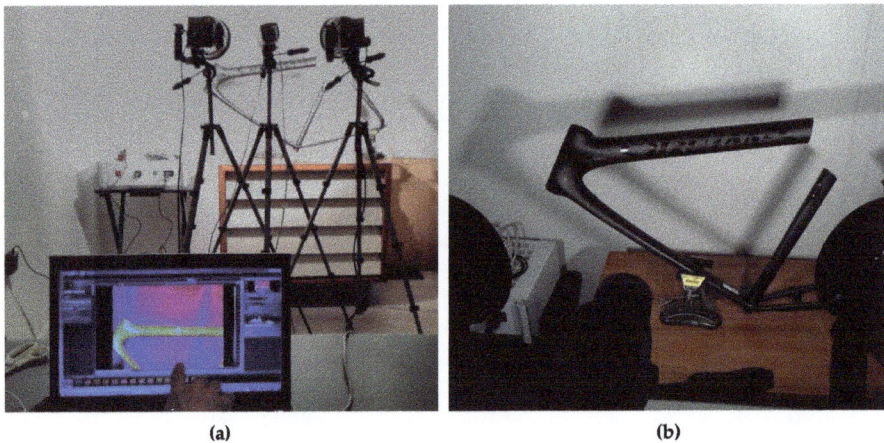

(a)   **(b)**

Figure 7. Inspection of the bicycle frame. (**a**) Complete view. (**b**) View from the camera position.

2.6. Post-Processing

The defect contrast caused by the thermal stimulus in infrared evaluation can be very subtle, almost inappreciable in some cases. Therefore, the signal levels associated with subsurface anomalies can be lost in the data noise. In those cases, the defects are undetectable from raw thermograms. Post-processing methods are algorithms that improve the visualization and the contrast of the defects. Moreover, these algorithms tend to remove harmful artefacts such as non-uniform illumination from the images, increasing the signal-to-noise ratio and greatly improving the defect detection rate. The post-processing methods used in this work are briefly described next.

2.6.1. Pulsed Phase Thermography

Pulsed phase thermography (PPT) is based on the calculation of FFT applied to the temperature-time history of every pixel in the acquired thermographic sequence. This operation approximates the temperature-time history by a sum of harmonic waves at different frequencies. The result of this operation describes the frequency response of the thermal stimulus. The operation is calculated using (1), where i is the imaginary number, n is the frequency increment and Re_n and Im_n are the real and imaginary parts of the DFT (discrete Fourier transform). The phase is of particular interest, as it contains relevant information about the structure of the material. It can be calculated using (2).

$$F_n = \sum_{k=1}^{N-1} T(k) e^{\frac{2\pi i k n}{N}} = Re_n + Im_n \tag{1}$$

$$\varnothing_n = \operatorname{atan}\left(\frac{Im_n}{Re_n}\right) \tag{2}$$

This method was originally proposed for pulse thermography [24], although it can also be applied to step heating inspections.

2.6.2. Principal Component Thermography

Principal component thermography (PCT) is based on principal component analysis, a statistical technique of information synthesis. The goal is to reduce the number of variables in a dataset, while losing the least amount of relevant information possible.

The calculation of PCT is based on singular value decomposition (SVD), which decomposes the thermal sequence into a series of statistic orthogonal functions known as empirical orthogonal functions (EOF) [25]. The first components provide a reduction of the data without removing useful information about the defects.

2.6.3. Polynomial Fit and Time Derivatives

In this method, the temperature-time history of every pixel in the thermal sequence is approximated by a polynomial. This method is usually called thermographic signal reconstruction (TSR) [26]. The evolution of the temperature is adjusted to an n degree polynomial as shown in (3).

$$T(t) = a_n t^n + a_{n-1} t^{n-1} + \ldots + a_1 t + a_0 \tag{3}$$

When this method is applied to pulse thermography, the signals are previously converted to the logarithmic domain. However, this conversion is not required when applying this method to step heating thermography [27].

The polynomial fitting provides the opportunity to filter noise and compress the thermal sequence, as only the coefficients of the polynomials are required to reconstruct the sequence. Moreover, it can also be used to calculate the time derivatives analytically. These derivatives have been proven to be one of the best methods to enhance the visualization of defects [28], especially the second derivative.

2.6.4. Partial Least Squares Thermography

Partial least squares thermography (PLST) is based on statistical correlation for the optimization of infrared inspection. The method decomposes the temperature-time history into a set of latent variables using partial least squares regression. In this method, non-relevant information is discarded, and only the most significant data are used in the regression. Non-uniform heating is removed while preserving the physical consistency [29].

3. Results and Discussion

3.1. Quantitative Evaluation

The analysis of the results in this work follows a quantitative approach. Therefore, it avoids subjective evaluation of the resulting images. In order to evaluate the results, the SNR metric is used to quantitatively assess the signal-to-noise ratio of the defects in the bicycle frames. The quantification of the defects is based on the definition of two regions in the images: the defect region and the reference or sound region [30]. The defect region encloses pixels in the images where the defect appears. This area is considered the signal. The sound region is an area in the image close to the defects, but outside the impact damage. This area is considered the noise. This approach to select the sound region close to the defect is usually referred to as the self-referencing method. This method is also recommended by the ASTM standards [31].

Prior information available about the defects is related to the approximate position in the bicycle frame. Thus, in this work, the defect and sound regions are calculated from the images obtained after the inspection. Some points in the images are selected as seed points. These points are placed in the centre of the defects. Using the intensity of these points as a reference, a similarity image is obtained using (4), where I_{Ref} is the intensity of the pixel selected as the seed. The resulting similarity image is then segmented using the fast marching method [32]. The results of the segmentation are the defect regions.

$$J = \frac{1}{\sqrt{I - I_{Ref}}} \tag{4}$$

The regions for the sound region are estimated around the defects using morphological dilatation applied to the binary image of the defects. Dilation is an operation used in mathematical morphology to expand the shapes of binary images. Two dilation operations are performed in the defect regions: a dilatation applied to the region of the defect to calculate a transitory region and a dilatation applied to this transitory region to calculate an extended region. The difference between these two regions is considered the reference or sound region. The diameter of the structuring element used in the morphological operation is the square root of the area of the region where the operation is applied, which follows the recommendation by ASTM [31].

Figure 8 shows the obtained regions using the proposed procedure. The region in red is the defect region, and the region in green is the sound region. In this case, an image obtained with good contrast of the defects is used as a reference.

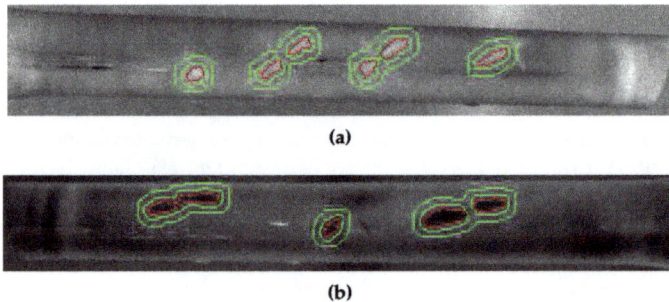

(a)

(b)

Figure 8. Defect and sound regions for the quantitative evaluation of the results. (a) Side A (from left to right, impact damage of 3, 4, 5 and 6 J). (b) Side B (from left to right, impact damage of 7, 8 and 9 J).

The SNR metric for the defects is calculated using (5), where μ_S is the arithmetic mean of all the pixels inside the defective area (signal), μ_N is the arithmetic mean of all the pixels inside the reference or sound area (noise) and σ_N is the standard deviation of the pixels inside the reference area. This is not the only definition of SNR, but it is widely used in infrared inspection [28,33].

$$SNR = 10 \log_{10} \frac{|\mu_S - \mu_N|^2}{\sigma_N^2} \tag{5}$$

3.2. Analysed Periods

Before applying the post-processing methods, it is necessary to define the periods of interest from the infrared inspections. Two periods are considered in the optical step heating stimulation: heating and cooling. The periods are represented in Figure 9, which shows the temperature evolution of a pixel in the image after the thermal stimulus. In this case, a 30-s pulse configuration was used. The images in these two periods are extracted and analysed independently. The number of images extracted from the cooling period is always the same. However, in the heating period, the number of images depends on the time of the pulse.

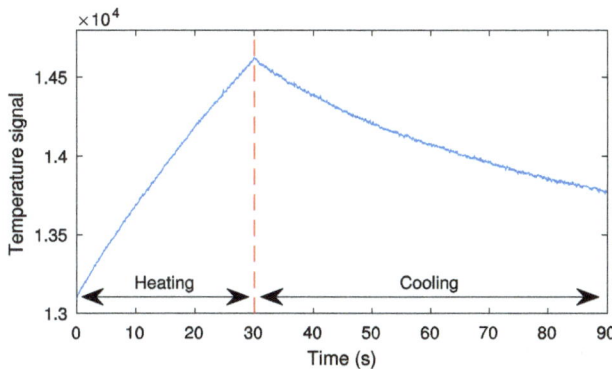

Figure 9. Periods of time considered for the analysis.

3.3. Comparative Results

The results of the experiments can be seen in Figure 10. The figure uses a colour scale to indicate the level of SNR: bright colours indicate high SNR and dark colours low SNR.

The results of the experiments show a huge difference between both sides of the bicycle frame. As expected, on Side B, the SNR is much higher because the impact energy of the defects is also higher. Thus, there is a correlation between SNR and impact energy. The SNR is higher as the energy of impact damage is increased, which makes defect detection easier.

The results also indicate that the heating period provides better information about the defects than the cooling period. In all the considered configurations (5-, 10- and 30-s pulses), the SNR results when processing the heating period are higher than the SNR results during cooling. Moreover, because the number of images during heating is significantly lower, the processing times are greatly reduced when processing the heating period. For example, in the 5-s pulse, only 250 images need to be processed. This is an important advantage, as the total time required to inspect the whole bicycle frame is reduced notably, providing the opportunity to design a fast nondestructive sensor.

There is a difference between applying a thermal stimulus of 1000 and 2000 W between 10 and 20%. The SNR improves as more energy is applied to the bicycle frame. A 2000-W stimulus seems to provide the best output, resulting in a high SNR. However, a 1000-W stimulus seems to be enough to detect the defects caused by impact damage in the bicycle frame.

Comparing the results of the three heating configurations (5-, 10- and 30-s pulses), it can be seen that the SNR results improve from a 5- to a 10-s pulse. However, in general, the results when the pulse is increased to 30 s are slightly worse. This indicates that the defects are very superficial, and the thermal contrast is generated at the beginning of the thermal stimulus. A 10-s pulse provides the best results, also with a reduced number of images to process.

| | Heating | | | | | | Cooling | | | | | |
| | 1000W | | | 2000W | | | 1000W | | | 2000W | | |
	5 s	10 s	30 s	5 s	10 s	30 s	5 s	10 s	30 s	5 s	10 s	30 s
Raw	-0.60	0.97	-0.55	0.76	0.22	1.72	-2.58	-3.35	-2.59	-1.79	-3.73	-5.76
Phase	2.31	-2.12	2.24	2.91	1.51	-0.34	-9.35	-8.86	-10.21	-7.32	-6.88	-5.99
PCT	5.31	5.63	4.01	6.62	7.89	5.88	-4.28	-2.81	-5.76	-2.88	-3.01	-3.56
Fit'	3.99	4.29	1.93	4.93	5.58	3.79	0.31	4.78	-0.70	1.94	0.63	0.53
Fit"	1.15	5.35	-1.17	4.57	7.45	7.11	-1.88	3.43	-3.70	-0.23	-1.60	-1.39
PLST	2.07	-0.80	3.06	5.10	0.43	-3.34	0.66	2.96	0.63	3.82	0.37	-2.78

(a)

| | Heating | | | | | | Cooling | | | | | |
| | 1000W | | | 2000W | | | 1000W | | | 2000W | | |
	5 s	10 s	30 s	5 s	10 s	30 s	5 s	10 s	30 s	5 s	10 s	30 s
Raw	8.58	7.31	7.82	9.63	10.06	9.65	4.28	-0.12	-3.10	4.03	0.18	-2.66
Phase	7.35	9.48	1.91	10.30	12.62	8.23	0.62	-0.18	-2.31	4.15	2.45	1.26
PCT	11.25	12.02	9.33	12.65	13.59	11.28	4.10	2.38	-0.70	7.92	7.29	5.60
Fit'	10.06	9.54	9.26	10.70	10.61	10.68	9.66	7.32	2.68	12.18	9.16	4.06
Fit"	6.52	9.53	11.56	11.57	13.54	15.53	9.20	7.36	5.20	11.89	9.41	7.17
PLST	10.27	9.35	0.29	13.69	12.34	-1.53	7.21	-0.09	1.36	8.62	3.18	0.81

(b)

Figure 10. Quantitative results of the inspection for the considered periods, power, stimulation time and post-processing methods. All values are expressed in dB. (**a**) Side A with impact damage from 1 to 6 J. (**b**) Side B with impact damage from 7 to 9 J. SNR calculated as the average of all the impacts from each side.

Figure 11 shows a comparison between the images obtained from the infrared inspection during heating, using a 10-s pulse and 2000 W. The first column shows the image where the best SNR was obtained for the raw sequence and the sequences resulting from image post-processing. The second and third columns show a contrast enhancement of the first image. The second column shows an image where the histogram has been stretched removing the intensity values below 2% and 98%. The third column shows the results of applying a contrast-limited adaptive histogram equalization [34]. In these images, most of the impacts can be clearly distinguished. The only impacts that cannot be identified are impacts with 1 and 2 J. All the others can be clearly appreciated in the images, especially defects with high impact energy. As can be seen, the pattern of defects is similar to a butterfly wings pattern, where the direction is affected by the alignment of the fibres in the composite used for manufacturing the bicycle frame. This pattern is commonly found in infrared inspections of impact damage [3,35,36].

Low energy impacts (1 and 2 J) are not detected in the resulting infrared inspection. Those low energy impacts may have simply not created any damage, in which case it is normal to not detect them, or the limit of detection was reached using this specific setup. These type of impacts may be detected using infrared cameras with higher sensitivity. However, further exploration should be performed to analyse the detectability and consequences of this type of impact in the bicycle frames, or if they are causing any integral damage at all.

The results in Figure 10 and the images in Figure 11 also show a huge difference between the raw sequence and the results of the post-processing methods. The raw images have a very low SNR, while the post-processing methods greatly increase the resulting SNR, providing images where the defects can be appreciated easily. Therefore, the selected post-processing methods are helpful in improving the visualization and localization of the defects caused by impact damage.

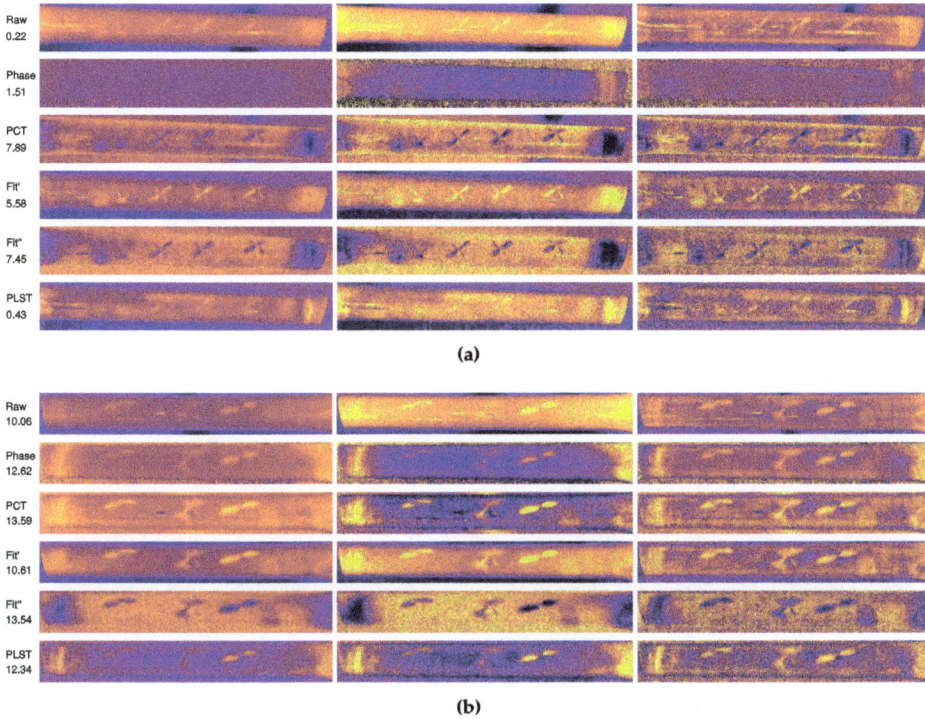

(a)

(b)

Figure 11. Images resulting from the infrared inspection during heating using a 10-second pulse and 2000 W. (a) Side A with impact damage from 1 to 6 J. (b) Side B with impact damage from 7 to 9 J.

All the post-processing methods tend to increase the SNR of the results. On average, the PCT method provides the best results, slightly above the second derivative of the polynomial fit. On the other hand, PLST provided the worse results, with irregular output across the experiments.

PCT also has a major advantage when compared with the other methods: the best SNR was always obtained in the third EOF, i.e., the third image in the resulting sequence of the PCT. This is an important advantage for inspection, as the technician can analyse this image without having to look at the entire sequence of images, as is normally the case with the polynomial fitting and derivatives.

The calculated SNR in these results is an average of the SNR in all defects for each side. In order to evaluate the SNR of single defects, new regions for the defects and for the sound areas have been defined using the same procedure described above. Figure 12 shows the binary images of the defect and sound regions for the defects with impact damage of 3, 6 and 9 J.

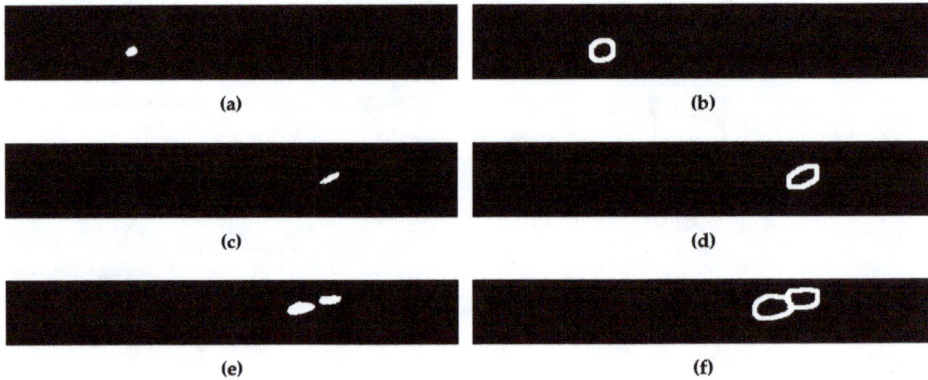

Figure 12. Defect and sound regions for defects with impact damage of 3, 6 and 9 J. (**a**) Defect region for 3 J impact. (**b**) Sound region for 3 J impact. (**c**) Defect region for 6 J impact. (**d**) Sound region for 3 J impact. (**e**) Defect region for 9 J impact. (**f**) Sound region for 9 J impact.

The SNR for these three defects is represented in Figure 13, which indicates that the SNR increases with the impact energy. This result is consistent with previous works on the topic, which indicated that the intensity in the images increases with the impact energy [14]. Figure 14 shows the images of these three defects. The images are on the same scale. Thus, this figure can be used to compare the size of the defects. The area of the defects is another feature that is correlated with the impact energy: defects caused by high impact energy are larger. The particular pattern of impact damage can be clearly observed in these images.

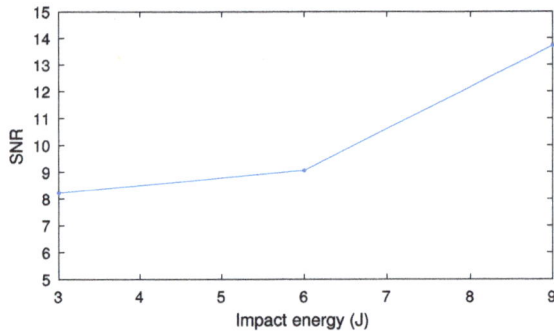

Figure 13. SNR for the there considered defects with impact energy of 3, 6 and 9 J.

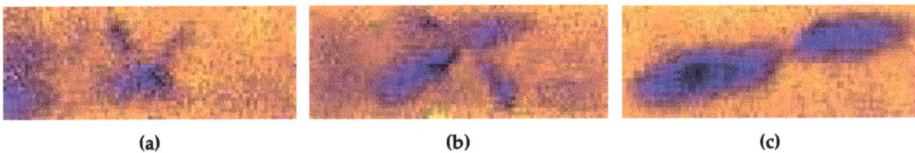

Figure 14. Images for the there considered defects. (**a**) Defect with 3 J impact energy. (**b**) Defect with 6 J impact energy. (**c**) Defect with 9 J impact energy.

4. Conclusions

Carbon-fibre composites are starting to be used in new and a wide variety of products. The resulting high-performance products are strong and also lightweight. However, these products are susceptible to damage caused by low energy impact loading not easily detected by visual inspection. Therefore, nondestructive evaluation is required in order to ensure safe long-term operation. High-performance carbon fibre bicycle frames are one of these products. In this case, impact damage can lead to dramatic consequences, exposing the rider to serious injury.

This work proposes a procedure to inspect carbon-fibre bicycle frames based on active infrared thermography. The proposed sensor is based on two halogen lamps and an infrared camera. The halogen lamps stimulate the bicycle frame thermally, and the infrared camera records the response. Different configurations have been tested for the inspection of the bicycle frames, including the power of the lamps and the time during which the lamps were turned on. These configurations were applied to a set a generated impact damage in a real carbon-fibre bicycle frame. The results were analysed quantitatively, including advanced image post-processing methods to improve the visualization of the localization of the defects.

The results indicate that lamps with a power of 1000 W turned on during only 5 s provide very good detection results. Most defects were clearly appreciable in the resulting infrared images. Moreover, when post-processing methods were applied, the SNR was greatly increased. The optimal configuration was a stimulation energy of 2000 W during 10 s. When combining the infrared sequence resulting from this configuration with principal component thermography, optimal results were obtained in terms of SNR. The resulting images clearly showed the size and intensity of the defects, which could be used to infer the impact energy of the damage.

The proposed inspection procedure in this work can be used to evaluate a carbon-fibre bicycle frame very quickly without damaging it. A similar approach is very likely to find potential applications in a number of different areas, where new products manufactured using carbon-fibre composites are starting to become more popular.

Acknowledgments: This research work has been partially funded by the Spanish National Program for Mobility of Professors and Researchers at an International Level, Reference PRX17/00031. The authors wish to thank Bicycle Record Inc., Quebec City, QC, Canada, for supplying carbon frames and parts damaged in real conditions of normal use, abusive use and accidents. We also thank Visiooimage Inc., Québec City, QC, Canada, for the facility, IRT equipment and testings.

Author Contributions: Alvaro Sanchez-Beato contributed to the research plan and some data in the paper. Rubén Usamentiaga wrote the paper and performed the data analysis and processing. Rubén Usamentiaga, Clemente Ibarra-Castanedo and Matthieu Klein performed the experiments. Clemente Ibarra-Castanedo, Matthieu Klein and Xavier Maldague designed the experiments. Clemente Ibarra-Castanedo and Jeroen Peeters reviewed the paper initially. All authors contributed to the final review.

Conflicts of Interest: The authors declare no conflict of interest.

References

1. Shull, P.J. *Nondestructive Evaluation: Theory, Techniques, and Applications*; CRC Press: Boca Raton, FL, USA, 2016.
2. Hellier, C. *Handbook of Nondestructive Evaluation*; McGraw-Hill: New York, NY, USA, 2001.
3. Maldague, X.P. *Nondestructive Evaluation of Materials by Infrared Thermography*; Springer Science & Business Media: Berlin, Germany, 2012.
4. Usamentiaga, R.; Venegas, P.; Guerediaga, J.; Vega, L.; Molleda, J.; Bulnes, F.G. Infrared thermography for temperature measurement and non-destructive testing. *Sensors* **2014**, *14*, 12305–12348.
5. Ibarra-Castanedo, C.; Maldague, X. Pulsed phase thermography reviewed. *Quant. Infrared Thermogr. J.* **2004**, *1*, 47–70.
6. Meola, C.; Boccardi, S.; Carlomagno, G.; Boffa, N.; Monaco, E.; Ricci, F. Nondestructive evaluation of carbon fibre reinforced composites with infrared thermography and ultrasonics. *Compos. Struct.* **2015**, *134*, 845–853.
7. Shrestha, R.; Park, J.; Kim, W. Application of thermal wave imaging and phase shifting method for defect detection in Stainless steel. *Infrared Phys. Technol.* **2016**, *76*, 676–683.

8. Tomić, L.D.; Jovanović, D.B.; Karkalić, R.M.; Damnjanović, V.M.; Kovačević, B.V.; Filipović, D.D.; Radaković, S.S. Application of pulsed flash thermography method for specific defect estimation in aluminium. *Therm. Sci.* **2015**, *19*, 1845–1854.

9. Pietrarca, F.; Mameli, M.; Filippeschi, S.; Fantozzi, F. Recognition of wall materials through active thermography coupled with numerical simulations. *Appl. Opt.* **2016**, *55*, 6821–6828.

10. Tran, Q.H.; Han, D.; Kang, C.; Haldar, A.; Huh, J. Effects of Ambient Temperature and Relative Humidity on Subsurface Defect Detection in Concrete Structures by Active Thermal Imaging. *Sensors* **2017**, *17*, 1718, doi:10.3390/s17081718.

11. Santiago Cintrón, M.; Montalvo, J.G.; Von Hoven, T.; Rodgers, J.E.; Hinchliffe, D.J.; Madison, C.; Thyssen, G.N.; Zeng, L. Infrared Imaging of Cotton Fiber Bundles Using a Focal Plane Array Detector and a Single Reflectance Accessory. *Fibers* **2016**, *4*, 27, doi:10.3390/fib4040027.

12. Chung, D.D.; Chung, D. *Carbon Fiber Composites*; Butterworth-Heinemann: Oxford, UK, 2012.

13. Sohn, M.; Hu, X.; Kim, J.K.; Walker, L. Impact damage characterisation of carbon fibre/epoxy composites with multi-layer reinforcement. *Compos. Part B Eng.* **2000**, *31*, 681–691.

14. Usamentiaga, R.; Venegas, P.; Guerediaga, J.; Vega, L.; López, I. Feature extraction and analysis for automatic characterization of impact damage in carbon fibre composites using active thermography. *NDT E Int.* **2013**, *54*, 123–132.

15. Viets, C.; Kaysser, S.; Schulte, K. Damage mapping of GFRP via electrical resistance measurements using nanocomposite epoxy matrix systems. *Compos. Part B Eng.* **2014**, *65*, 80–88.

16. Bowkett, M.; Thanapalan, K.; Williams, J. Review and analysis of failure detection methods of composites materials systems. In Proceedings of the 2016 22nd International Conference on Automation and Computing (ICAC), Colchester, UK, 7–8 September 2016; pp. 138–143.

17. Angelidis, N.; Irving, P. Detection of impact damage in CFRP laminates by means of electrical potential techniques. *Compos. Sci. Technol.* **2007**, *67*, 594–604.

18. Castaings, M.; Singh, D.; Viot, P. Sizing of impact damages in composite materials using ultrasonic guided waves. *NDT E Int.* **2012**, *46*, 22–31.

19. Chady, T.; Lopato, P.; Szymanik, B. Terahertz and thermal testing of glass-fibre reinforced composites with impact damages. *J. Sens.* **2012**, *2012*.

20. Meola, C.; Carlomagno, G.M. Infrared thermography to evaluate impact damage in glass/epoxy with manufacturing defects. *Int. J. Impact Eng.* **2014**, *67*, 1–11.

21. Kozak, M.W. *Radiation Protection and Safety in Industrial Radiography*; International Atomic Energy Agency Safety Series Number 13; International Atomic Energy Agency: Vienna, Austria, 2000.

22. American Section of the International Association for Testing Materials (ASTM E1933-97). *Standard Test Methods for Measuring and Compensating for Emissivity Using Infrared Imaging Radiometers*; ASTM International: West Conshohocken, PA, USA, 1997.

23. Thyng, K.M.; Greene, C.A.; Hetland, R.D.; Zimmerle, H.M.; DiMarco, S.F. True colors of oceanography: Guidelines for effective and accurate colormap selection. *Oceanography* **2016**, *29*, 9–13.

24. Maldague, X.; Marinetti, S. Pulse phase infrared thermography. *J. Appl. Phys.* **1996**, *79*, 2694–2698.

25. Rajic, N. Principal component thermography for flaw contrast enhancement and flaw depth characterisation in composite structures. *Compos. Struct.* **2002**, *58*, 521–528.

26. Shepard, S.M.; Lhota, J.R.; Rubadeux, B.A.; Wang, D.; Ahmed, T. Reconstruction and enhancement of active thermographic image sequences. *Opt. Eng.* **2003**, *42*, 1337–1342.

27. Usamentiaga, R.; Venegas, P.; Guerediaga, J.; Vega, L.; López, I. Automatic detection of impact damage in carbon fibre composites using active thermography. *Infrared Phys. Technol.* **2013**, *58*, 36–46.

28. Usamentiaga, R.; Venegas, P.; Guerediaga, J.; Vega, L.; López, I. A quantitative comparison of stimulation and post-processing thermographic inspection methods applied to aeronautical carbon fibre reinforced polymer. *Quant. InfraRed Thermogr. J.* **2013**, *10*, 55–73.

29. Lopez, F.; Ibarra-Castanedo, C.; de Paulo Nicolau, V.; Maldague, X. Optimization of pulsed thermography inspection by partial least-squares regression. *NDT E Int.* **2014**, *66*, 128–138.

30. Balageas, D.L. Balageas, D.L. Defense and Illustration of Time-Resolved Pulsed Thermography for NDE. In Proceedings of the SPIE 8013, Thermosense: Thermal Infrared Applications XXXIII, Orlando, FL, USA, 10 May 2011, doi: 10.1117/12.882967.

31. American Section of the International Association for Testing Materials (ASTM E2737). *Standard Practice for Digital Detector Array Performance Evaluation and Long-Term Stability*; ASTM International: West Conshohocken, PA, USA, 2010.

32. Sethian, J.A. Level set methods and fast marching methods. *J. Comput. Inf. Technol.* **2003**, *11*, 1–2.

33. Madruga, F.J.; Ibarra-Castanedo, C.; Conde, O.M.; López-Higuera, J.M.; Maldague, X. Infrared thermography processing based on higher-order statistics. *NDT E Int.* **2010**, *43*, 661–666.

34. Reza, A.M. Realization of the contrast limited adaptive histogram equalization (CLAHE) for real-time image enhancement. *J. VLSI Signal Process.* **2004**, *38*, 35–44.

35. Thompson, D.O.; Chimenti, D.E. *Review of Progress in Quantitative Nondestructive Evaluation*; Springer Science & Business Media: Berlin, Germany, 2012; Volume 18.

36. Derusova, D.A.; Vavilov, V.P.; Pawar, S.S. Evaluation of equivalent defect heat generation in carbon epoxy composite under powerful ultrasonic stimulation by using infrared thermography. In *IOP Conference Series: Materials Science and Engineering*; IOP Publishing: Bristol, UK, 2015; Volume 81, p. 012084.

sensors

MDPI

Review

Superconducting Quantum Interferometers for Nondestructive Evaluation

M. I. Faley [1,*] [ID]**, E. A. Kostyurina [2,3], K. V. Kalashnikov [2,3], Yu. V. Maslennikov [3], V. P. Koshelets [3] and R. E. Dunin-Borkowski [1]** [ID]

1   Peter Grünberg Institute, Forschungszentrum Jülich GmbH, 52428 Jülich, Germany;
     r.dunin-borkowski@fz-juelich.de
2   Moscow Institute of Physics and Technology, Moscow 141700, Russia; kostyurina.katya@gmail.com (E.A.K.);
     kalashnikovkv@gmail.com (K.V.K.)
3   Kotel'nikov Institute of Radio Engineering & Electronics RAS, Moscow 125009, Russia;
     cryoton@inbox.ru (Y.V.M.); valery@hitech.cplire.ru (V.P.K.)
*   Correspondence: m.faley@fz-juelich.de

Received: 8 October 2017; Accepted: 29 November 2017; Published: 6 December 2017

Abstract: We review stationary and mobile systems that are used for the nondestructive evaluation of room temperature objects and are based on superconducting quantum interference devices (SQUIDs). The systems are optimized for samples whose dimensions are between 10 micrometers and several meters. Stray magnetic fields from small samples (10 μm–10 cm) are studied using a SQUID microscope equipped with a magnetic flux antenna, which is fed through the walls of liquid nitrogen cryostat and a hole in the SQUID's pick-up loop and returned sidewards from the SQUID back to the sample. The SQUID microscope does not disturb the magnetization of the sample during image recording due to the decoupling of the magnetic flux antenna from the modulation and feedback coil. For larger samples, we use a hand-held mobile liquid nitrogen minicryostat with a first order planar gradiometric SQUID sensor. Low-T_c DC SQUID systems that are designed for NDE measurements of bio-objects are able to operate with sufficient resolution in a magnetically unshielded environment. High-T_c DC SQUID magnetometers that are operated in a magnetic shield demonstrate a magnetic field resolution of ~4 fT/$\sqrt{\text{Hz}}$ at 77 K. This sensitivity is improved to ~2 fT/$\sqrt{\text{Hz}}$ at 77 K by using a soft magnetic flux antenna.

Keywords: magnetic analysis; magnetic sensors; nondestructive testing; scanning probe microscopy; SQUIDs

1. Introduction

Nondestructive evaluation (NDE) describes the characterization of the structure and/or functionality of an object without compromising its usability. The recording of magnetic fields is a non-invasive contactless method that provides a direct view of magnetic features and/or electrical currents deep in the object. For an NDE technique that involves magnetic field measurement, it is challenging to construct a magnetic sensor that has high magnetic field sensitivity, high dynamic range and a broad frequency bandwidth that allows high sampling rates. Superconducting quantum interference devices (SQUIDs) provide unprecedented sensitivity down to the sub-fT/$\sqrt{\text{Hz}}$ range, a broad frequency range of >1 MHz and a dynamic range of up to ~120 dB [1]. SQUID-based NDE systems have been developed for the investigation of objects that have dimensions of nanometers (nanoSQUID microscopes [2]) to kilometers (nondestructive archeology or geomagnetic evaluation [3,4]). Related scanning methods vary from 3D piezo stages to airborne systems transported by planes or helicopters. Successful applications of SQUID-based NDE systems from the last thirty years for monitoring materials and structures have been described and assessed elsewhere [5].

The disadvantages of such systems include their operation at cryogenic temperatures and, hence, the expense of performing routine measurements. In spite of the relatively high price of cryogenic equipment and technical difficulties, SQUID-based systems are employed when the required efficiency cannot be reached using alternative NDE techniques [6]. SQUID-based NDE systems have been developed and employed for the detection of defects in steel plates [7], the study of stress–strain states in ferromagnetic materials [8], the detection of ruptures in steel ropes on bridge structures [9], and the detection of cracks in turbine blades of aircraft engine turbine blades [10]. Here, we briefly review stationary and mobile low-T_c and high-T_c SQUID systems that have been developed in Forschungszentrum Jülich (FZJ) and the Kotel'nikov Institute of Radio Engineering and Electronics (IRE) for the NDE of room temperature objects, in the context of those developed elsewhere.

2. Basic Principle of Operation and Important Features of SQUIDs

A direct current SQUID (DC SQUID) is essentially a loop of superconductor interrupted by two Josephson junctions (JJs) that have non-hysteretic current-voltage characteristics and, in an ideal case, identical critical currents I_c and normal state resistances R_n (see [1,11,12] and references therein). The operation of SQUIDs is based on the dependence of the phase shift of the wave-function of Cooper pairs on the magnetic flux passing through the SQUID loop, similar to the phase shift of the wave-function of a charged particle in the Aharonov-Bohm effect. Both effects result from the fundamental dependence of the canonical momentum of a charged particle $\vec{p} = m\vec{v} + q\vec{A}$ on the magnetic vector potential $\vec{A}$ and represent a particular case of the presence of a geometric phase shift (Berry phase) in the wave function of a charged particle after its adiabatic evolution around a closed path in the parameter space of magnetic vector potentials [13]. A DC SQUID is sensitive to the magnetic flux Φ that passes through its loop, leading to spatial variations in the phase of the wave function of Cooper pairs in the superconducting electrodes. These spatial variations lead to phase shifts $\Delta\varphi_1$ and $\Delta\varphi_2$ at the Josephson junctions and, as a result, to a voltage signal. At an optimal bias current of $I_B \cong 2I_c$, the DC voltage V on a DC SQUID depends periodically on the magnetic flux Φ that passes through the SQUID loop according to the expression [14]

$$V \approx \frac{R_n I_B}{2}\sqrt{1 - \left(\frac{2I_c}{I_B}\cos\frac{\pi\Phi}{\Phi_0}\right)^2},\tag{1}$$

where the modulation period is equal to the magnetic flux quantum $\Phi_0 \approx 2.07 \times 10^{-15}$ T·m^2. The periodic dependence of the SQUID voltage on magnetic field can be linearized by implementing a dynamic range higher than 120 dB and a slew rate larger than 1 MΦ_0/s using the DC SQUID control electronics, providing a digital negative feedback signal within each period and counting the periods when the magnetic flux exceeds Φ_0 [15].

According to Equation (1), a SQUID is sensitive to the magnetic flux Φ that penetrates through its loop. For sensitive measurements of magnetic fields, the SQUID should be equipped with a superconducting flux transformer that collects the magnetic flux in a pickup loop from a relatively large area and concentrates it into the SQUID loop using a multiturn input coil. The magnetic field sensitivity B_N of a DC SQUID magnetometer with an inductively coupled superconducting flux transformer can be estimated according to the equation

$$B_N = \frac{L_{pu} + L_i}{kA_{pu}\sqrt{L_iL_S}}S_\Phi^{1/2},\tag{2}$$

where S_Φ is the magnetic flux noise of the high-T_c DC SQUID, L_{pu} and A_{pu} are the inductance and the area of the pickup loop, respectively, k is the coupling coefficient between the input coil and the SQUID loop, L_i is the inductance of the input coil and L_S is the inductance of the SQUID loop.

3. Low-T_c vs. High-T_c JJs and DC SQUIDs: Technologies and Properties

Currently, the most sensitive detector for subtle magnetic field measurements is a DC SQUID magnetometer based on low-T_c superconducting polycrystalline Nb films and planar JJs. A magnetic field resolution below 1 fT/$\sqrt{Hz}$ at 4.2 K has been demonstrated [16]. Thin film JJs based on Nb films are widely implemented in superconducting electronics, including low-T_c DC SQUID magnetometers. The noise and signal characteristics of such magnetometers depend directly on the quality of the JJs. High quality JJs with a small spread of parameters over the substrate and between batches are vitally important for the development of low-noise sensors that are suitable for NDE applications. Several methods for the fabrication of shunted JJs have been developed. These methods include the use of double-barrier junctions with an additional normal layer between two conventional JJs [17,18] and Nb/αSi/Nb structures with a doped Si layer [19]. However, the most widely used and best-developed method involves the use of Nb/Al-AlO$_x$/Nb tunnel junctions [20,21] with an additional external resistive shunt made from Mo (Figure 1). The Mo shunt resistor is highlighted in green in Figure 1.

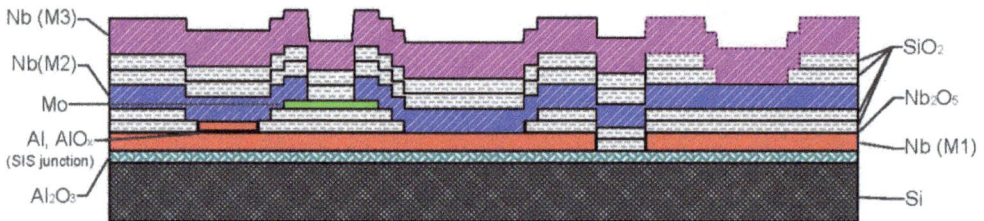

Figure 1. Schematic representation of a Nb-based low-T_c Josephson junction developed at IRE.

One of the factors that results in a reduction in the quality of Nb-based junctions is the presence of internal mechanical stress in the thin superconducting Nb films, which can lead to destruction of the tunnel barrier and junction degradation. The surface roughness of the bottom electrode caused by the internal stress increases Al diffusion at the Nb-Al boundary and can lead to micro-shortcuts. These micro-shortcuts typically result in increased noise levels of the JJs and SQUIDs. In order to minimize tension in Nb films prepared using DC magnetron sputtering, the operating modes of the magnetron have been investigated. Experimental studies of the dependence of internal tension on magnetron power level and Ar pressure have shown that the optimal deposition of Nb films is realized at a power of ~600 W for a target area of ~122 cm^2 and an Ar pressure of ~10^{-2} mbar.

The typical capacitance of the Nb/AlO$_x$/Nb JJs that are used in SQUID sensors is ~0.5 pF at a critical current density of the JJs of ~200 A/cm^2 and an area of 3.2 μm × 3.2 μm [22]. Up to ~100 low-T_c DC SQUID structures with integrated input coils can be produced simultaneously on a single large-area Si wafer. Pick-up loops of superconducting flux transformers made from thin Nb wires can be used to measure the magnetic field or field gradient and to transfer it, in the form of an induced superconducting current, into the multiturn thin film input coil, which concentrates the magnetic flux into the SQUID loop, which is integrated on the same substrate. The SQUID sensor is placed in a superconducting shield, in order to isolate it from the parasitic influence of external electromagnetic interference. Standard highly sensitive low-T_c SQUIDs are available from commercial companies (see, for example, [23]). Special SQUID sensors that are intended for NDE experiments have been developed and produced in small quantities at IRE (see Figure 2). The primary advantage of using such self-made low-T_c SQUID sensors is the possibility to adapt their design to a particular NDE system, in order to reduce the coupling of parasitic background signals to the SQUID. The current design of a SQUID loop includes 4 balanced slots that are coupled gradiometrically to two input coils, one modulation coil and one feedback coil. The sensors are encapsulated inside a Nb shield together with screw contacts that are machined from Nb and provide a superconducting connection to the Nb wire of the gradiometric pick-up loops. The Nb contact pads on the SQUID chip are connected to the Nb screw contacts using a 25-μm-diameter Nb wire.

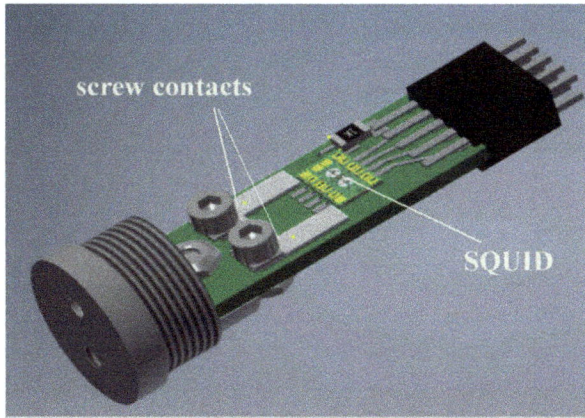

Figure 2. Schematic representation of Nb-based low-T_c DC SQUID sensor developed at IRE. The cylindrical superconducting (Nb) shield has been removed for clarity.

High-T_c JJs and SQUIDs are based on epitaxial films of the high-T_c superconductor YBa$_2$Cu$_3$O$_{7-x}$ (YBCO). The much shorter and highly anisotropic coherence length in YBCO ($\xi_{ab} \approx 2$ nm, $\xi_c \approx 0.4$ nm), as well as the d-wave symmetry of the superconducting order parameter and the strong dependence of the order parameter on the local strain and oxygen content in YBCO, when compared to the isotropic coherence length $\xi \approx 38$ nm and s-wave symmetry of the superconducting order parameter in polycrystalline Nb films, results in a completely different technology for high-T_c JJs. Grain boundaries can play the role of weak links in YBCO, whereas they do not significantly suppress the superconducting order parameter in Nb. High-T_c JJs are based mainly on grain boundary weak links, which can be realized by the epitaxial growth of YBCO films on bicrystal substrates [24,25] or on sharp steps etched on the surfaces of single crystal substrates [26–31]. Step-edge JJs can be placed on any part of a substrate, allowing the more efficient use of the substrate surface to design more efficient SQUID structure(s) with grain boundaries that are located exclusively at the JJ (see Figure 3). Newly-developed high-T_c step-edge JJs are based on the presence of two synchronously operating 45° [100]-tilted grain boundaries and possess optimal parameters for operation in high-T_c DC SQUIDs: critical current $I_c \approx 40$ μA, capacitance $C \approx 10$ fF, normal state resistance $R_n \approx 20$ Ω and characteristic voltage $I_c R_n \approx 800$ μV at 77 K [28–31]. The 50 times smaller capacitance of high-T_c JJs when compared to the capacitance of low-T_c JJs is advantageous for the low noise properties of high-T_c DC SQUIDs based on high-T_c JJs. In comparison to high-T_c step-edge JJs on SrTiO$_3$ (STO) and LaAlO$_3$ (LAO) substrates, such buffered 45° [100]-tilted step edge JJs on MgO substrates demonstrate better reproducibility and have lower noise values, also because of the absence of multiple low-angle grain boundaries at the bottom corner of the step.

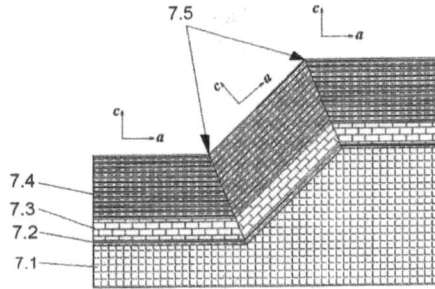

Figure 3. Schematic representation of a step-edge high-T_c Josephson junction developed at FZJ [28–30]. (7.1) Textured MgO substrate with a step height of ~400 nm; (7.2, 7.3) Graphoepitaxial buffer layers; (7.4) YBCO film; (7.5) Grain boundaries.

Only a few high-T_c SQUIDs can be produced simultaneously on the relatively small single crystal substrates of STO, LAO and MgO materials that are used for deposition of the epitaxial high-T_c films and heterostructures. The sensitivity of a high-T_c SQUID is typically improved by using a thin film pick-up loop that is connected directly to the SQUID loop or inductively coupled to it via a multiturn input coil. Low noise high-T_c superconducting flux transformers are made from epitaxial films because of the absence of sufficiently flexible and thin high-T_c superconducting wires. Thin film 20-mm multilayer superconducting flux transformers based on heterostructures with YBCO films are used to concentrate magnetic flux into the loop of the high-T_c SQUID to achieve a magnetic field resolution of ~4 fT/$\sqrt{Hz}$ at 77 K [25,31]. Further improvements in the magnetic field resolution of flip-chip high-T_c SQUID magnetometers down to ~2 fT/$\sqrt{Hz}$ at 77 K have recently been achieved by using a soft magnetic flux antenna in addition to the 20-mm multilayer superconducting flux transformer [32].

High-T_c SQUIDs demonstrate low noise properties up to temperatures of ~80 K, which can easily be reached by cooling using relatively cheap liquid nitrogen or energy-efficient cryocoolers. A wide variety of high-T_c SQUID sensors have been developed for specific NDE applications. Typically, they are vacuum-tight-encapsulated in fiberglass capsules together with a heater and feedback coil. The propensity of YBCO films and MgO substrates to degrade in the presence of humidity or corrosive contaminants in the air results in the need for vacuum-tight encapsulation or passivation, which is required for long-term stability of the high-T_c SQUID sensors.

4. Low-T_c and High-T_c SQUID NDE Systems

A wide variety of NDE systems equipped with specific SQUID sensors have been developed to study objects with different requirements. The measurement of magnetic fields generated by remote objects in magnetically unshielded environments during nondestructive archeological or geomagnetic surveys can be performed to a first approximation using room temperature magnetometers such as fluxgates, induction coils or optically pumped magnetometers. A low-T_c SQUID gives the best results for apparent resistivity at both shallow and deep regions simultaneously because it covers a larger response time interval than conventional coils during transient electromagnetic measurements [33], which require frequency-independent sensitivity at the level of several fT/$\sqrt{Hz}$. During transient electromagnetic measurements, electromagnetic fields are induced by transient pulses of electric current through a large loop of wire and the subsequent decay response from currents induced in underground layers can be measured. As a result of their superior sensitivity at low frequencies, only SQUID systems are currently able to resolve changes in the electrical conductivity of underground layers with sufficient sensitivity for depths exceeding ~500 m. Both low-T_c and high-T_c mobile systems have been demonstrated for the recording of magnetic anomalies during movement of the systems in the Earth's magnetic field [3,33–35]. High-T_c SQUID magnetometers or gradiometers with directly coupled 8-mm pick-up loops that are inductively coupled to first-order single-layer superconducting gradiometers, as well as low-T_c SQUID gradiometers with integrated multilayer gradiometric flux transformers [22], are currently the most suitable low-T_c SQUID sensors for mobile geomagnetic and archeological NDE.

The nondestructive monitoring of ion beam currents in particle accelerators is performed by the non-invasive measurement of magnetic fields generated by moving charged elementary particles. By using a Cryogenic Current Comparator (CCC) based on a low-T_c SQUID with a ferromagnetic Vitrovac core in the pick-up loop, a resolution of ~6 pA/$\sqrt{Hz}$ at 4.2 K and 2 kHz with a system 10-kHz frequency bandwidth has been achieved for monitoring accelerated electrons or $^{20}Ne^{10+}$ ions [36]. The sensor part of the CCC was optimized for the lowest possible noise-limited current resolution, in combination with a high system bandwidth of ~200 kHz, without compromising the resolution [37]. The ferromagnetic core was made from NANOPERM® with different annealing recipes by the company MAGNETEC. The fine structure of a beam could be observed. The CCC could also be used for the calibration of different devices, such as a secondary electron monitor. By using a ferromagnetic-core-free monitor based on a high-T_c DC SQUID gradiometer with a multilayer flux transformer operating at 77 K, fabricated at FZJ, the intensity of a 1 μA beam of $^{132}Xe^{20+}$ (50 MeV/u) ions could be measured non-invasively with 100 nA resolution [38].

In "traditional" NDE, high-T_c DC SQUID systems have demonstrated their superior capabilities for the inspection of metal plates, aircraft wheels and fuselage and pre-stressed concrete bridges [5,6,39–42]. The chosen measurement scheme depends on the NDE application: an eddy current excitation scheme and a narrowband lock-in readout scheme are used for the investigation of metal plates and aircraft parts, while measurements of static magnetic fields are efficient for monitoring magnetic flux leakage from ferromagnetic objects such as the pre-stressed steel tendons of concrete bridges. Deeper defects can be detected using SQUIDs at lower excitation frequencies, when compared to the conventional eddy current technique based on induction coils, because the sensitivity of coils decreases strongly with frequency.

Figure 4a shows a nonmagnetic ~200 mL cryostat with fiberglass walls that is able to hold liquid nitrogen for up to ~4 h while operating in different orientations (see Figure 4b). It was held by hand or fixed on the robotic arm of an automatic scanner during NDE measurements. A high-T_c DC SQUID first order planar gradiometer produced on a 1 cm² LAO substrate with a [110] orientation of its edges was fixed on a sapphire rod in the vacuum part of the cryostat, which was cooled by liquid nitrogen and placed ~1 mm from the outer surface of the bottom of the cryostat. Such gradiometers are able to operate in industrial environments, while providing a high sensitivity of ~50 fT/cm$\sqrt{Hz}$ at 77 K to the magnetic field gradient $\partial B_z/\partial x$.

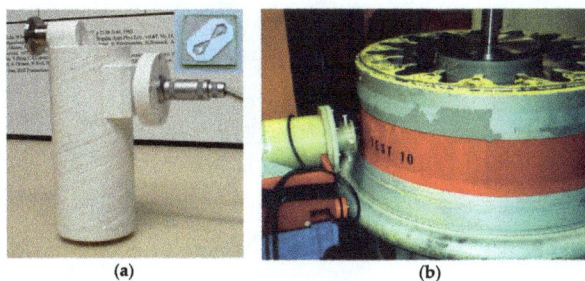

(a)   (b)

Figure 4. (a) Liquid nitrogen minicryostat used for the operation of a high-T_c DC SQUID gradiometer in an NDE system. The inset shows a photograph of the directly coupled high-T_c DC SQUID first order planar gradiometer, which was produced on a 1 cm² LAO substrate and installed in the cryostat; (b) Scan of an airplane wheel rim using the high-T_c DC SQUID gradiometer system. The robotic arm scanner moves the cryostat along the outer surface of the wheel rim, while the wheel is rotated around its axis.

An interesting application of high-T_c SQUIDs for the NDE of non-magnetic Al pipes involves the use of a magnetostrictive transmitter and sensor based on the use of pre-magnetized thin Ni plates to generate ultrasonic waves in the pipes and to convert the ultrasonic waves that are reflected from defects into magnetic signals, which can be measured contactlessly using a high-T_c SQUID gradiometer [43,44]. Another prospective application of high-T_c SQUIDs is a multi-channel system intended for the detection of magnetic metallic contaminants in packaged food [45].

At IRE, a low-T_c DC SQUID-based NDE system for operation in a magnetically unshielded environment was developed. The measurement probe in this system is based on fiberglass tubes and consists of the following elements: a first-order axial gradiometer as an input magnetic flux transformer, the low-T_c DC SQUID sensor CE2blue (a product of Supracon AG) with a low-T_c DC SQUID and input coil, connecting wires with a LEMO connector and a filling port for liquid He (see Figure 5). Low-T_c DC SQUID sensors developed at IRE are intended for the replacement of commercial sensors in future NDE systems.

Figure 5. Photograph of a single-channel low-T_c DC SQUID-based gradiometer system with a liquid He cryostat and a measurement probe. The first-order gradiometer was made of insulated Nb wire with a diameter of 0.05 mm using a "1:1" configuration (one lower and one upper turn) on a textolite rod. The diameter of the pick-up loops of the gradiometer is 4 mm and the base line of the gradiometer is 40 mm. The initial unbalance of the gradiometer is below 1%. The gradiometer ends are fixed mechanically on the Nb lamella of the SQUID sensor for connection to the SQUID input coil.

The single-channel low-T_c NDE system includes a liquid He cryostat, as shown in Figure 5. The inner diameter of the neck and inner tail of the cryostat is 22 mm. The distance between the outer and inner surfaces in the tail in the cooled system is no greater than 10 mm. The working time of the cryostat, which is cooled by 1.2 L of liquid He, is more than 2 days. The parameters of the liquid helium cryostat are as follows: outer diameter 110 mm; length 500 mm; outer diameter of the tail 45 mm; outer length of the tail 85 mm; inner diameter of the neck 22 mm; inner diameter of the cryogenic volume 80 mm; weight of the empty cryostat 2.2 kg. As the cryostat volume is relatively small, the filling procedure is relatively simple and takes several minutes. The small volume of He and the presence of a relief valve result in safety of the cryostat if the vacuum conditions in the space between the inner and outer walls are violated.

In tests of the gradiometer in such a configuration, the transfer coefficient of the input magnetic field B_{in} into magnetic flux Φ_e in the SQUID was measured to be ~9.5 nT/Φ_0, corresponding to an equivalent sensitivity of the gradiometer with respect to the magnetic field of ~30 fT/$\sqrt{\text{Hz}}$ at a SQUID intrinsic noise level of 3 $\mu\Phi_0/\sqrt{\text{Hz}}$. Such a sensitivity is sufficient for applications of SQUID-based gradiometers in NDE systems.

The DC SQUID electronics of the NDE system prototype are mounted on an Al box of size 117 mm $\times$ 62 mm $\times$ 19 mm located close to the cryostat and connected to the measurement probe using a cable of length 70 cm. The low-T_c DC SQUID electronics contain analog and digital components. The analog part contains a conventional modulation circuit of a null detector and a circuit of negative feedback with respect to magnetic flux. The analog components allow tuning of the low-T_c DC SQUID operating parameters. The digital components make it possible to switch the tuning and working regimes of the low-T_c DC SQUID gradiometer and system control using a personal computer. The low-T_c DC SQUID electronics are connected to the control unit by a 5-m-long cable. The preamplifier of the electronics unit is based on a Toshiba K-369 low-noise field effect transistor (FET) in the cascade circuit. The intrinsic noise of the preamplifier, without a transformer between the SQUID and the transistor, is <0.7 nV/$\sqrt{\text{Hz}}$. The transformer improves this value by approximately a factor of 10.

A single-pole integrator generates a feedback signal, which is fed to the modulation coil via a feedback resistor. The voltage across the feedback resistor is used as the output signal of the gradiometer. The DC SQUID electronics operate at a fixed feedback coefficient of ~1 V/Φ_0. The bandwidth of the system is approximately 0–16 kHz. The control unit of the NDE system contains stabilized power supply sources and a data acquisition system based on a 24-bit ADC.

The elements described above were used to construct a working prototype of a DC SQUID-based gradiometer. The prototype was tested under laboratory conditions without additional magnetic shielding and the main working parameters were studied. The Stanford Research low-frequency spectrum analyzer was used to study the noise characteristics of the output signal of the DC SQUID-based gradiometer prototype.

Noise spectra were registered over a frequency interval of 1–1000 Hz at a feedback coefficient of K_{FB} = 1 V/Φ_0. The measured transfer coefficient of the external magnetic field into the magnetic flux in the SQUID of ~9.5 nT/Φ_0 corresponds to an equivalent noise level with respect to the magnetic field of ~30 fT/$\sqrt{\text{Hz}}$. Such noise levels of the DC SQUID-based gradiometer indicate sufficient balancing and confirm that such devices can be employed in NDE systems. DC SQUID-based gradiometers can be used to develop multichannel DC SQUID-based systems. The prototype of the single-channel DC SQUID-based gradiometer shows stable operation in unshielded laboratory conditions and can be used for the development of multichannel gradiometric DC SQUID-based systems for the NDE of defects in metal structures and materials.

One of the important elements of a SQUID-based NDE system is the XY-scanner used to scan samples under a stationary liquid helium cryostat. The developed XY-scanner was equipped with two computer-controlled stepper motors (5RK60GE-CW2TE, ORIENTAL MOTOR), in order to move samples in the X and Y directions. The scanned area was 300 $\times$ 300 mm, with an accuracy for sample positioning of ~0.3 mm. In order to avoid external magnetic noise from magnetic components, the sample holder was fabricated using non-metallic and non-magnetic materials, such as fiberglass and plexiglass.

5. High-T$_c$ SQUID Microscope System with a Ferromagnetic Flux Antenna for NDE

A scanning SQUID microscope (SSM) is a powerful noninvasive tool for fundamental and applied research (see for example [2,46,47] and references therein). The high-T$_c$ DC SQUID microscope developed at FZJ for studies of room temperature objects is based on a high-T$_c$ DC SQUID with a magnetic flux antenna and was described in detail in our previous publications [48–50] (see Figure 6). Here, we review it briefly, report new results obtained with the system and provide an outlook for further developments.

The principle of operation of the microscope is shown in Figure 6b. An amorphous metallic soft magnetic 25 μm thick foil Vitrovac 6025X (Vacuumschmelze GmbH, Hanau, Germany) was used to guide magnetic flux from an object at room temperature through the pick-up loop of the high-T$_c$ SQUID and to return the flux back to the object. 2-mm-wide stripes were cut using scissors in a direction normal to the rolling direction of the foil, in order to reduce Barkhausen noise from the ferromagnetic foil. The tip of the flux antenna was first formed at a 50° angle using scissors and the end of the tip was then sharpened to a radius of ~200 nm using 0.3 μm diamond polishing sheets.

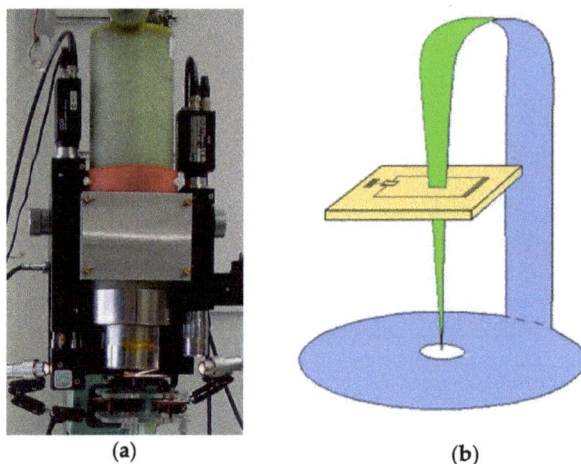

(a)   (b)

Figure 6. (a) Photograph of a high-T$_c$ DC SQUID microscope with a fiberglass cryostat that can support 0.8 L of liquid nitrogen; (b) Schematic diagram of a high-T$_c$ DC SQUID with a magnetic flux antenna made of soft magnetic foil penetrating the directly coupled pick-up loop [49].

The SQUID was fixed using vacuum grease on a sapphire rod together with the modulation coil and the low temperature part of the flux antenna (see Figure 7a) in the vacuum part of the cryostat. The sapphire rod was cooled using liquid nitrogen through the inner wall of the fiberglass cryostat. The cryostat contains ~0.8 L of liquid nitrogen when it is completely filled and provides 2 days of SQUID operation at a temperature of ~78 K. The room temperature parts of the flux antenna were vacuum-sealed using epoxy in the outer wall of the cryostat and connected to their cooled counterparts (see Figure 7b). Commercial ac-bias electronics was used for SQUID operation in flux-locked loop mode (Cryoton Co. Ltd., Moscow, Russia).

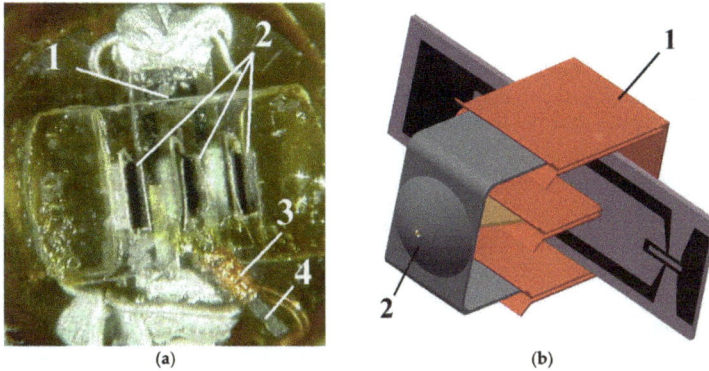

(a)   (b)

Figure 7. (**a**) Photograph of a high-T_c DC SQUID (1) assembled on a sapphire rod, showing parts of the magnetic flux antenna (2) and the modulation coil (3) on ferromagnetic wires (4); (**b**) Sketch of a DC SQUID with a directly coupled pick-up loop assembled together with low temperature (1) and room temperature (2) parts of the flux antenna.

This system was used to perform measurements of the magnetic field distribution over a US $1 bill, for a qualitative comparison of the device with SQUID microscope systems made by other groups [51,52]. The magnetic signal originates from the black ink used for printing banknotes, which contains a small quantity of magnetite (Fe_3O_4) nanoparticles. The measurements were nondestructive. Such a system can also be used for the detection of magnetic ink on old bills, which can result in false alarm signals in the detection of counterfeit notes using conventional magnetic ink testers.

The nondestructive evaluation of magnetic features in stainless steel X5CrNi18-10 (German grade 1.4301, AISI 304) samples caused by welding and wear-out was performed. Although this corrosion-resisting austenitic steel is not magnetic, heat treatment or wear [53] partially transform non-magnetic austenite to ferromagnetic α-martensite that is brittle and less resistant to corrosion. The detection of magnetic signals at weld seams provides valuable information about the quality of the welding. An example of a magnetic image of a weld seam made by laser welding of 1.4301 stainless steel plates is shown in Figure 8. The magnetic signal measured along such a weld seam is relatively weak compared to the more than 10 times stronger magnetic field above seams made using wolfram-inert-gas (WIG) welding of the same steel plates.

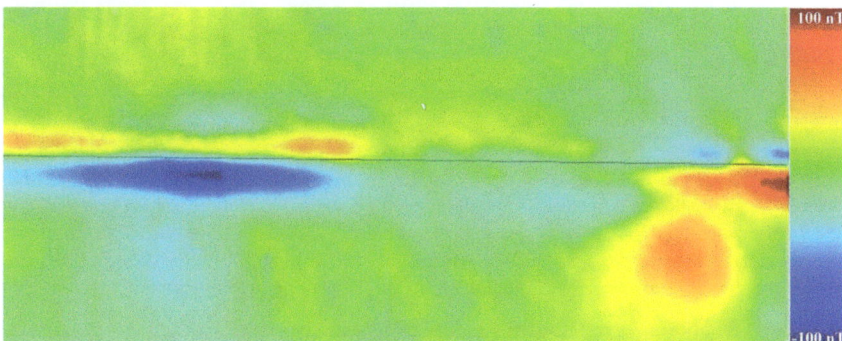

Figure 8. 3D color-scale image of the magnetic field distribution measured over a weld seam (indicated by a black line) made by laser welding. The range of color-scale values is from −100 nT (blue) to 100 nT (red). The scanned area is 30 mm × 10 mm.

The wear-out of stainless steel plates was simulated by scratching [50] the plates using a diamond tip or engraving by a diamond drill. The measured magnetic signal originates from inclusions of the ferromagnetic α-martensite form of the steel crystalline structure, which appear as a result of the plastic deformation of austenite in the contact area due to tribological stressing [53].

A SQUID microscope has been used for the investigation of the magnetization states of thin magnetic films and heterostructures intended for magneto-electronic devices and recording media. Bit patterns of information stored ferromagnetically on old floppy disks and hard disks have been evaluated. Changes in the distributions of magnetic stray fields in the $Co/Al_2O_3/Co$-tunnel junctions of tunneling magneto-resistive devices during their magnetization have been measured. The dependence of magnetic domain structure in thin Fe films on the thicknesses of $(SiGe)_n$ barrier layers between them has been reported [54]. Magnetic stray fields originating from 30-nm-thick Co films fabricated using electron beam lithography on 50-nm-thick SiN membranes have been registered [50]. A measurement of the latter structure after demagnetization is shown in Figure 9. Measurements of stray magnetic fields using a SQUID microscope were performed in the frequency range 1–10 Hz and did not result in observable changes in magnetization.

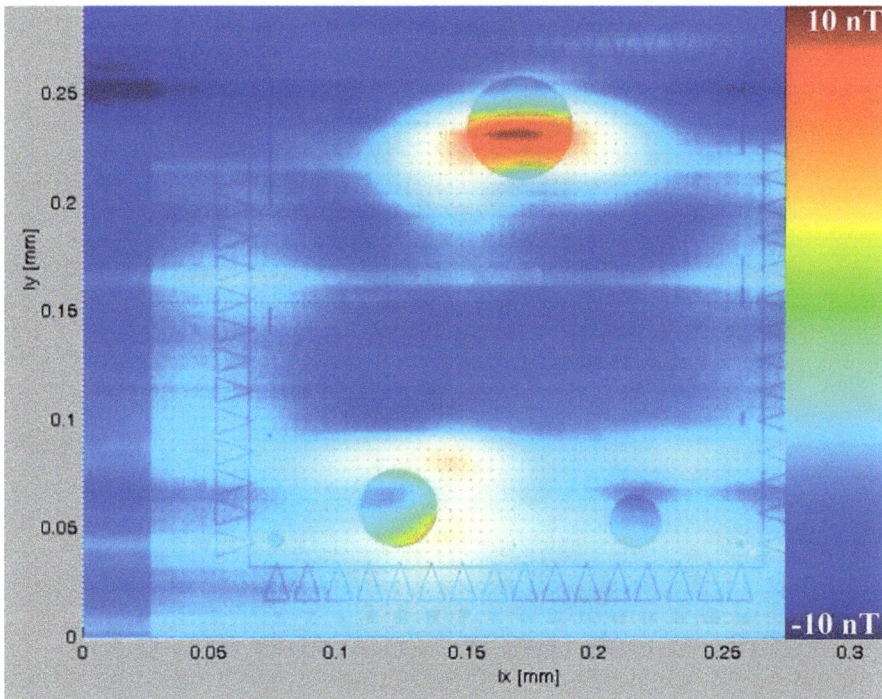

Figure 9. Magnetic field distribution of the demagnetized state of a 30-nm-thick Co film (contours showing the Co pattern have been added to the picture) prepared on a 50-nm-thick SiN membrane. The color scale represents magnetic fields of between -10 nT (blue) and 10 nT (red). Signals recorded from the magnetic domain structure of 40 μm, 30 μm and 20 μm dots are observable.

The spatial resolution of the SQUID microscope of ~10 μm was limited primarily by the shape of the ferromagnetic tip of the magnetic flux antenna and the tip-to-sample separation. Additional sharpening of the tip by focused ion beam milling and the implementation of a tuning fork for controlling the tip-to-sample distance would improve the spatial resolution. The resulting thinning of the tip would deteriorate the magnetic field sensitivity. A possible solution involves optimization of the shape and material of the magnetic flux antenna. For example, Nanoperm M033 may result in better magnetic field sensitivity of the sensor [55].

Replacement of the direct-coupled pick-up loop by a multilayer flux transformer improves transfer of the magnetic flux from the pick-up loop to the loop of the high-T_c DC SQUID (see [56] and references therein). For a 20-mm flip-chip magnetometric high-T_c SQUID sensor, a magnetic field resolution of ~4 fT/$\sqrt{Hz}$ at 77 K was measured in magnetically shielded conditions [25,31]. This sensitivity was further improved to 2 fT/$\sqrt{Hz}$ at 77 K by using an extremely soft magnetic flux antenna made from ferromagnetic Vitrovac 6025 foil [32]. In order to provide low values of Barkhausen and Johnson noise of the sensor, the magnetic flux antenna was assembled from ~250 pieces of 2-mm-wide 3.5-cm-long strips, which were cut in a direction perpendicular to the rolling direction of the foil and insulated on both sides by ~200-nm-thick insulating Al_2O_3 film. An example of noise measurement of the 20-mm flip-chip magnetometric high-T_c SQUID sensor with such a soft magnetic flux antenna in a magnetic shield is shown in Figure 10. The 20 mm sensors were initially developed for human magnetoencephalography [57] and other noninvasive noncontact investigations of biological objects. A composite ferromagnetic antenna can be prolonged through the walls of the cryostat in the future to measure the strongest component of the magnetic field in the nearest vicinity of the object under investigation. The combination of a superconducting flux transformer with a ferromagnetic flux antenna will also be useful for other NDE applications, such as improving the magnetic field resolution of a SQUID microscope or continuous non-invasive current monitoring of a high energy ion beam in a particle accelerator using a high-T_c SQUID sensor operating at temperature of up to 80 K.

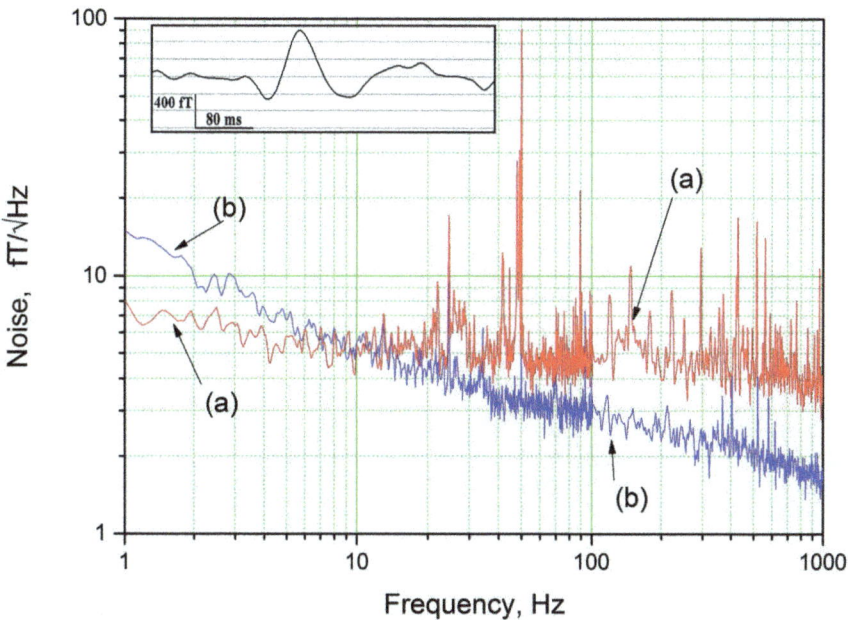

Figure 10. Noise spectra of a 20 mm high-T_c DC SQUID magnetometer measured at 77 K in a magnetic shield: (**a**) without a ferromagnetic antenna and (**b**) with a ferromagnetic antenna. The inset shows a measurement of human magnetoencephalography performed using a high-T_c DC SQUID magnetometer that has a sensitivity in the femto-Tesla range at low frequencies.

Low-T_c DC SQUIDs with sizes of below 1 μm ("nanoSQUIDs") have been fabricated on sharp tips of pulled quartz tubes and have demonstrated unprecedented spin sensitivities of ~0.38 μB/√Hz [58] with spatial resolutions of ~20 nm [2]. The implementation of an electrically tunable multi-terminal SQUID configuration [59] provided optimal flux bias conditions by the direct injection of flux modulation and feedback current into the SQUID loop, thereby avoiding the need for the application of bias fields as high as ~0.4 T in the case of a 40-nm loop of a nanoSQUID. Such nanoSQUIDs can potentially be used for the nondestructive measurement of distributions of stray fields of magnetic nanoparticles and nanostructures, as well as for the nondestructive readout of the final states of superconducting flux qubits after their protection by sufficiently high potential barriers. The self-biasing of SQUIDs using YBCO-Nb JJs has also been realized [60]. NanoSQUIDs based on YBCO films and step-edge or bicrystal JJs should be able to operate at liquid nitrogen temperature or have a large $I_c R_n$ product at lower temperatures [27,61].

Acknowledgments: The research leading to these results has received funding from the European Research Council under the European Union's Seventh Framework Programme (FP7/2007-2013)/ERC grant agreement number 320832. This work was supported in part by the Russian Science Foundation under Grant No. 15-19-00206. The authors gratefully acknowledge valuable discussions with U. Poppe and the technical assistance of R. Speen.

Conflicts of Interest: The authors declare no conflicts of interest. The funding sponsors had no role in the design of this study, in the collection, analysis or interpretation of data, in the writing of the manuscript, or in the decision to publish the results.

References

1. Clarke, J.; Braginski, A.I. (Eds.) Fundamentals and technology of SQUIDs and SQUID systems. In *The SQUID Handbook*; WILEY-VCH Verlag GmbH&Co. KgaA: Weinheim, Germany, 2004; Volume 1, ISBN 3-52740229-2.
2. Anahory, Y.; Reiner, J.; Embon, L.; Halbertal, D.; Yakovenko, A.; Myasoedov, Y.; Rappaport, M.L.; Huber, M.E.; Zeldov, E. Three-junction SQUID-on-tip with tunable in-plane and out-of-plane magnetic field sensitivity. *Nano Lett.* **2014**, *14*, 6481–6487. [CrossRef] [PubMed]
3. Schultze, V.; Linzen, S.; Schüler, T.; Chwala, A.; Stolz, R.; Schulz, M.; Meyer, H.-G. Rapid and sensitive magnetometer surveys of large areas using SQUIDs—The measurement system and its application to the Niederzimmern Neolithic double-ring ditch exploration. *Archaeol. Prospect.* **2008**, *15*, 113–131. [CrossRef]
4. Schiffler, M.; Queitsch, M.; Stolz, R.; Chwala, A.; Krech, W.; Meyer, H.-G.; Kukowski, N. Calibration of SQUID vector magnetometers in full tensor gradiometry systems. *Geophys. J. Int.* **2014**, *198*, 954–964. [CrossRef]
5. Krause, H.-J.; Michael Mück, M.; Tanaka, S. SQUIDs in Nondestructive Evaluation. In *Applied Superconductivity: Handbook on Devices and Applications*; Seidel, P., Ed.; Wiley: Weinheim, Germany, 2015; Volume 2.
6. Krause, H.-J.; von Kreutzbruck, M. Recent developments in SQUID NDE. *Physics C* **2002**, *368*, 70–79. [CrossRef]
7. Bain, R.J.P.; Donaldson, G.B.; Evanson, S.; Hayward, G. Design and operation of SQUID-based planar gradiometers for NDT of ferromagnetic plates. *IEEE Trans. Magn.* **1987**, *23*, 473–476. [CrossRef]
8. Hahlbohm, H.D.; Lubbig, H. *SQUID '85 Superconducting Quantum Interference Devices and Their Applications: Proceedings of the Third International Conference on Superconducting Quantum Devices*; Walter de Gruyter & Co.: Berlin, Germany, 1985; p. 843.
9. Sawade, G.; Straub, J.; Krause, H.J.; Bousack, H.; Neudert, G.; Ehrlich, R. Signal analysis methods for remote magnetic examination of prestressed elements. In *Proceedings of the International Symposium on Non Destructive Testing in Civil Engineering (NDT-CE), Berlin, Germany, 26–28 September 1995*; Schickert, G., Wiggenhauser, H., Eds.; DGZfP: Berlin, Germany, 1995; Volume II, pp. 1077–1084.
10. Tavrin, Y.; Siegel, M.; Hinken, J.-H. Standard Method for Detection of Magnetic Defects in Aircraft Engine Discs using a HTS SQUID Gradiometer. *IEEE Trans. Appl. Supercond.* **1999**, *9*, 3809–3812. [CrossRef]
11. Kleiner, R.; Buckel, W. Applications of Superconductivity. In *Superconductivity: An Introduction*, 3rd ed.; Wiley-VCH Verlag GmbH & Co. KgaA: Weinheim, Germany, 2016; Chapter 7, ISBN 978-3-527-41162-7.
12. Faley, M.I. Epitaxial oxide heterostructures for ultimate high-T_c quantum interferometers. In *Applications of High-T_c Superconductivity*; Luiz, A., Ed.; InTech: Rijeka, Croatia, 2011; ISBN 978-953-307-308-8.
13. Berry, M. Die geometrische Phase. *Spektrum der Wissenschaft*, February 1989; pp. 74–81.

14. Tinkham, M. *Introduction to Superconductivity*, 2nd ed.; McGraw-Hill Inc.: New York, NY, USA, 1996; p. 225, ISBN 0-07-064878-6.
15. Ludwig, C.; Kessler, C.; Steinfort, A.J.; Ludwig, W. Versatile high performance digital SQUID electronics. *IEEE Trans. Appl. Supercond.* **2001**, *11*, 1122–1125. [CrossRef]
16. Drung, D.; Bechstein, S.; Franke, K.-P.; Scheiner, M.; Schurig, T. Improved Direct-Coupled dc SQUID Read-out Electronics with Automatic Bias Voltage Tuning. *IEEE Trans. Appl. Supercond.* **2001**, *11*, 880–883. [CrossRef]
17. Kupriyanov, M.Y.; Brinkman, A.; Golubov, A.A.; Siegel, M.; Rogalla, H. Double-barrier Josephson structures as the novel elements for superconducting large-scale integrated circuits. *Physics C* **1999**, *326–327*, 16–45. [CrossRef]
18. Brinkman, A.; Cassel, D.; Golubov, A.A.; Kupriyanov, M.Y.; Siegel, M.; Rogalla, H. Double-barrier Josephson junctions: Theory and experiment. *IEEE Trans. Appl. Supercond.* **2001**, *11*, 1146–1149. [CrossRef]
19. Gudkov, A.L.; Kupriyanov, M.Y.; Samus, A.N. Properties of planar Nb/α-Si/Nb Josephson junctions with various degrees of doping of the α-Si layer. *J. Exp. Theor. Phys.* **2012**, *114*, 818–829. [CrossRef]
20. Gurvitch, M.; Washington, W.A.; Huggins, H.A. High quality refractory Josephson tunnel junctions utilizing thin aluminium layers. *Appl. Phys. Lett.* **1983**, *42*, 472–474. [CrossRef]
21. Shiota, T.; Imamura, T. Fabrication of high quality Nb/AlO$_x$-Al/Nb Josephson junctions: III-Annealing stability of AlO$_x$ tunneling barriers. *IEEE Trans. Appl. Supercond.* **1992**, *2*, 222–227. [CrossRef]
22. Stolz, R.; Fritzsch, L.; Meyer, H.-G. LTS SQUID sensor with a new configuration. *Supercond. Sci. Technol.* **1999**, *12*, 806–808. [CrossRef]
23. Supracon AG, Jena, Germany. Available online: http://www.supracon.com/de/kontakt.html (accessed on 30 November 2017).
24. Dimos, D.; Chaudhari, P.; Mannhart, J.; LeGoues, F.K. Orientation Dependence of Grain-Boundary Critical Currents in YBa$_2$Cu$_3$O$_{7-\delta}$ Bicrystals. *Phys. Rev. Lett.* **1988**, *61*, 219. [CrossRef] [PubMed]
25. Faley, M.I.; Jia, C.L.; Poppe, U.; Houben, L.; Urban, K. Meandering of the grain boundary and d-wave effects in bicrystal Josephson junctions. *Supercond. Sci. Technol.* **2006**, *19*, S195–S199. [CrossRef]
26. Daly, K.P.; Dozier, W.D.; Burch, J.F.; Coons, S.B.; Hu, R.; Platt, C.E.; Simon, R.W. Substrate step-edge YBa$_2$Cu$_3$O$_7$ rf SQUIDs. *Appl. Phys. Lett.* **1990**, *58*, 543. [CrossRef]
27. Mitchell, E.E.; Foley, C.P. YBCO step-edge junctions with high $I_c R_n$. *Supercond. Sci. Technol.* **2010**, *23*, 65007. [CrossRef]
28. Faley, M.I. Reproducible Step-Edge Josephson Junction. U.S. Patent 9,666,783 B2, 30 May 2017.
29. Faley, M.I. Reproduzierbarerer Stufen-Josephson-Kontakt für ein Bauelementder supraleitenden Elektronik und Herstellverfahren dafür. DE Patent 102,012,006,825 B4, 26 February 2015.
30. Faley, M.I. Reproduzierbarerer Stufen-Josephson-Kontakt. EP Patent 2,834,860 B1, 30 December 2015.
31. Faley, M.I.; Dammers, J.; Maslennikov, Y.V.; Schneiderman, J.F.; Winkler, D.; Koshelets, V.P.; Shah, N.J.; Dunin-Borkowski, R.E. High-T$_c$ SQUID biomagnetometers. *Supercond. Sci. Technol.* **2017**, *30*, 83001. [CrossRef]
32. Faley, M.I.; Maslennikov, Y.V.; Koshelets, V.P.; Dunin-Borkowski, R.E. Flip-Chip High-T$_c$ DC SQUID Magnetometer with a Ferromagnetic Flux Antenna. *IEEE Trans. Appl. Supercond.* **2017**.
33. Ji, Y.; Du, S.; Xie, L.; Chang, K.; Liu, Y.; Zhang, Y.; Xie, X.; Wang, Y.; Lin, J.; Rong, L. TEM measurement in a low resistivity overburden performed by using low temperature SQUID. *J. Appl. Geophys.* **2016**, *135*, 243–248. [CrossRef]
34. Supracon AG. JESSY STAR. Available online: http://www.supracon.com/de/star.html (accessed on 30 November 2017).
35. Keenan, S.T.; Young, J.A.; Foley, C.P.; Du, J. A high-T$_c$ flip-chip SQUID gradiometer for mobile underwater magnetic sensing. *Supercond. Sci. Technol.* **2010**, *23*, 25029. [CrossRef]
36. Peters, A.; Vodel, W.; Koch, H.; Neubert, R.; Reeg, H.; Schroeder, C.H. A cryogenic current comparator for the absolute measurement of nA beams. *AIP Conf. Proc.* **1998**, *451*, 163–180. [CrossRef]
37. Tympel, V.; Golm, J.; Neubert, R.; Seidel, P.; Schmelz, M.; Stolz, R.; Zakosarenko, V.; Kurian, F.; Schwickert, M.; Sieber, T.; Stöhlker, T. The Next Generation of Cryogenic Current Comparators for Beam Monitoring. In Proceedings of the International Beam Instrumentation Conference (IBIC 2016), Barcelona, Spain, 11–15 September 2016; pp. 441–444.
38. Watanabe, T.; Fukunishi, N.; Kase, M.; Kamigaito, O.; Inamori, S.; Kon, K. Beam Current Monitor with a High-T$_c$ Current Sensor and SQUID at the RIBF. *J. Supercond. Nov. Magn.* **2013**, *26*, 1297–1300. [CrossRef]

39. Krause, H.-J.; Hohmann, R.; Soltner, H.; Lomparski, D.; Grüneklee, M.; Banzet, M.; Schubert, J.; Zander, W.; Zhang, Y.; Wolf, W.; et al. Mobile HTS SQUID system for eddy current testing of aircraft. In *Review of Progress in Quantitative Nondestructive Evaluation*; Thompson, D.O., Chimenti, D.E., Eds.; Plenum Publishing: New York, NY, USA, 1997; Volume 16, pp. 1053–1060, ISBN 0-306-45597-8.

40. Grüneklee, M.; Krause, H.-J.; Hohmann, R.; Maus, M.; Lomparski, D.; Banzet, M.; Schubert, J.; Zander, W.; Zhang, Y.; Wolf, W.; et al. HTS SQUID System for Eddy Current Testing of Airplane Wheels and Rivets. In *Review of Progress in Quantitative Nondestructive Evaluation*; Thompson, D.O., Chimenti, D.E., Eds.; Springer: Boston, MA, USA, 1998; Volume 17, pp. 1075–1082. [CrossRef]

41. Krause, H.-J.; Hohmann, R.; Grüneklee, M.; Maus, M.; Zhang, Y.; Lomparski, D.; Soltner, H.; Wolf, W.; Banzet, M.; Schubert, J.; et al. Aircraft Wheel and Fuselage Testing with Eddy Current and SQUID. *NDTnet* **1998**, *3*. Available online: http://www.ndt.net/article/ecndt98/aero/043/043.htm (accessed on 30 November 2017).

42. Krause, H.-J.; Wolf, W.; Glaas, W.; Zimmermann, E.; Faley, M.I.; Sawade, G.; Mattheus, R.; Neudert, G.; Gampe, U.; Krieger, J. SQUID Array for Magnetic Inspection of Prestressed Concrete Bridges. *Physica C Supercond. Appl.* **2002**, *368*, 91–95. [CrossRef]

43. Masutani, N.; Teranishi, S.; Masamoto, K.; Kanenaga, S.; Hatsukade, Y.; Adachi, S.; Tanabe, K. Defect Detection of Pipes using Guided Wave and HTS SQUID. *J. Phys. Conf. Ser.* **2017**, *871*, 12074. [CrossRef]

44. Hatsukade, Y.; Kobayashi, T.; Nakaie, S.; Masutani, N.; Tanaka, Y. Novel Remote NDE Technique for Pipes Combining HTS-SQUID and Ultrasonic Guided Wave. *IEEE Trans. Appl. Supercond.* **2017**, *27*, 1600104. [CrossRef]

45. Tanaka, S.; Ohtani, T.; Krause, H.-J. Prototype of Multi-Channel High-T_c SQUID Metallic Contaminant Detector for Large Sized Packaged Food. *IEICE Trans. Electron.* **2017**, *E100-C*, 269–273. [CrossRef]

46. Kirtley, J.R.; Wikswo, J.P. Scanning SQUID microscopy. *Annu. Rev. Mater. Sci.* **1999**, *29*, 117–148. [CrossRef]

47. Cui, Z.; Kirtley, J.R.; Wang, Y.; Kratz, P.A.; Rosenberg, A.J.; Watson, C.A.; Gibson, G.W., Jr.; Ketchen, M.B.; Moler, K.A. Scanning SQUID sampler with 40-ps time resolution. *Rev. Sci. Instrum.* **2017**, *88*, 83703. [CrossRef] [PubMed]

48. Poppe, U.; Faley, M.I.; Breunig, I.; Speen, R.; Urban, K.; Zimmermann, E.; Glaas, W.; Halling, H. HTS dc-SQUID Microscope with soft-magnetic Flux Guide. *Supercond. Sci. Technol.* **2004**, *17*, S191–S195. [CrossRef]

49. Faley, M.I.; Zimmermann, E.; Poppe, U.; Urban, K.; Halling, H.; Soltner, H.; Jungbluth, B.; Speen, R.; Glaas, W. Magnetic Flow Sensor Comprising a Magnetic Field Conductor and a Hole Diaphragm. U.S. Patent 7,221,156 (B2), 22 May 2007.

50. Faley, M.I.; Kostyurina, E.A.; Diehle, P.; Poppe, U.; Kovacs, A.; Maslennikov, Y.V.; Koshelets, V.P.; Dunin-Borkowski, R.E. Nondestructive Evaluation Using a High-T_c SQUID Microscope. *IEEE Trans. Appl. Supercond.* **2017**, *27*, 1600905. [CrossRef]

51. Wellstood, F.C.; Mathai, A.; Song, D.; Black, R.C. Method and Apparatus for Imaging Microscopic Spatial Variations in Small Currents and Magnetic Fields. U.S. Patent 5,491,411, 13 February 1996.

52. Faley, M.I.; Pratt, K.; Reineman, R.; Schurig, D.; Gott, S.; Sarwinski, R.E.; Paulson, D.N.; Starr, T.N.; Fagaly, R.L. HTS dc-SQUID Micro-Susceptometer for Room Temperature Objects. *Supercond. Sci. Technol.* **2004**, *17*, S324–S327. [CrossRef]

53. Assmus, K.; Hübner, W.; Pyzalla, A.; Pinto, H. Structure transformations in CrNi steels under tribological stressing at low temperatures. *Tribotest* **2006**, *12*, 149–159. [CrossRef]

54. Gareev, R.R.; Weides, M.; Schreiber, R.; Poppe, U. Resonant tunneling magnetoresistance in antiferromagnetically coupled Fe-based structures with multilayered Si/Ge spacers. *Appl. Phys. Lett.* **2006**, *88*, 172105. [CrossRef]

55. Geithner, R.; Heinert, D.; Neubert, R.; Vodel, W.; Seidel, P. Low temperature permeability and current noise of ferromagnetic pickup coils. *Cryogenics* **2013**, *54*, 16–19. [CrossRef]

56. Faley, M.I.; Poppe, U.; Urban, K.; Paulson, D.N.; Starr, T.; Fagaly, R.L. Low noise HTS dc-SQUID flip-chip magnetometers and gradiometers. *IEEE Trans. Appl. Supercond.* **2001**, *11*, 1383–1386. [CrossRef]

57. Dammers, J.; Chocholacs, H.; Eich, E.; Boers, F.; Faley, M.; Dunin-Borkowski, R.E.; Shah, N.J. Source localization of brain activity using helium-free interferometer. *Appl. Phys. Lett.* **2014**, *104*, 213705. [CrossRef]

58. Vasyukov, D.; Anahory, Y.; Embon, L.; Halbertal, D.; Cuppens, J.; Neeman, L.; Finkler, A.; Segev, Y.; Myasoedov, Y.; Rappaport, M.L.; et al. A scanning superconducting quantum interference device with single electron spin sensitivity. *Nat. Nanotechnol.* **2013**, *8*, 639–644. [CrossRef] [PubMed]

59. Uri, A.; Meltzer, A.Y.; Anahory, Y.; Embon, L.; Lachman, E.O.; Halbertal, D.; HR, N.; Myasoedov, Y.; Huber, M.E.; et al. Electrically Tunable Multiterminal SQUID-on-Tip. *Nano Lett.* **2016**, *16*, 6910–6915. [CrossRef] [PubMed]

60. Smilde, H.-J.H.; Ariando, R.H.; Hilgenkamp, H. Bistable superconducting quantum interference device with built-in switchable $\pi/2$ phase shift. *Appl. Phys. Lett.* **2004**, *85*, 4091. [CrossRef]

61. Poppe, U.; Divin, Y.Y.; Faley, M.I.; Wu, J.S.; Jia, C.L.; Shadrin, P.; Urban, K. Properties of $YBa_2Cu_3O_7$ Thin Films Deposited on Substrates and Bicrystals with Vicinal Offcut and Realization of High I_cR_n Junctions. *IEEE Trans. Appl. Supercond.* **2001**, *11*, 3768–3771. [CrossRef]

sensors

MDPI

Article

Construction Condition and Damage Monitoring of Post-Tensioned PSC Girders Using Embedded Sensors

Kyung-Joon Shin [1], Seong-Cheol Lee [2], Yun Yong Kim [1], Jae-Min Kim [3], Seunghee Park [4] and Hwanwoo Lee [5,*]

1   Department of Civil Engineering, Chungnam National University, Daejeon 34134, Korea; kjshin@cnu.ac.kr (K.-J.S.); yunkim@cnu.ac.kr (Y.Y.K.)
2   Department of NPP Engineering, KEPCO International Nuclear Graduate School, Ulsan 45014, Korea; sclee@kings.ac.kr
3   Department of Marine and Civil Engineering, Chonnam National University, Yeosu 59626, Korea; jm4kim@jnu.ac.kr
4   School of Civil, Architectural Engineering and Landscape Architecture, Sungkyunkwan University, Suwon 16419, Korea; shparkpc@skku.edu
5   Department of Civil Engineering, Pukyong National University, Busan 48513, Korea
*   Correspondence: hwanwoo@pknu.ac.kr; Tel.: +82-51-629-6073

Received: 25 July 2017; Accepted: 7 August 2017; Published: 10 August 2017

Abstract: The potential for monitoring the construction of post-tensioned concrete beams and detecting damage to the beams under loading conditions was investigated through an experimental program. First, embedded sensors were investigated that could measure pre-stress from the fabrication process to a failure condition. Four types of sensors were installed on a steel frame, and the applicability and the accuracy of these sensors were tested while pre-stress was applied to a tendon in the steel frame. As a result, a tri-sensor loading plate and a Fiber Bragg Grating (FBG) sensor were selected as possible candidates. With those sensors, two pre-stressed concrete flexural beams were fabricated and tested. The pre-stress of the tendons was monitored during the construction and loading processes. Through the test, it was proven that the variation in thepre-stress had been successfully monitored throughout the construction process. The losses of pre-stress that occurred during a jacking and storage process, even those which occurred inside the concrete, were measured successfully. The results of the loading test showed that tendon stress and strain within the pure span significantly increased, while the stress in areas near the anchors was almost constant. These results prove that FBG sensors installed in a middle section can be used to monitor the strain within, and the damage to pre-stressed concrete beams.

Keywords: Keywords: load cell; optical fiber sensor; pre-stressed concrete; monitoring; pre-stress; damage

1. Introduction

For the last half century, post-tensioned pre-stressed concrete (PSC) girders have been constructed due to the advantages and effectiveness of their structural behavior. To secure structural safety in post-tensioned PSC girders, it is very important to know the effective pre-stress force in the tendon [1–3]. Of the many studies and design specifications that have been proposed, there are several methods that can predict the pre-stress in a tendon of a PSC girder. However, because there are too many uncertainties related to the loss of pre-stress, there is always a huge gap between predictions and measurements.

Although it is important to know the effective pre-stress force in PSC girders, it is not easy to measure this during their years of service using conventional methods. Electrical strain gauges are difficult to install on tendons located inside the concrete and are easy to lose during the construction process [4,5]. Load cells embedded in the anchors of tendons are not available to measure the effective pre-stress force in the middle of PSC girders when the tendons are bonded. Therefore, information regarding effective pre-stress force is too limited. Generally, only the jacking force measured from the hydraulic unit during tensioning is available.

Various nondestructive test (NDT) methods have been studied to estimate the force of the pre-stressing tendons during the construction and service stages. However, most studies are limited to lab-scale applications. Specifically, there is practically no example in which an economically efficient estimation of the pre-stress force has been realized for the bonded pre-stressing tendons applied in existing pre-stressed concrete bridges [6].

The studies related to pre-stress measurement in bonded tendons can be classified based on the theory applied. The most famous approach is to use a guided wave or stress wave. However, this approach is limited to being applied to bonded pre-stressing steel filled with grout. The guided wave passing through the concrete cannot be measured practically due to its attenuation in concrete [7,8]. As an alternative to overcome the drawbacks of ultrasonic and stress waves, a technique using a magnetic field was studied. Wang et al. [9,10] developed magneto-elastic sensors that can monitor the stress in a multistrand cable for cable stayed or cable suspension bridges. However, most applications have been for cables and tendons not embedded inside the concrete [11,12].

Recently, optical fiber sensors have been developed and applied to monitor the pre-stress of tendons because of several advantages, such as high accuracy and electromagnetic interference resistance [13–22]. Previous studies reported that it is possible to measure the pre-stressing force of PSC structures by attaching an optical sensor to the surface of one of the strands used in such structures [16–20]. Distributed optical fiber sensing technology was also used to monitor the pre-stress of a tendon [21,22]. Fiber optic sensors can be attached directly to such strands, or in combination with a material similar to that of the strands.

As an application method of Fiber Bragg Grating (FBG) sensors, smart tendons with embedded FBG sensors have been developed to measure tendon strain directly [23,24]. By encapsulating the FBG sensor inside a seven-wire strand, the effective pre-stress force can be measured during construction and even during the service life of the product. In the literature, it was found that the effective pre-stress force can be accurately measured through FBG sensors, not only when tendons are directly exposed to the air [25,26], but also when tendons are embedded in concrete [27].

For this paper, the feasibility of monitoring the pre-stress of tendons during beam construction and of detecting damage to PSC girders that are in service was investigated experimentally. First, several kinds of sensors were tested and their appropriateness were verified. Next, PSC girders were fabricated with the sensors embedded. During their construction, the condition of the pre-stressing tendons was measured by the embedded sensors. Finally, girder stress was monitored while the applied load was increased until failure.

2. Validation Test of Sensors

2.1. Experimental Program

As a first step, the sensors available with current technology were investigated, and it was verified that they could be used to measure the pre-stress of a tendon in a PSC girder. It should be noted that even though the alignment of the hydraulic jack, anchor head, and steel wire is adjusted carefully, the individual strands sustain difference prestress forces [12], which can lead to eccentric force on the anchor head. This eccentric effect cannot be avoided when a typical hydraulic jacking device is used. Four types of sensors were tested, as shown in Figure 1. Three of them were strain gauge-based force transducers, and the other was an FBG sensor.

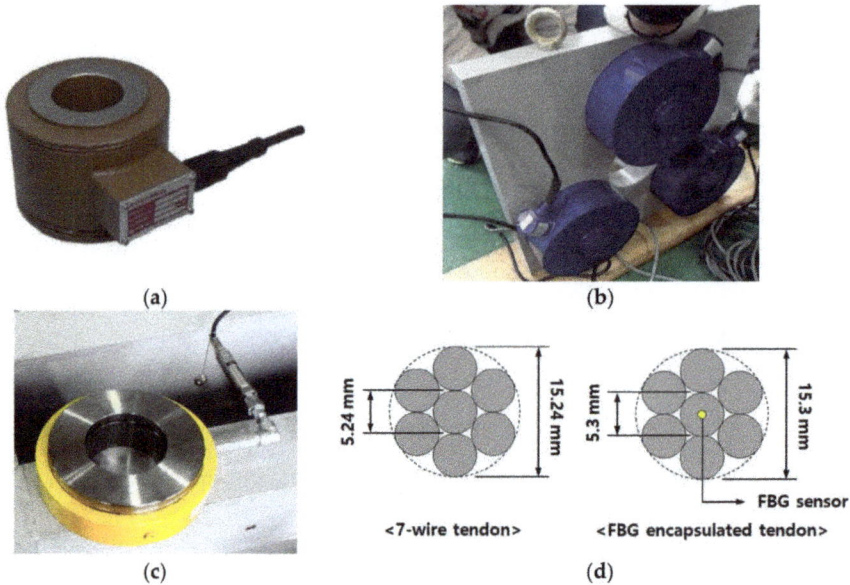

Figure 1. Sensors tested in this study. (**a**) Center-hole load cell; (**b**) Tri-sensor loading plate; (**c**) Hydraulic load cell; (**d**) Fiber Bragg Grating (FBG) sensor encapsulated seven-wire steel tendon.

2.1.1. Center-Hole Load Cell

Center-hole load cells have been used widely to measure the pre-stress force directly. Because the sensor has a hole in its center, it is optimal for measuring the pre-stress of the tendon at the anchorage. However, it has been reported that sometimes the measured value is not as consistent or as precise as given in the specifications when an eccentric load is applied [25]. Therefore, this center-hole load cell needed verification as to whether it is appropriate for the measurement of pre-stress in a PSC girder. In this study, the effect of eccentricity on typical center-hole load cells was tested and investigated. The result indicates that such eccentricity leads to error in the measurements.

2.1.2. Tri-Sensor Loading Plate

Because a center-hole load cell has high sensitivity to eccentricity, a new type of load measuring device was developed and tested. Three load cells were attached to a loading plate with a hole. The applied load on the plate could be measured by summing the loads of the three load cells. This loading plate was designed to minimize the effect of eccentricity based on mechanical theory. Even though there is eccentricity, the applied load can be measured precisely due to the statically determined structural characteristic of the loading plate. Figure 1b shows a schematic of the proposed loading plate with the load cells. The proposed tri-sensor loading plate was verified by experiment.

2.1.3. Hydraulic Load Cell

Hydraulic load cells are force-balance devices that measure load as a change in the pressure of the internal filling fluid. The load, when applied to the surface area of the piston, causes a pressure increase in the hydraulic fluid. This pressure is transferred to the attached pressure transducer for measurement. Because the load is measured through hydrostatic pressure, it can be inferred that the measurements are unaffected by the eccentricity. Thus, a hydraulic load cell was tested as a candidate for the measurement of pre-stress with eccentricity.

2.1.4. FBG Sensor

A Fiber Bragg Grating (FBG) sensor is a type of distributed Bragg reflector constructed in a short segment of optical fiber that reflects particular wavelengths of light and transmits all others. This is achieved by creating a periodic variation in the refractive index of the fiber core, which generates a wavelength-specific dielectric mirror. An FBG can therefore be used as an inline optical filter to block certain wavelengths, or as a wavelength-specific reflector. Based on these principles, an FBG sensor can be used to measure a strain profile of a tendon. In this study, an FBG sensor was tested for possible application to pre-stress measurement. In order to measure the strain of a tendon, a special tendon with an embedded FBG sensor was made in the lab [23].

2.2. Test Method and Equipment

The sensors listed in Section 2.1 measure load using different principles and mechanisms. To compare the various types of sensors for the purpose of measuring pre-stress, the sensors should be calibrated on the same basis. Therefore, all the sensors used in the experiment were calibrated and compared in the same test machine. The piece of equipment used for the calibration was a fatigue material testing machine manufactured by MTS. The sensors were tested and calibrated with the load cell built into the MTS testing machine. The load cells were calibrated based on compressive load while tensile force was applied on the FBG sensors.

In order to measure all of the sensors at once with the same loading condition, a small-scale girder was prepared. Instead of using a PSC girder that has friction between tendons and sheath, a steel girder was made with two I-shaped steel girders connected together, as shown in Figure 2. The girder had a large interior space so that no friction would occur when the tendons were installed inside the girder. As shown in Figure 2, the size of the model was 560 × 400 × 5000 mm and it had two anchor plates, each with a circular hole in which to place the two ends of the pre-stressing tendon. At the ends of the girder, the sensors needed for verification were installed, as shown in Figure 3. All of the sensors were calibrated in advance to improve the credibility of the results.

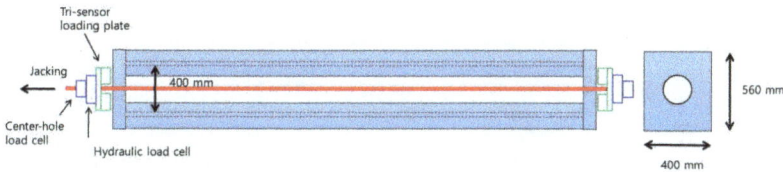

Figure 2. Steel frame of a girder for validation test (not to scale, stiffeners not shown).

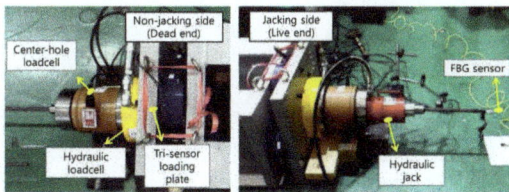

Figure 3. Installation of sensors at the ends of the steel frame.

After the test setup was completed, one pre-stressing tendon was installed. Then, the pre-stress was applied to the tendon with a hydraulic jack until the stress reached about 70% of maximum strength. The maximum load applied to a single tendon was about 160 kN.

The load was increased from zero to a target value in stages. During this loading process, the sensors installed on the girder were monitored. The sensors based on the strain gauge principle were monitored continuously during the entire loading process using a National Instruments (NI) data acquisition system. Also, the strain was calculated from the wavelength change of the FBG sensor measured using an interrogator. The temperature of the specimens was recorded and used for the consideration of the temperature effect of FBG sensors. The coefficients and constants for the FBG sensors were the same as those of Kim et al. [24].

2.3. Test Results

2.3.1. General Behavior

During the pre-stressing process, the load applied to the specimen was monitored using a center-hole load cell, a tri-sensor loading plate, and a hydraulic load cell. Among these three load cells, the measurement from the tri-sensor loading plate was selected as the reference load. This is because the mechanical behavior of this load cell is determinate and stable so that the eccentricity of the load cannot influence its measurement of the total pre-stress. During the entire test, the reference load showed values consistent with the pressure of the hydraulic jack.

Figure 4 shows a result for one tendon. In the figure, the displacement means the movement of the hydraulic pre-stressing jack. Figure 4a shows the results measured at the jacking side (live end), while Figure 4b shows the results at the opposite side (dead end). The results show that the measurements from the dead end had relatively lower variation than those from the live end. The figures show that the hydraulic load cell underestimated the pre-stress, while the center-hole load cell overestimated it.

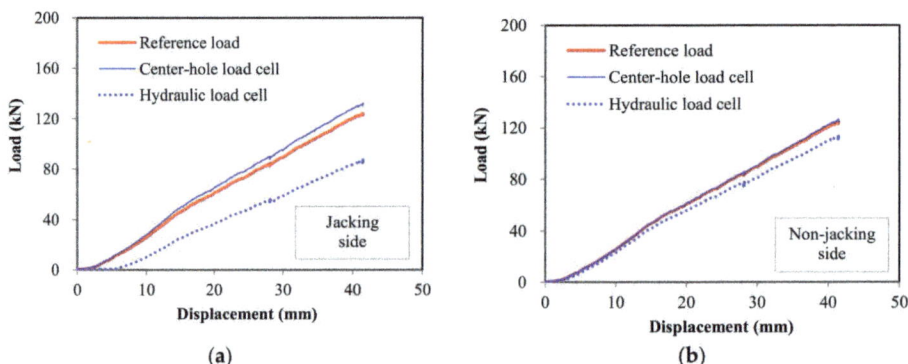

Figure 4. Test results of one tendon in the steel frame. (**a**) Jacking side; (**b**) Non-jacking side.

The R-square values were above 0.997 when linear regressions were conducted for each measurement. However, the slope of the fitted line varied with the load cell used. This means that the data from all of the tested load cells can be considered linear, but the accuracy varies with the type of load cell.

2.3.2. Center-Hole Load Cell

When the center-hole load cells were calibrated for a centric load, the calibration results were sufficient to provide the linearity and repeatability specified by the manufacturer. However, the results shown in Figure 4 show different patterns. The linearity of the results was sufficient, but the measurements were overestimated, as shown in Figure 5. This trend coincides with previous reports [25] that the measured value is not consistent with that given in the specifications when an eccentric load is applied to a center-hole load cell.

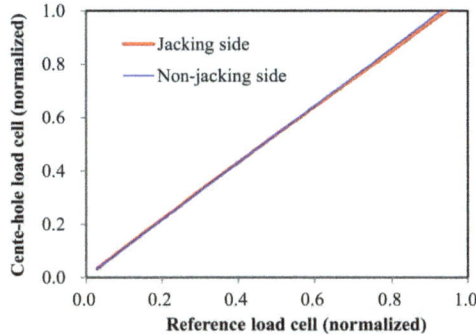

Figure 5. Relationship of the measurements between reference and center-hole load cells.

Therefore, in order to investigate this characteristic intensively, the influence of eccentricity on the measurement was tested. First, the load was applied to the center-hole load cell without eccentricity. Then, the load was applied to the same load cell with various eccentric configurations. The directions of loading point varied from the center of the load cell, while the distance from the central point was maintained as shown in Figure 6.

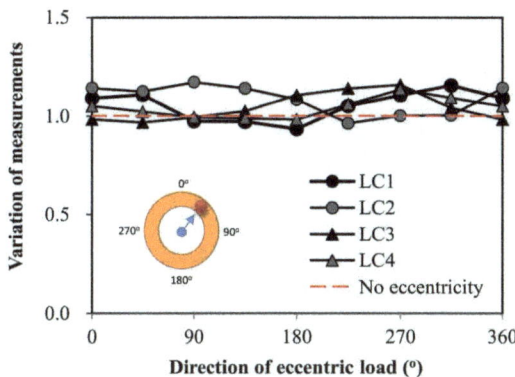

Figure 6. Variations of measurements in relation to the direction of eccentric loads.

Four load cells were tested and the test results are shown in Figure 6. The variations are shown as the ratios between the measurement from the load cell and the applied load on the load cell. It needs to be noted that the loads applied by the MTS test machine were almost the same values.

For a centric load, the ratio is equal to '1'. However, Figure 6 indicates that the eccentric measurement varied as the location of the load changed, even though the same load was applied. Most of the eccentric measurements were larger than the applied load, while some were smaller. This can be interpreted to mean that the measurements from the center-hole load cell were overestimated or underestimated when an eccentric load was applied. This trend can be considered not limited to this product, but a typical characteristic of a center-hole load cell. In the center-hole load cell, the load is supposed to be transferred through the whole body. However, the load is typically evaluated using only a few strain gauges attached to the body, so errors can be induced when the strain on the body is not uniform. Therefore, special attention needs to be paid when a prestress is measured using a center-hole-type load cell.

In brief, the center-hole load cell is the most commonly used transducer for the measurement of pre-stress. However, it generates errors when an eccentric load is applied. The magnitude of eccentricity and error did not show any particular tendency, but in most cases, the value indicated by the center-hole load cell was larger than the reference load. Therefore, careful attention should be paid when using a center-hole load cell.

2.3.3. Hydraulic Load Cell

As done for the center-hole load cell, the hydraulic load cells were calibrated first for a centric load, and then tested for validation. The calibration result showed linearity and repeatability for a centric loading condition. However, Figure 4 shows that the measured value from the hydraulic load cell was different from the expected value.

Figure 7 shows the characteristics of the hydraulic load cell. It can be observed that the loads measured from either the jacking or non-jacking side were less than the reference load. In addition, these differences were much larger on the jacking side. In pre-stressed concrete structures, the eccentricity of the pre-stress cannot be avoided due to the misalignment of the hydraulic jack, anchor head and steel wire, and due to variation of each tendon's stress [12]. In addition, this eccentricity is much larger on the jacking side than on the non-jacking side. Thus, it can be concluded that the eccentricity of the applied load influenced the characteristics of the hydraulic load cell due to the inherent structural characteristics of the sensor.

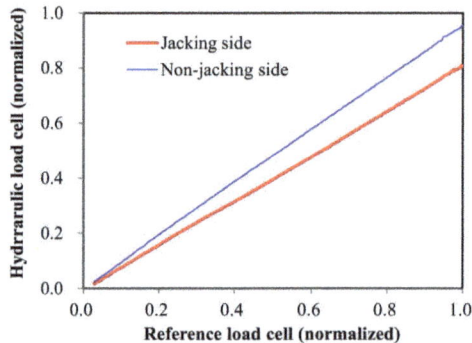

Figure 7. Relationship of the measurements between reference and hydraulic load cells.

A pressure-loaded load cell has clearance for its piston to move, and there are interior oil rings to prevent the leakage of oil. Therefore, when an eccentric load is applied, it could be expected that the hydraulic piston would rotate a little within its clearance, thereby causing loss due to the frictional force between the piston and the outer wall. In particular, the pressure-type load cell manufactured in this study was considered to be more vulnerable to eccentricity, because the axial length of the piston was short (low height) so that the probability that rotational deformation would occur was high. Thus, it can be concluded that pressure-type load cells without sufficient height to prevent rotation of the cylinder are not suitable for the measurement of pre-stressing force.

2.3.4. Tri-Sensor Loading Plate

The tri-sensor loading plates were calibrated and tested. Three individual compressive load cells attached to the tri-sensor loading plate were calibrated first. Then, the tri-sensor loading plates were assembled and tested for validation. The results show that the measures from each load cell varied in relation to the eccentricity. As the eccentricity increased, the standard deviations of the measurements from each load cell increased. However, the summation of the measurements did not change regardless of the eccentricity. Thus, it was deduced that the tri-sensor loading plate could provide a reference load not influenced by eccentricity.

2.3.5. FBG Sensor

Figure 8 shows the results of the FBG sensor measurement. In this case, the strain is calculated from changes in the wavelength from a sensor. The result shows that the strain increased in relation to the increase of the tensile force of the pre-stressing tendons. The results prove that the strain increased proportionally as the applied pre-stress increased. Thus, it can be concluded that the applied load can be estimated from the measured strain using the relationship between the strains measured and the forces applied. The greatest advantage of this FBG sensor is that the strain at any location inside the tendon can be measured, while the other load cells can only measure the stress at the end of a PSC beam.

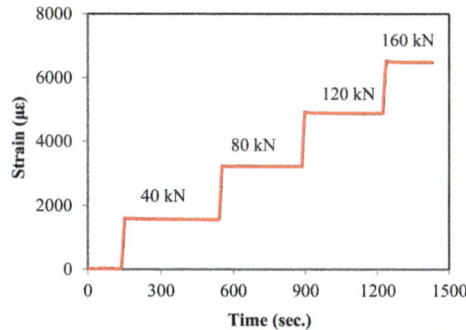

Figure 8. Strain measured by FBG sensor.

3. Construction Process and Damage Monitoring of PSC Girders

3.1. Experimental Program with PSC Girders

In order to evaluate the feasibility of monitoring the pre-stress during the construction stage and monitoring the damage to the PSC girder during a service stage, a structural beam member intended to represent the behavior of a typical PSC girder was designed and fabricated [28]. For monitoring purposes, load cells and FBG sensors were installed as shown in Figure 9.

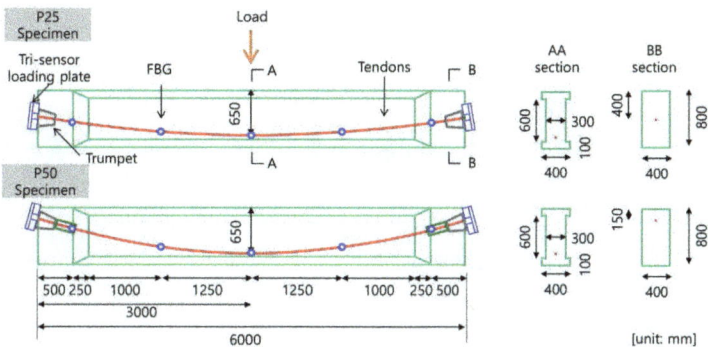

Figure 9. Details about the post-tensioned girder specimens and sensors installed. P25, eccentricity of tendon profile set to be 250 mm; P50, eccentricity of tendon profile set to be 500 mm.

Tri-sensor loading plates were installed at both (live and dead) ends of the beam in order to measure the pre-stressing forces. The live anchor refers to the end where the pre-stressing is applied, and the dead anchor means the opposite end where pre-stress is not applied. Because the pre-stress inside the beam cannot be measured using typical load cells, a smart tendon with an FBG sensor was installed to monitor the strain distribution of the tendon.

Two post-tensioned girders were fabricated. The dimensions of the girders (height and length) were 0.8 and 6 m, respectively. The web thickness was 300 mm. The test variable was the tendon profile: the eccentricity of the tendon profile was set to be 250 mm (P25) or 500 mm (P50). Each specimen contained three conventional 15.2 mm seven-wire strands and one smart tendon, in which the FBG sensor was embedded. Reinforcing bars of 1.28% longitudinal reinforcement ratio and 0.23% shear reinforcement ratio were embedded. The compressive strength determined in a loading test was 45 MPa. To measure tendon strains during the test, five measuring points were set along the FBG sensors in each specimen. Figure 9 shows details about the dimensions of the specimens, tendon profiles, and locations of the FBG sensors.

The specimens were made using the same procedures as those for constructing the PSC girders at the site. First, the assembly of reinforcing bars and sheaths for the entire specimens was completed. Concrete was cast and cured. Then, the seven-wire strands were tensioned using a hydraulic jack and a pump. The pre-stress was applied to the live anchor. During these construction procedures, the pre-stress was monitored in real time using the installed sensors.

After the specimens were stored for four months, the specimens were loaded using a 3-point loading method with a hydraulic actuator of which the capacity was 3000 kN, as presented inFigure 9. The pure span was set to be 4.5 m to observe any difference in tendon strain between the damaged zone within the pure span and the undamaged zone beyond the pure span. During the test, the tendon strain was measured throughout using the embedded FBG sensors, and the pre-stress force was measured through the load cells.

3.2. Test Results: Construction Stage

3.2.1. Pre-Stressing Stage

During the construction process, pre-stress was applied to the tendon and transferred to the concrete beam. The pre-stress at the live anchor, pre-stress at the dead anchor, and elongation of the tendon were measured during the jacking process, as shown in Figure 10. As seen in Figure 10, the pre-stress at the dead end and elongation increased with an increase in the pre-stress applied to the live end. There were regions where the measured values were constant, because the pre-stressing was applied in several steps.

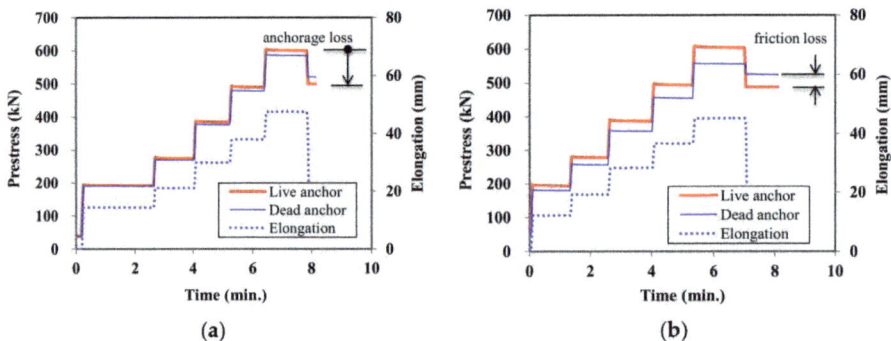

Figure 10. Pre-stress and elongation measured at the anchorage. (**a**) P25 specimen; (**b**) P50 specimen.

The pre-stress measured at the live end coincided well with the applied pressure measured at the hydraulic jack. The measured load at the dead end was smaller than that at the live end due to friction losses that occurred in the tendon. The gap between loads at the live and dead anchors varies with respect to the test specimens. As the curvature of the tendon profile increased, the difference increased too. From Figure 10, it was calculated that the measured friction loss through the tendon was 4.2% and 7.7% for the P25 and P50 specimens, respectively.

After applying pre-stress to the tendon, the pre-stress was released and transferred to the anchorage and concrete. During this process, a pre-stress loss occurred due to the anchorage-seating. From the measurements, it was observed that the anchorage-seating loss of live anchor was 17% and 19% for the P25 and P50 specimens, respectively.

From the load cell measurements, it was shown that the pre-stressing force at both ends of the beam can be measured so that the construction process and quality of a PSC girder can be monitored successfully. The pre-stress losses, which occur during the construction process, can be evaluated from exterior measurements. However, the stress or strain inside the beam cannot be measured directly using the load cells.

Figure 11 shows the measurements from the FBG sensors during the process of jacking the tendon. Figure 11a shows that the strains along the tendon increase as the applied load increases, as expected. However, Figure 11b shows that the strain near the live anchor decreased in the last stage due to unexpected slippage of the tendon. This kind of small change cannot be monitored using only a load cell. This figure proves that the strain profile of a tendon located inside a concrete beam can be monitored successfully using FBG sensors. Even local changes of strain can be monitored.

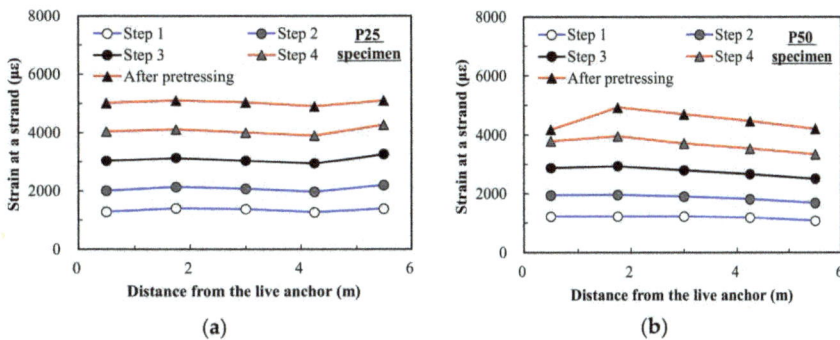

Figure 11. Strain measured by embedded FBG sensors. (**a**) P25 specimen; (**b**) P50 specimen.

Figures 10 and 11 prove that the load cells are able to measure the pre-stress of a tendon at the anchorage, and that FBG sensors can measure the strain at any location of the tendon where FBG sensors are embedded. The pre-stress of tendons during prestressing stage can be monitored successfully with these load cells and FBG sensors.

3.2.2. Yard Storage Stage and Pre-Stress Losses

After the beams were fabricated and pre-stressed, they were stored for four months. After this, the beams were tested for structural performance and damage monitoring. After pre-stressing was applied to the tendon, the pre-stress on the tendon varied with time due to several causes, such as elastic shortening, creep, and shrinkage of the concrete. Calculation of the effective pre-stress that remains in a tendon plays a major part in the design stage of PSC beams, because the amount of applied pre-stress is a key factor governing the structural performance of a PSC beam. The pre-stress losses can be evaluated using well-known theories or design codes [1–3]. However, it is not easy to know the actual pre-stress applied to the tendon in the construction or in-service stages.

Table 1 shows the summarized pre-stresses that were monitored using the embedded load cells and FBG sensors. In the table, LL and LD represent the load cell at the live anchor and dead anchor, respectively. This table shows that the actual pre-stress on the tendon can be monitored rationally with embedded sensing technology. The calculated losses can be used to prove (or improve) the design process and construction quality of PSC beams.

Table 1. Results from monitoring pre-stress and corresponding pre-stress loss.

P25 Specimen	LL (kN)	FBG1 (με)	FBG2 (με)	FBG3 (με)	FBG4 (με)	FBG5 (με)	LD (kN)
Initial (A)	0	0	0	0	0	0	0
Jacking (B)	600	5009	5091	5030	4894	5092	585
After anchoring (C)	500	4153	4292	4342	4255	4555	520
Before loading test (D)	467	3682	3532	3824	3753	3645	479
Jacking loss (1-B/C)	16.7	17.1	15.7	13.7	13.0	10.6	11.1
Long term loss (1-D/C)	6.6	11.3	17.7	11.9	11.8	20.0	7.9
P50 Specimen	**LL (kN)**	**FBG1 (με)**	**FBG2 (με)**	**FBG3 (με)**	**FBG4 (με)**	**FBG5 (με)**	**LD (kN)**
Initial (A)	0	0	0	0	0	0	0
Jacking (B)	603	4172	4919	4687	4459	4195	555
After anchoring (C)	487	3038	3900	3933	4153	4000	524
Before loading test (D)	468	2790	3348	3517	4000	3253	464
Jacking loss (1-B/C)	19.2	27.2	20.7	16.1	6.9	4.7	5.6
Long term loss (1-D/C)	3.9	8.2	14.2	10.6	3.7	18.7	11.5

LL: load cell at live anchor; LD: load cell at dead anchor.

3.3. Test Results: Loading Stage

3.3.1. General Behavior

As the applied load increased, the deflection increased, as shown in Figure 12. The crack patterns were recorded and checked at various loading steps until the specimens failed. During the test, flexural cracks occurred at the bottom of the center, and then flexural-shear cracks were induced as the applied load increased. Subsequently, web-shear cracks were formed. As a result, all of the specimens showed shear failures with a dominant diagonal crack. Detailed crack patterns of the specimens can be found in the study of Lee et al. [28].

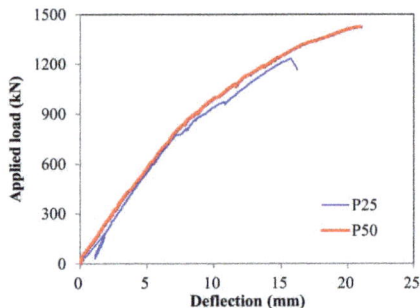

Figure 12. Load-center deflection response.

3.3.2. Test Results: Pre-Stress Forces and Damage

During the loading process, the pre-stress of the tendon was measured through the load cells. However, the pre-stress forces measured at the ends of the beam changed very little; the changes measured were less than 1% of the initial values.

Generally, the tendons in a PSC girder are grouted for several reasons, such as protection against corrosion and improvement of structural integrity. After tendons are grouted and bonded, the tendon and concrete beam behave as a single structure, so that there is no increment of strain at the ends of a beam where external force has no influence.

Thus, it can be concluded that load cells, which can measure the pre-stress only at the ends of a beam, are not appropriate for evaluation of damage to PSC girders, because no noticeable differences were measured using such sensors.

3.3.3. Test Results: Tendon Strain and Damage

The tendon strain measured through the FBG sensors during the test is presented over time in Figure 13. At the beginning of the loading test, the tendon strain was around 0.003–0.004, which decreased by about 0.0007 from the value measured just after post-tensioning. This was due to the effect of concrete creep and shrinkage. As the load was applied at the center of the girders, the tendon strains at FBG1 and FBG5 (located near the anchors) showed very little variation, even under failure. This indicates that damage caused by flexural cracks in PSC girders cannot be detected with sensors embedded at the anchors.

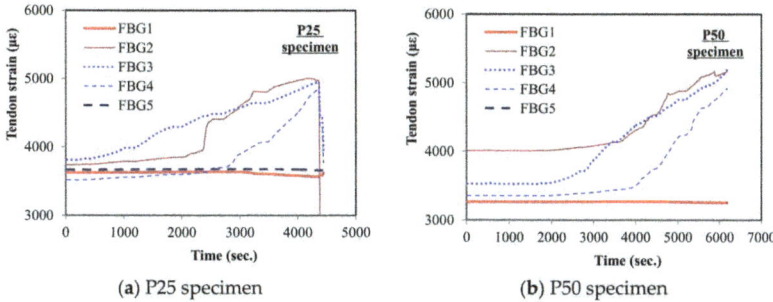

(a) P25 specimen (b) P50 specimen

Figure 13. Tendon strain of specimens in relation to time.

On the other hand, tendon strains at the sensing points within the pure span exhibited significant increase as the applied load increased and cracks occurred. Because flexural cracks were formed at the bottom of the center, the tendon strain at FBG3 (near the first flexural crack) started increasing first while the others remained almost constant. Then, the tendon strains at FBG2 and FBG4 increased later as the number of cracks increased and the damage increased. This indicates that damage due to flexural cracks in PSC girders can be detected by the strain increment monitored by FBG sensors near cracks. In addition, it can be inferred that FBG sensors installed in appropriate regions would make it possible to evaluate how much a PSC girder is damaged.

4. Conclusions

Pre-stressing force has not been managed after construction, even though it is a very important factor for maintaining the structural safety of PSC girder bridges. Usually, the pre-stressing force is measured only during construction using a jacking device, and after that it cannot be managed practically. For this reason, in this study, the measurement of pre-stress was investigated using embedded sensors that are currently available, with the goal of proposing 'smart' pre-stressed concrete girders that could monitor the pre-stress of the tendon and damage of a beam from its birth to the end of its life.

Four types of sensors were installed on a small steel frame, and the applicability and accuracy of those sensors were tested while pre-stress was applied to the frame. The results show that the hydraulic load cells have a tendency to underestimate the pre-stress when it is loaded with eccentricity. The center-hole load cell shows irregular error in the presence of the eccentricity. The tri-sensor loading plate and FBG sensors measure the pre-stress successfully.

In order to investigate the feasibility of monitoring the pre-stress of tendons in construction and service stages, 6 m-long post-tensioned PSC girders were fabricated. During the construction process, the pre-stress applied to the tendons was monitored. The results prove that the pre-stressing history and construction quality can be monitored precisely using embedded sensors. Pre-stress losses can be evaluated using the monitored results.

Through a 3-point loading test for the PSC girders, tendon stress and strain were measured by embedded sensors. The measurements prove that the stress and strain of tendons near the anchors change only a little, even after failure. However, it was observed that the FBG sensors at the middle section showed significant increases due to flexural deformation and cracks within the pure span. Consequently, it can be concluded that damage due to flexural deformation and cracks in PSC girders can be detected by FBG sensors at the location where the damage occurs.

Acknowledgments: This research was supported by a grant (16CTAP-C078425-03) from Technology Advancement Research Program (TARP) funded by Ministry of Land, Infrastructure and Transport of Korean government.

Author Contributions: Hwanwoo Lee supervised the whole work; Kyung-Joon Shin, Seong-Cheol Lee, Yun Yong Kim, Jae-Min Kim, and Seunghee Park were principal investigators.

Conflicts of Interest: The authors declare no conflict of interest.

References

1. Nilson, A.H. *Design of Prestressed Concrete*; Wiley: Hoboken, NJ, USA, 1987.
2. American Concrete Institute (ACI) Committee 318. *Building Code Requirements for Structural Concrete (ACI 318–14)*; American Concrete Institute (ACI): Farmington Hills, MI, USA, 2014.
3. Euro-International Committee for Concrete (CEB). *CEB-FIP Model Code*; Thomas Telford Services Ltd.: Lausanne, Switzerland, 1993.
4. Barr, P.J.; Kukay, B.M.; Halling, M.W. Comparison of prestress losses for a prestress concrete bridge made with high-performance concrete. *J. Bridge Eng.* **2008**, *13*, 468–475. [CrossRef]
5. Jain, S.K.; Goel, S.C. Discussion of "prestress force effect on vibration frequency of concrete bridges". *J. Struct. Eng.* **1996**, *122*, 458–460. [CrossRef]
6. Joh, C.; Lee, J.W.; Kwahk, I. Feasibility study of stress measurement in prestressing tendons using villari effect and induced magnetic field. *Int. J. Distrib. Sens. Netw.* **2013**, *9*. [CrossRef]
7. Chaki, S.; Bourse, G. Guided ultrasonic waves for nondestructive monitoring of the stress levels in prestressed steel strands. *Ultrasonics* **2009**, *49*, 162–171. [CrossRef] [PubMed]
8. Salamone, S.; Bartoli, I.; Phillips, R.; Nucera, C.; Scalea, F.L. Health monitoring of prestressing tendons in posttensioned concrete bridges. *J. Transp. Res. Board* **2011**, *2220*, 21–27. [CrossRef]
9. Wang, M.L.; Chen, Z.L.; Koontz, S.S. Magnetoelastic method of stress monitoring in steel tendons and cables. In Proceedings of the SPIE—Nondestructive Evaluation of Highways, Utilities, and Pipelines IV, Newport Beach, CA, USA, 7–9 March 2000; Volume 395, pp. 492–500.
10. Zhao, Y.; Wang, M.L. Non-destructive condition evaluation of stress in steel cable using magnetoelastic technology. In Proceedings of the SPIE—The International Society for Optical Engineering, San Diego, CA, USA, 27 February–2 March 2006; Volume 6178, pp. 1–7.
11. Sumitro, S.; Jarošević, A.; Wang, M.L. Elasto-magnetic sensor utilization on steel cable stress measurement. In Proceedings of the 1st Fib Congress: Concrete Structures in the 21th Century, Osaka, Japan, 13–19 October 2002.
12. Cho, K.; Park, S.Y.; Cho, J.-R.; Kim, S.T.; Park, Y.-H. Estimation of prestress force distribution in themulti-strand system of prestressed concrete structures. *Sensors* **2015**, *15*, 14079–14092. [CrossRef] [PubMed]
13. Li, H.-N.; Li, D.-S.; Song, G.-B. Recent applications of fiber optic sensors to health monitoring in civil engineering. *Eng. Struct.* **2004**, *26*, 1647–1657. [CrossRef]
14. Ye, X.; Su, Y.; Han, J. Structural health monitoring of civil infrastructure using optical fiber sensing technology: A comprehensive review. *Sci. World J.* **2014**, *2014*. [CrossRef] [PubMed]
15. Bao, X.; Chen, L. Recent progress in Brillouin scattering based fiber sensors. *Sensors* **2011**, *11*, 4152–4187. [CrossRef] [PubMed]
16. Maaskant, R.; Alavie, T.; Measures, R.; Tadros, G.; Rizkalla, S.; Guha-Thakurta, A. Fiber-optic Bragg grating sensors for bridge monitoring. *Cem. Concr. Compos.* **1997**, *19*, 21–33. [CrossRef]
17. Moyo, P.; Brownjohn, J.; Suresh, R.; Tjin, S. Development of fiber Bragg grating sensors for monitoring civil infrastructure. *Eng. Struct.* **2005**, *27*, 1828–1834. [CrossRef]

18. Majumder, M.; Gangopadhyay, T.K.; Chakraborty, A.K.; Dasgupta, K.; Bhattacharya, D.K. Fibre Bragg gratings in structural health monitoring—Present status and applications. *Sens. Actuators A Phys.* **2008**, *147*, 150–164. [CrossRef]

19. Zhou, Z.; Graver, T.W.; Hsu, L.; Ou, J. Techniques of advanced FBG sensors: Fabrication, demodulation, encapsulation and their application in the structural health monitoring of bridges. *Pac. Sci. Rev.* **2003**, *5*, 116–121.

20. Kim, S.T.; Park, Y.; Park, S.Y.; Cho, K.; Cho, J.-R. A sensor-type PC strand with an embedded FBG sensor for monitoring prestress forces. *Sensors* **2015**, *15*, 1060–1070. [CrossRef] [PubMed]

21. Zhou, Z.; He, J.P.; Ou, J.P. Long-term monitoring of a civil defensive structure based on distributed Brillouin optical fiber sensor. *Pac. Sci. Rev.* **2008**, *8*, 1–6.

22. Lan, C.; Zhou, Z.; Ou, J. Full-scale prestress loss monitoring of damaged RC structures using distributed optical fiber sensing technology. *Sensors* **2012**, *12*, 5380–5394. [CrossRef] [PubMed]

23. Kim, J.M.; Kim, H.W.; Park, Y.H.; Yang, I.H.; Kim, Y.S. FBG sensors encapsulated into 7-wire steel strand for tension monitoring of a prestressing tendon. *Adv. Struct. Eng.* **2012**, *15*, 907–917. [CrossRef]

24. Kim, J.M.; Kim, C.M.; Choi, S.Y.; Lee, B.Y. Enhanced strain measurement range of an FBG sensor embedded in seven-wire steel strands. *Sensors* **2017**, *17*. [CrossRef] [PubMed]

25. Shin, K.J.; Park, Y.U.; Lee, S.C.; Kim, Y.Y.; Lee, H.W. Experimental evaluation of prestress force in tendons for prestressed concrete girders using sensors. *J. Comput. Struct. Eng. Inst. Korea* **2015**, *28*, 715–722. [CrossRef]

26. Kim, J.K.; Kim, J.M.; Lee, H.W.; Park, S.H.; Choi, S.Y.; Shin, K.J. ANN based tensile force estimation for pre-stressed tendons of PSC girders using FBG/EM hybrid sensing. *Struct. Control Health Monitor.* **2016**, in press.

27. Lee, S.C.; Choi, S.Y.; Shin, K.J.; Kim, J.M.; Lee, H.W. Measurement of transfer length for a seven-wire strand with FBG sensors. *J. Comput. Struct. Eng. Inst. Korea* **2015**, *28*, 707–714. [CrossRef]

28. Lee, S.C.; Shin, K.J.; Kim, J.M.; Lee, H.W. Damage detection with FBG sensors for pre-stress concrete girders. *Key Eng. Mater.* **2017**, *737*, 454–458. [CrossRef]

sensors

MDPI

Article

Development of Embedded EM Sensors for Estimating Tensile Forces of PSC Girder Bridges

Junkyeong Kim [1] [ID]**, Ju-Won Kim [2], Chaggil Lee [1] and Seunghee Park [2,***

[1] Department of Civil & Environmental System Engineering, Sungkyunkwan University 2066, Seobu-ro, Jangan-gu, Suwon-si, Gyonggi-do 16419, Korea; junk135@nate.com (J.K.); tolck81@gmail.com (C.L.)
[2] School of Civil & Architectural Engineering, Sungkyunkwan University 2066, Seobu-ro, Jangan-gu, Suwon-si, Gyonggi-do 16419, Korea; malsi@nate.com
*   Correspondence; shparkpc@skku.edu; Tel.: +82-031-290-7525

Received: 21 July 2017; Accepted: 28 August 2017; Published: 30 August 2017

Abstract: The tensile force of pre-stressed concrete (PSC) girders is the most important factor for managing the stability of PSC bridges. The tensile force is induced using pre-stressing (PS) tendons of a PSC girder. Because the PS tendons are located inside of the PSC girder, the tensile force cannot be measured after construction using conventional NDT (non-destructive testing) methods. To monitor the induced tensile force of a PSC girder, an embedded EM (elasto-magnetic) sensor was proposed in this study. The PS tendons are made of carbon steel, a ferromagnetic material. The magnetic properties of the ferromagnetic specimen are changed according to the induced magnetic field, temperature, and induced stress. Thus, the tensile force of PS tendons can be estimated by measuring their magnetic properties. The EM sensor can measure the magnetic properties of ferromagnetic materials in the form of a B (magnetic density)-H (magnetic force) loop. To measure the B-H loop of a PS tendon in a PSC girder, the EM sensor should be embedded into the PSC girder. The proposed embedded EM sensor can be embedded into a PSC girder as a sheath joint by designing screw threads to connect with the sheath. To confirm the proposed embedded EM sensors, the experimental study was performed using a down-scaled PSC girder model. Two specimens were constructed with embedded EM sensors, and three sensors were installed in each specimen. The embedded EM sensor could measure the B-H loop of PS tendons even if it was located inside concrete, and the area of the B-H loop was proportionally decreased according to the increase in tensile force. According to the results, the proposed method can be used to estimate the tensile force of unrevealed PS tendons.

Keywords: tensile force estimation; embedded EM sensor; PS Tendon; B-H loop measurement; PSC girder

1. Introduction

Recently, civil structures and their behaviors have become more complicated due to the development of materials, design, and construction technology. Also, the evaluation and maintenance of these structures have become very important. To inspect these structures, nondestructive tests (NDTs) have become a solution for evaluating structural health [1–7]. The main benefit of such non-destructive evaluation systems is that a structure does not need to be altered while being monitored. The condition of a structure can be assessed on site, while the information derived from a non-destructive evaluation can be instrumental in making engineering decisions concerning the fate of a structure. This ensures better judgment when determining whether or not a structure is safe, thus avoiding the construction, labor, and social costs of replacing a structure that actually does not require replacement. In addition, NDT techniques can be applied to new structures as part of a monitoring scheme, which leads to a better understanding of the behavior and performance during the construction and servicing of a structure.

Since the first post-tensioned concrete bridge was built in 1936, many PSC (pre-stressing concrete) bridges have been constructed globally [8]. However, after the sudden collapse of a number of post-tensioned concrete bridges, it was found that the post-tension system has long-term risks, such as corrosion of tendons caused by ingress of water and chloride ions into partially grouted ducts [9,10]. The tensile forces in the pre-stressing strands can vary due to a variety of losses including instantaneous losses such as elastic shortening, friction, and anchorage set occurring at the time of transfer of the pre-stressing force, as well as time-dependent losses due to steel relaxation and the concrete creep and shrinkage that occur after transfer of the pre-stressing force and during the life of the member. Accordingly, the measurement of the tensile force in a tendon becomes very important for long-term maintenance of such bridges, as well as for the purpose of design [11–15].

Various NDT methods have been studied to estimate tensile forces of tendons or cables. In addition, field measurements have been carried out by attaching sensors, such as a strain measuring gauges (Tensmeg), directly to the outside of the strand, or indirectly, by sensing the strain near the strand using an electrical strain gauge and vibrating wire strain gauge (VWSG) installed in the concrete or on a rebar near the duct [16]. More recently, various NDT methods for measuring pre-stressing forces have been studied using methods based on guided stress waves [17,18], a system identification technique based on modal parameters [19], an impedance method applied to an anchorage plate [20], and use of the in-strand encapsulated fiber Bragg grating (FBG) sensor [21,22].

Especially, a magnetic sensor is a reliable method for stress estimation of steel specimens, due to its outstanding superiorities including corrosion resistance, actual-stress measurement, nondestructive monitoring, and long service life [23]. The elasto-magnetic (EM) sensor is usually applied to measure the stress of ferromagnetic members and it consists of a primary excitation coil and a secondary induction coil. It can measure the permeability to estimate the stress and it has been used to monitor the stress in exposed steel cables on field more than ten years [24–26]. The many types of EM sensors were invested using coil type [27], yoke-shaped electromagnet with hall sensors [28], and coil type electromagnet with magneto-electric-laminated composite sensor [23].

However, the previous magnetic sensors should located closely to specimen and it cannot be applied actual PSC girder that the PS (pre-stressing) tendons did not expose to outside of girder. To overcome this limitation, this research proposed an embedded EM sensor that can embed into the PSC girder and measure the magnetic responses of internal PS tendons to improve field applicability. For this purpose, embedded EM sensors are developed for measuring the magnetic hysteresis loop variations in pre-stressing tendons and a test was carried out using a 6 m down-scaled PSC girder beam in order to verify the proposed method.

2. Development of the Embedded EM Sensor

This study applied an EM sensor to estimate the tensile force of PS tendons. The EM sensor can measure the induced stress of ferromagnetic specimen by the elasto-magnetic effect [29]. The EM sensor should be located inside a PSC girder to measure the magnetic responses of the PS tendons because the tendons are built into the PSC girder and could not approach after casting concrete. To apply an EM sensor to an actual PSC girder, an embedded EM sensor was developed. The embedded EM sensor consists of a cylindrical bobbin with screw threads on both ends to connect with the sheath, a primary coil to generate a magnetic field, a secondary coil to measure the magnetic response of the PS tendon, and an external cover to protect the coils, as shown in Figure 1. The ends of the primary and secondary coils are connected to coaxial cables (Figure 2a), and a PVC (poly vinyl chloride) pipe is used as the external cover (Figure 2b) to protect the EM sensor from impact and water from the concrete.

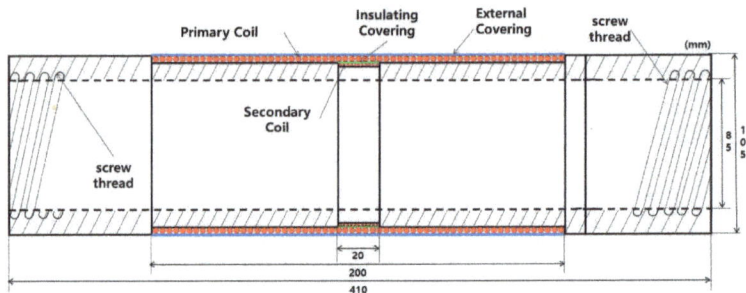

Figure 1. Schematic of the embedded EM sensor.

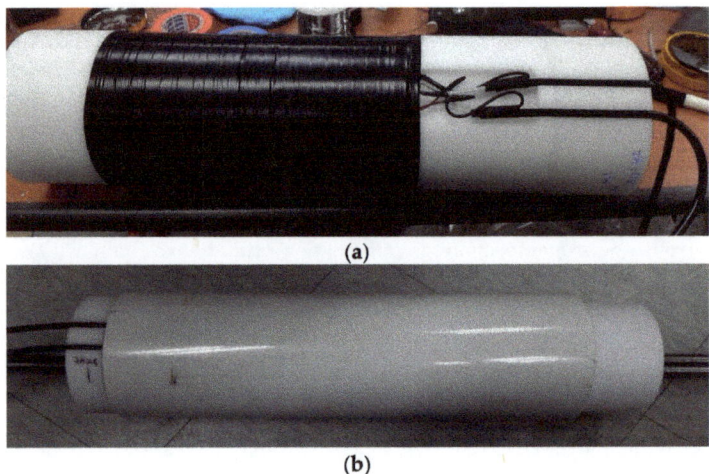

Figure 2. Embedded EM sensor. (**a**) Connection with coils and cable, (**b**) External protection cover.

The embedded EM sensor is installed between the sheaths as a sheath joint, as shown in Figure 3. Both sides of the EM sensor have screw threads, allowing the sensor to be installed at any position along the sheath line. Also, the joints are sealed using tape to prevent water permeation into the sheath.

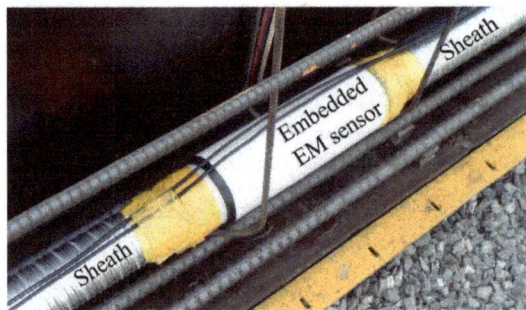

Figure 3. Installation of an embedded EM sensor.

3. Experimental Study

3.1. Experimental Setup and Test Procedure

To confirm the proposed tensile force monitoring method, two down-scaled PSC girder specimens were constructed as shown in Figure 4. The span of the girders was 6 m, and the height was 1 m. The girders had 1 sheath line in each specimen (specimen 1 had a straight sheath, and specimen 2 had a curved sheath), and the embedded EM sensors were installed at the left and right anchorage parts and the maximum eccentric part as shown in Figures 4 and 5.

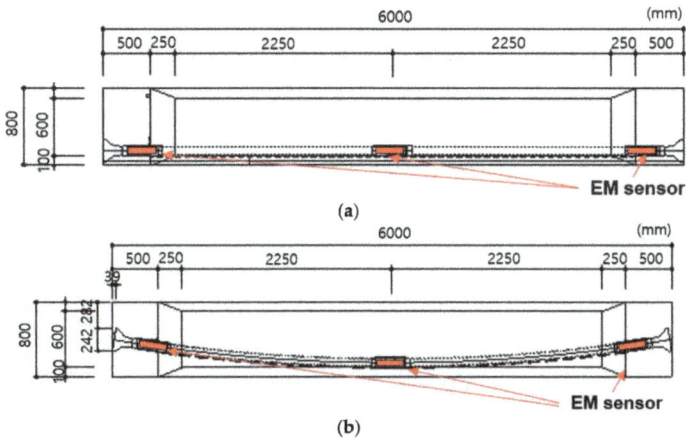

Figure 4. Schematic diagram of the test specimens. (**a**) Specimen 1: Straight sheath; (**b**) Specimen 2: Curved sheath.

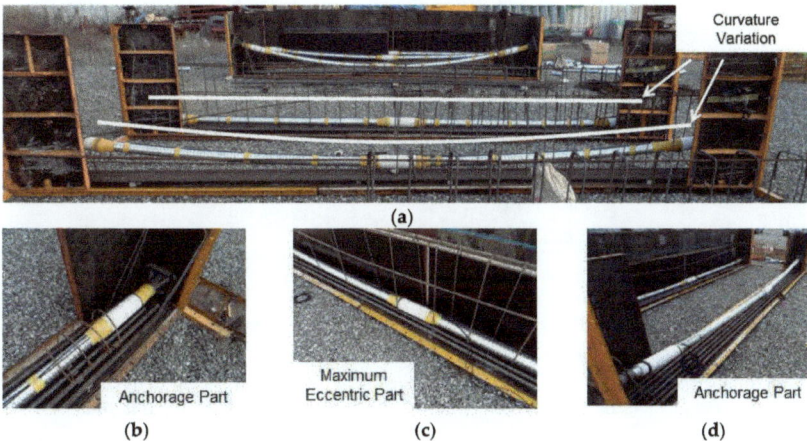

Figure 5. Installation of embedded EM sensors in a PSC girder. (**a**) EM sensor installed PSC girder, (**b**) Right anchorage part, (**c**) Maximum eccentric part, (**d**) Left anchorage part.

After installation of the EM sensors, the concrete was cast, and the PS tendons were arranged through the sheath lines. The nominal cross-section of each PS tendon was 158.8 mm, and four PS tendons were placed in each sheath line. To measure the actual tensile force, a three-point load cell was installed as shown in Figure 6.

Figure 6. Installation of a load cell.

The tensile force on the PS tendons was induced by a multi-tendon hydraulic jacking machine. The jacking step was divided into 6 steps, and the details are listed in Table 1.

Table 1. Jacking steps of specimens.

Jacking Step	Tensile Force (kN)	
	Specimen 1 (Straight Sheath)	Specimen 2 (Curved Sheath)
1	0	0
2	189	194
3	272	275
4	379	386
5	486	492
6	595	602

3.2. Results of EM Measurement

The B-H loops of PS tendons were measured every jacking step using the embedded EM sensors. Figures 7 and 8 show the results of B-H loop measurement of specimens 1 and 2.

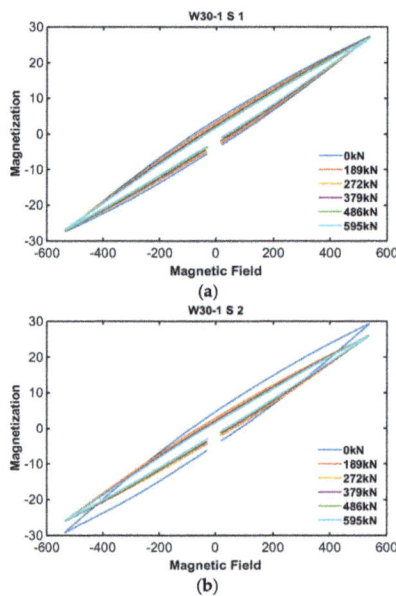

(a)

(b)

Figure 7. *Cont.*

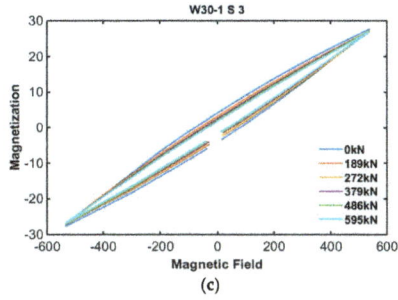

Figure 7. B-H loop of specimen 1: (**a**) Sensor no. 1-1; (**b**) Sensor no. 1-2; (**c**) Sensor no. 1-3.

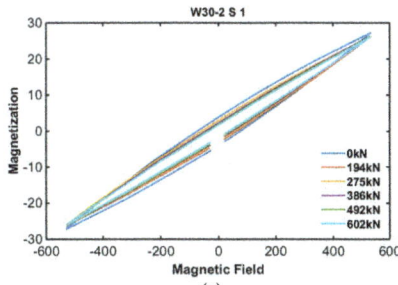

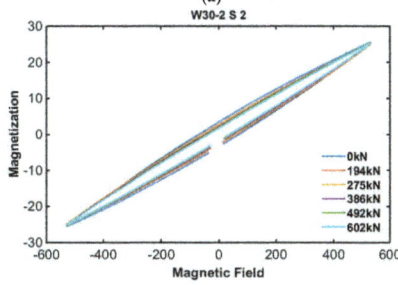

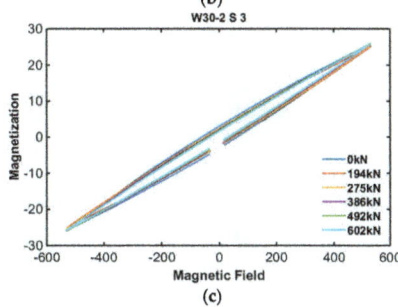

Figure 8. B-H loop of specimen 2: (**a**) Sensor no. 2-1; (**b**) Sensor no. 2-2; (**c**) Sensor no. 2-3.

According to the measurement results, proposed embedded EM sensor can measure the B-H loop of unrevealed PS tendon even if it located inside of concrete.

To quantify the B-H loop variation due to induced tensile force, the area of the B-H loop was extracted. Figure 9 shows the variations in area of the B-H loop at each measurement of specimen 1. The results confirm that the area of the B-H loop decreased with increase in tensile force.

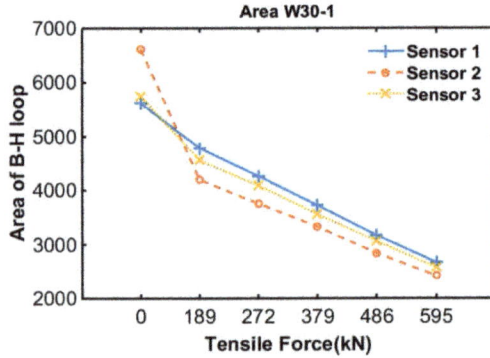

Figure 9. B-H loop of specimen 1.

Figure 10 shows the result of specimen 2. The results are similar to that of sheath no. 2; the area of the B-H loop decreased with increased tensile force.

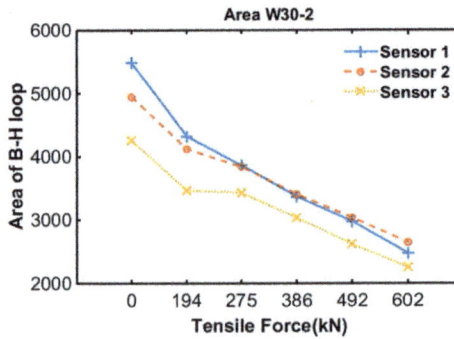

Figure 10. B-H loop of specimen 2.

To quantify the area variation due to tensile force, the area ratio was proposed as follows:

$$A_r = \frac{A_i - A_c}{A_i} \tag{1}$$

where A_i is the measured area of the B-H loop at a tensile force of 0 (step 1 of each specimen) and A_c is the measured area of the B-H loop at the current step.

Figure 11 shows the area ratio results of specimen 1. The area ratio increased with an increase in tensile force. The area ratio of Sensor no. 1-2 was larger than those of Sensor no. 1-1 and 1-3, indicating that the tensile force was concentrated at the center of the PS tendon. It caused by the sheath of specimen 1 was straight and there was no friction loss.

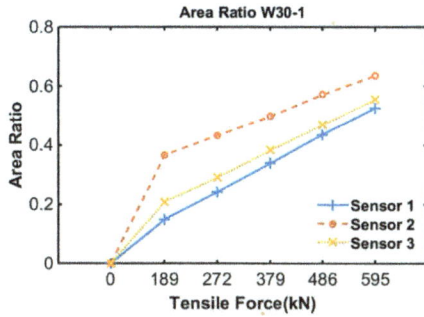

Figure 11. Area ratio of specimen 1.

Figure 12 shows the result of the area ratio calculation from the area of the B-H loop of specimen 2. The result shows a similar pattern to that of the result of specimen 1. However, friction loss was observed with the sensors because of the curved sheath of specimen 2.

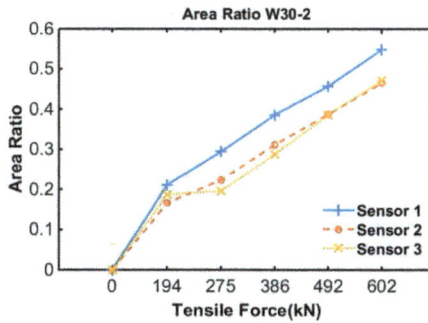

Figure 12. Area ratio of specimen 2.

Figure 13 shows the relationship between area ratio of Sensor no. 1 of each specimen and reference tensile force measured by the load cell at the left side of each specimen. As shown in the figure, the area ratio and tensile force had a linear relationship, and the tensile force of a PSC girder can be estimated using Equation (2).

$$Tensile\ Force(kN) = 1065 \times A_r \tag{2}$$

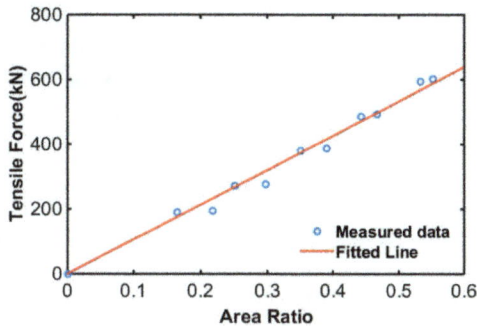

Figure 13. Relationship between area ratio and reference stress.

According to the results, the embedded EM sensors could estimate the tensile forces in PS tendons with acceptable error by tracking the area ratio variations. Furthermore, the friction losses and the tensile force distribution in the PS tendons could be monitored using the embedded EM sensor.

4. Conclusions

An embedded EM sensor based tensile force monitoring for PSC girders was proposed in this research. The B-H loop of ferromagnetic material is affected by the induced tensile force in the specimen. To measure the B-H loop of PS tendons in a PSC girder, an embedded EM sensor was developed as a sheath joint. To validate the proposed method, a down-scaled PSC girder test was performed using two PSC girder specimens that had 1 sheath line in each specimen. The embedded EM sensors were installed at the left and right anchorage parts and at the maximum eccentric part through the sheath before casting concrete. After casting and curing concrete, tensile force was induced in steps, and the B-H loop was measured at every tensile force step. Also, the reference tensile forces were measured from the load cell at the left anchorage of each specimen. The area of the B-H loop decreased with tensile force increase. To quantify the area variation, the area ratio was calculated, and the equation for estimating tensile force was derived by comparing the area ratio and reference tensile force. According to the results, the embedded EM sensor can measure the magnetic responses of unrevealed PS tendon even if it located inside of concrete and the tensile force could be estimated based on the area ratio variations of the PS tendon using the embedded EM sensors in the field environment.

Acknowledgments: This research was supported by a grant (14CTAP-C078424-01#) from the Infrastructure and Transportation Technology Promotion Research Program funded by the Ministry of Land, Infrastructure, and Transport of Korean government, the Development of Design and Construction Technology for Double Deck Tunnel in Great Depth Underground Space (14SCIP-B088624-01) from the Construction Technology Research Program funded by the Ministry of Land, Infrastructure and Transport of Korean government, a grant (2017-MPSS31-001) from the Supporting Technology Development Program for Disaster Management funded by the Ministry of Public Safety and Security (MPSS) of the Korean government, and the National Research Foundation of Korea (NRF) grant funded by the Korean government (MSIP) (No. NRF-2017R1A2B3007607, NRF-2017-R1D1A1B03033399, NRF-2014R1A2A1A11054299).

Author Contributions: Junkyeong Kim and Ju-Won Kim conceived and designed the experiments; Junkyeong Kim performed the experiments; Junkyeong Kim and Changgil Lee analyzed the data; Junkyeong Kim and Seunghee Park wrote the paper. In addition, Junkyeong Kim and Seunghee Park are responsible for the implementation the proposed scheme.

Conflicts of Interest: The authors declare no conflict of interest.

References

1. Rens, K.L.; Wipf, T.J.; Klaiber, F.W. Review of nondestructive evaluation techniques of civil infrastructure. *J. Perform. Const. Facil.* **1997**, *11*, 152–160. [CrossRef]
2. Kim, J.W.; Kim, J.; Park, S.; Oh, T.K. Integrating embedded piezoelectric sensors with continuous wavelet transforms for real-time concrete curing strength monitoring. *Struct. Infrastruct. E* **2015**, *11*, 897–903. [CrossRef]
3. Lee, C.; Kim, J.; Park, S.; Kim, D.H. Advanced Fatigue Crack Detection using Nonlinear Self-Sensing Impedance Technique for Automated NDE of Metallic Structures. *Res. Nondestruct. Eval.* **2015**, *26*, 107–121. [CrossRef]
4. Lee, C.; Park, S. Damage visualization of pipeline structures using laser-induced ultrasonic waves. *Struct. Health Monit.* **2015**, *14*, 475–488. [CrossRef]
5. Lee, C.; Park, S. Flaw Imaging Technique for Plate-Like Structures Using Scanning Laser Source Actuation. *Shock Vib.* **2014**, 725030. [CrossRef]
6. Lee, C.; Kang, D.; Park, S. Visualization of Fatigue Cracks at Structural Members using a Pulsed Laser Scanning System. *Res. Nondestruct. Eval.* **2015**, *26*, 123–132. [CrossRef]
7. Park, S.; Kim, J.W.; Lee, C.; Lee, J.J. Magnetic Flux Leakage Sensing-Based Steel Cable NDE Technique. *Shock Vib.* **2014**, 929341. [CrossRef]

8. Weiher, H.; Zilch, K. Condition of post-tensioned concrete bridges-assessment of the German stock by a spot survey of damages. In Proceedings of the First International Conference on Advances in Bridge Engineering, London, UK, 26–28 June 2006; Brunel University: London, UK, 2006; pp. 26–28.

9. Youn, S.G.; Kim, E.K. Deterioration of bonded post-tensioned concrete bridges and research topics on the strength evaluation in ISARC. In Proceedings of the JSCE-KSCE Joint Seminar on Maintenance and Management Strategy of Infrastructure in Japan and Korea, Shiga, Japan, 20 September 2006; Sakai, K., Ed.; JSCE: Tokyo, Japan, 2006; pp. 49–63.

10. Bruce, S.M.; McCarten, P.S.; Freitag, S.A.; Hasson, L.M. Deterioration of prestressed concrete bridge beams. In *Land Transport New Zealand Research Report*; Land Transport New Zealand: Wellington, New Zealand, 2008; Volume 337, p. 72.

11. Shenoy, C.V.; Frantz, G.C. Structural tests of 27-year-old prestressed concrete bridge beams. *PCI J.* **1991**, *36*, 80–90. [CrossRef]

12. Aalami, B.O. Time-dependent analysis of post-tensioned concrete structures. *Prog. Struct. Eng. Mater.* **1998**, *1*, 384–391. [CrossRef]

13. Pantelides, C.P.; Saxey, B.W.; Reaveley, L.D. Posttensioned tendon losses in a spliced-girder bridge, Part 1: Field measurements. *PCI J.* **2007**, *52*, 1–15.

14. Lakshmanan, N.; Saibabu, S.; Murthy, A.R.C.; Ganapathi, S.C.; Jayaraman, R. Experimental, numerical and analytical studies on a novel external prestressing technique for concrete structural components. *Comput. Concr.* **2009**, *6*, 41–57. [CrossRef]

15. Im, S.; Hurlebaus, S.; Trejo, D. Inspections of voids in external post-tensioned tendons. In *Transportation Research Board*; Business Office: Washington, DC, USA, 2010.

16. Onyemelukwe, O.; Kunnath, S. *Field Measurement and Evaluation of Time-Dependent Losses in Prestressed Concrete Bridges*; Research Report: Project Number: WPI-0510735; Florida Department of Transportation: Tallahassee, FL, USA, 1997.

17. Chen, H.L.; Wissawapaisal, K. Measurement of tensile forces in a seven-wire prestressing strands using stress waves. *J. Eng. Mech.* **2001**, *127*, 599–606. [CrossRef]

18. Washer, G.A.; Green, R.E.; Pond, R.B. Velocity constants for ultrasonic stress measurement in prestressing tendons. *Res. Nondestruct. Eval.* **2002**, *14*, 81–94. [CrossRef]

19. Kim, J.T.; Yun, C.B.; Ryu, Y.S.; Cho, H.M. Identification of prestress-loss in PSC beams using modal information. *Struct. Eng. Mech.* **2003**, *17*, 467–482. [CrossRef]

20. Kim, J.T.; Park, J.H.; Hong, D.S.; Cho, H.M.; Na, W.B.; Yi, J.H. Vibration and impedance monitoring for prestress-loss prediction in PSC girder bridges. *Smart Struct. Syst.* **2009**, *5*, 81–94. [CrossRef]

21. Kim, J.M.; Kim, H.W.; Park, Y.H.; Yang, I.H.; Kim, Y.S. FBG sensors encapsulated into 7-wire steel strand for tension monitoring of a prestressing tendon. *Adv. Struct. Eng.* **2012**, *15*, 907–917. [CrossRef]

22. Lan, C.; Zhou, Z.; Ou, J. Monitoring of structural prestress loss in RC beams by inner distributed brillouin and fiber Bragg grating sensors on a single optical fiber. *Struct. Control Health Monit.* **2014**, *21*, 317–330. [CrossRef]

23. Duan, Y.F.; Zhang, R.; Zhao, Y.; Or, S.W.; Fan, K.Q.; Tang, Z.F. Smart elasto-magneto-electric (EME) sensors for stress monitoring of steel structures in railway infrastructures. *J. Zhejiang Univ.-Sci. A* **2011**, *12*, 895–901. [CrossRef]

24. Cho, S.; Yim, J.; Shin, S.W.; Jung, H.; Yun, C.; Wang, M.L. Comparative Field Study of Cable Tension Measurement for a Cable-Stayed Bridge. *J. Bridg. Eng.* **2012**, *18*, 748–757. [CrossRef]

25. Zhao, Y.; Wang, M.L. Fast EM stress sensors for large steel cables. In Proceedings of the SPIE 6934, Nondestructive Characterization for Composite Materials, Aerospace Engineering, Civil Infrastructure, and Homeland Security 2008, San Diego, CA, USA, 9 March 2008.

26. Wang, M.L.; Chen, Z.; Koontz, S.S.; Lloyd, G.D. Magneto-elastic permeability measurement for stress monitoring. In Proceedings of the SPIE 7th Annual Symposium on Smart Structures and Materials, Health Monitoring of the Highway Transportation Infrastructure, Newport Beach, CA, USA, 6–9 March 2000; Volume 3995, pp. 492–500.

27. Holst, A.; Wichmann, H.-J.; Hariri, K.; Budelmann, H. Monitoring of tension members of civil structures—New concepts and testing. In Proceedings of the 3rd European workshop on SHM 2006, Granada, Spain, 5–7 July 2006; pp. 117–125.

28. Joh, C.; Lee, J.W.; Kwahk, I. Feasibility study of stress measurement in prestressing tendons using Villari effect and induced magnetic field. *Int. J. Distrib. Sens. Netw.* **2013**, *9*, 1–8. [CrossRef]

29. Duan, Y.F.; Zhang, R.; Zhao, Y.; Or, S.W.; Fan, K.Q. Steel stress monitoring sensor based on elasto-magnetic effect and using magneto-electric laminated composite. *J. Appl. Phys.* **2012**, *111*, 07E516. [CrossRef]

sensors

MDPI

Article

Magnetic Flux Leakage Sensing and Artificial Neural Network Pattern Recognition-Based Automated Damage Detection and Quantification for Wire Rope Non-Destructive Evaluation

Ju-Won Kim [ID] and **Seunghee Park** *

School of Civil, Architectural Engineering and Landscape Architecture, Sungkyunkwan University, Suwon 16419, Korea; malsi@nate.com
* Correspondence: shparkpc@skku.edu; Tel.: +82-031-290-7525

Received: 30 November 2017; Accepted: 27 December 2017; Published: 2 January 2018

Abstract: In this study, a magnetic flux leakage (MFL) method, known to be a suitable non-destructive evaluation (NDE) method for continuum ferromagnetic structures, was used to detect local damage when inspecting steel wire ropes. To demonstrate the proposed damage detection method through experiments, a multi-channel MFL sensor head was fabricated using a Hall sensor array and magnetic yokes to adapt to the wire rope. To prepare the damaged wire-rope specimens, several different amounts of artificial damages were inflicted on wire ropes. The MFL sensor head was used to scan the damaged specimens to measure the magnetic flux signals. After obtaining the signals, a series of signal processing steps, including the enveloping process based on the Hilbert transform (HT), was performed to better recognize the MFL signals by reducing the unexpected noise. The enveloped signals were then analyzed for objective damage detection by comparing them with a threshold that was established based on the generalized extreme value (GEV) distribution. The detected MFL signals that exceed the threshold were analyzed quantitatively by extracting the magnetic features from the MFL signals. To improve the quantitative analysis, damage indexes based on the relationship between the enveloped MFL signal and the threshold value were also utilized, along with a general damage index for the MFL method. The detected MFL signals for each damage type were quantified by using the proposed damage indexes and the general damage indexes for the MFL method. Finally, an artificial neural network (ANN) based multi-stage pattern recognition method using extracted multi-scale damage indexes was implemented to automatically estimate the severity of the damage. To analyze the reliability of the MFL-based automated wire rope NDE method, the accuracy and reliability were evaluated by comparing the repeatedly estimated damage size and the actual damage size.

Keywords: magnetic flux leakage; steel wire rope inspection; signal processing; damage quantification; artificial neural network

---

1. Introduction

Recently, elevators for convenience of movement and transportation have become essential facilities inside of buildings, particularly with the development of high-rise buildings, and the installation of elevators has been rapidly increasing all over the world. In addition, cranes and lifts are essential pieces of equipment for transporting materials on construction sites.

In various industrial facilities, steel wire rope, which has a high strength and high flexibility, is a key mechanical element used for power transmission and is widely used because it has the advantages of reliability and efficiency.

Wire ropes fully support the load of structures or cargo, so damage to a wire rope can lead to great risks. However, local cross-sectional damage in a wire rope can occur due to aging caused by long-term use, corrosion caused by the external environment, damage due to unexpected mechanical movement, and local defects due to friction with peripheral devices, etc. These small defects can expand quickly because of the tension in the wire rope, which can lead to the lifting structure falling apart or other structural failure. However, such damage is not easily detected due to certain properties of the wire rope, such as its complicated cross section and long length; and thus some wire ropes are being used in very dangerous conditions in situ [1,2].

Additional, due to the features mentioned above, the remaining service life of wire ropes cannot be accurately predicted, so the ropes are regularly replaced. In particular, it is estimated that more than 70% of ropes have been replaced despite the fact that there is no problem with the strength, and there is a recurring economic loss as a result. Thus, it is very important to detect initial defects in wire ropes at the early stages to prevent both accidents and avoid economic losses.

Currently, wire rope inspection relies almost entirely on manual inspection methods along with visual inspection. This requires a lot of time and money, and there is a high probability of human error because it is diagnosed is based on an individual's decision. Visual inspection also has a fatal disadvantage in that it cannot inspect local flaws or internal corrosion.

To overcome these drawbacks, non-destructive testing has recently been introduced to check the state of wires, but only up to a limit. Techniques such as radiographic testing and ultrasonic testing (referred to as representative non-destructive testing methods) have the potential to reduce the risk of radiation exposure, but they are associated with low inspection efficiency and problems with wave propagation due to the geometry of the complicated wire ropes [2].

Therefore, magnetic testing can be a very useful method for detecting defects in wire ropes because such wire ropes are ferromagnetic materials that are easily magnetized, and most non-destructive testing methods for wire rope inspections primarily use magnetic sensors. In this method, a magnetic field is applied to a wire, and a magnetic sensor detects any changes in the magnetic flux or leakages of the magnetic field from a defective portion of the wire [3–5].

Magnetic sensors have the advantages of excellent reliability and reproducibility, and excellent diagnostic performance can be expected when combined with signal processing technology [6–9].

Magnetic sensors vary in type and are used in accordance with the characteristics of the target structure considering any possible damage [10]. Wire ropes are ferromagnetic and continuous, and have a complicated cross section in which several wires are twisted. In addition, the type of damage that occurs in the wire rope is characterized by disconnection, corrosion, and collapse of the shape, most of which occurs locally. Further, corrosion and abrasion damage, which can occur globally, the degree of degeneration of the corrosion and abrasion damage is progressed to leads to disconnection of the wire.

Considering these characteristics, the magnetic flux leakage (MFL) method was selected as a suitable test method for wire rope in this study. This is because the MFL method has been effectively applied to diagnose local damage in continuous ferromagnetic structures such as pipes and rails, and it has the advantage of high-speed non-contact diagnosis [11–13].

Currently, most wire rope diagnostic techniques using the MFL technique employ a one-dimensional raw magnetic flux signal. However, there is a restriction in that the damage must be judged via subjective signal analysis by a professional, and thus such methods are not widely used.

To overcome these limitations, this study aimed to develop an automatic damage assessment method based on MFL that quantitatively assesses the degree of damage by objectively determining whether or not the damage has occurred automatically without expert intervention.

To accomplish this, magnetic flux data measured from 8-channel MFL sensor heads were processed by filtering and the Hilbert transform method, outlier analysis was performed by comparing the results with a threshold using the GEV distribution, and the existence of damage was objectively determined.

Assuming damage was identified, damage indexes were created to quantify the MFL signals by extracting the magnetic characteristics of the MFL signals. Finally, an artificial neural network

(ANN)-based pattern recognition technique using extracted damage indexes was developed, and the size of the damage in the wire rope was automatically and objectively estimated.

2. Theoretical Background

2.1. Magnetic Flux Leakage-Based Damage Detection Technique

Any magnetized ferromagnetic material can be considered as a magnet. The magnetic field spreads out when it encounters a small air gap created by the defect, as the air cannot support as much magnetic field per unit volume when compared to the magnet. When the field spreads out, it starts to leak out of the material, and this is called magnetic flux leakage. Before measuring the magnetic flux leakage, specimens must reach saturated magnetization conditions by applying a field large enough to cause the magnetic flux to effectively leak out. To establish the magnetic flux in the material to be inspected, a strong permanent magnet is used to magnetize the specimen in this study. When no damage is present, the magnetic flux in the specimen remains uniform, as illustrated in Figure 1a. In contrast, flux leakage occurs when damage due to local defects has occurred, as shown in Figure 1b.

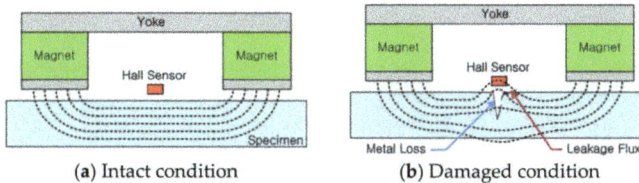

(a) Intact condition (b) Damaged condition

Figure 1. Schematic of the MFL method [9].

Magnetic flux leaks out of the metal specimen near the damaged areas. Sensors that can detect this flux leakage are placed between the poles of the magnet. The sensors then generate a voltage signal proportional to the magnetic flux leakage [10,14]. In this study, Hall sensors that operate based on the Hall effect were used to capture the MFL signal, as illustrated in Figure 2.

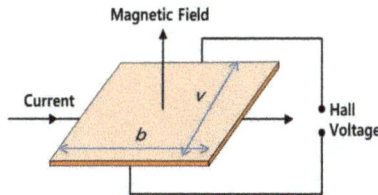

Figure 2. Schematic of the Hall effect [10].

When a magnetic field (B) is applied to a plate, an electron moving through the magnetic field experiences a force, known as the Lorentz force, which is perpendicular to both to the direction of motion and the direction of the field. The response to this force then creates a Hall voltage [10]. This Hall voltage can be measured using a data acquisition (DAQ) system and can then be used to examine the condition of the target structure.

2.2. Signal Processing for Improving Signal Quality

Signal processing techniques, such as low-pass filtering and offset correction, were performed to improve signal resolution after measuring the magnetic flux. After the de-noising process was performed, the enveloping process was carried out to determine flux leakage to improve the accuracy of damage detection using the characteristics of the magnetic flux signal [9]. Samples of the raw MFL signal and the enveloped MFL signal are shown in Figure 3.

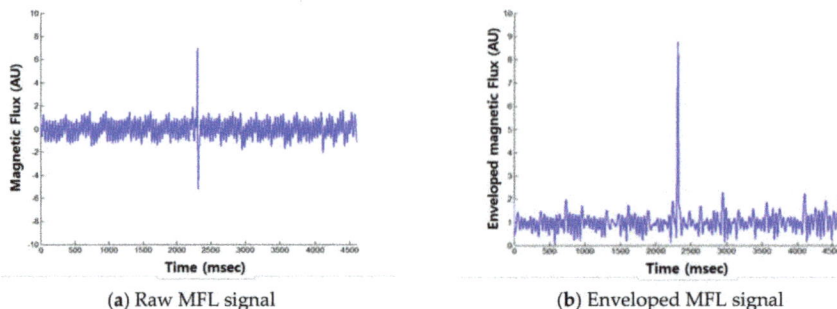

| (a) Raw MFL signal | (b) Enveloped MFL signal |

Figure 3. Effect of the enveloping process [9].

This enveloping process based on the Hilbert transform was performed to bring out the MFL signal due to damage [15]. The envelope using the Hilbert transform can be obtained based on the instantaneous amplitude, and it is useful for analyzing abnormal signals generated by defects in a time series signal. This enveloping process can help reveal important information about the signal by reducing meaningless information, which is used to improve damage detection. In addition, it is helpful for comparing the damage with a threshold value for decision making during damage detection, and additional damage indexes can be extracted from envelope signals to quantify the damage level.

2.3. Establishing Threshold Levels Using the GEV Distribution for Damage Detection

After obtaining the magnetic flux signal, the appropriate threshold that distinguishes between the normal and damaged conditions needs to be established. In this study, a 99.99% confidence level threshold of the normal condition was established using the generalized extreme value (GEV) distribution. The GEV distribution is the limit distribution of the properly normalized maxima of a sequence of independent and identically distributed random variables according to the extreme value theorem [16]. The GEV distribution was therefore used as an approximation to model the maxima of long sequences of random variables. The generalized extreme value distribution has a cumulative distribution function, that is shown in Equation (1):

$$F(x; \mu, \sigma, \xi) = \exp\{-[1 + \xi(\frac{x - \mu}{\sigma})]^{-1/\xi}\} \qquad (1)$$

for $1 + \xi(x - \mu)/\sigma > 0$, where $\mu \in R$ is the location parameter, $\sigma > 0$ is the scale parameter, and $\xi \in R$ is the shape parameter [16]. When the magnetic flux signal exceeds the calculated threshold value, the signal is determined to be within a damaged range.

2.4. Damage Quantification Using MFL Signal Based Damage Indexes

Typically, two kinds of damage indexes that can be extracted from raw MFL signals have been used to quantify the MFL signal to estimate the amount of damage [17,18].

The peak to peak value (P-P value: $P\text{-}P_V$) shown in Figure 4a is used to represent the y-component (amplitude) of an MFL signal and is known to represent the depth of damage. On the other hand, the x-component (width) of the MFL signal is represented by the peak to peak width (P-P width: $P\text{-}P_W$), as shown in Figure 4b. In this study, four types of new damage indexes were extracted from the relationship between the enveloped MFL signal and the threshold, which are utilized to quantify the damage level [9].

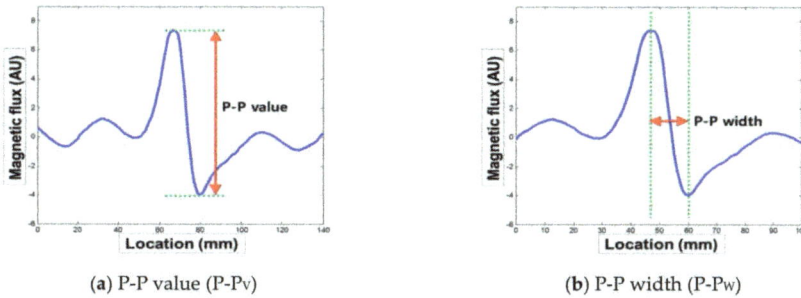

(a) P-P value (P-Pv)

(b) P-P width (P-Pw)

Figure 4. Common damage index for an MFL signal [9].

First, the maximum peak of the enveloped signal that exceeds the threshold was extracted and was named the 'peak value of envelope (E_P)' as shown in Figure 5a. This measurement more effectively represents the level of damage when the MFL signal is saturated; this is because the saturated part of an MFL signal can be restored using the enveloping process since the trends in the remaining signal are reflected in the enveloped signal.

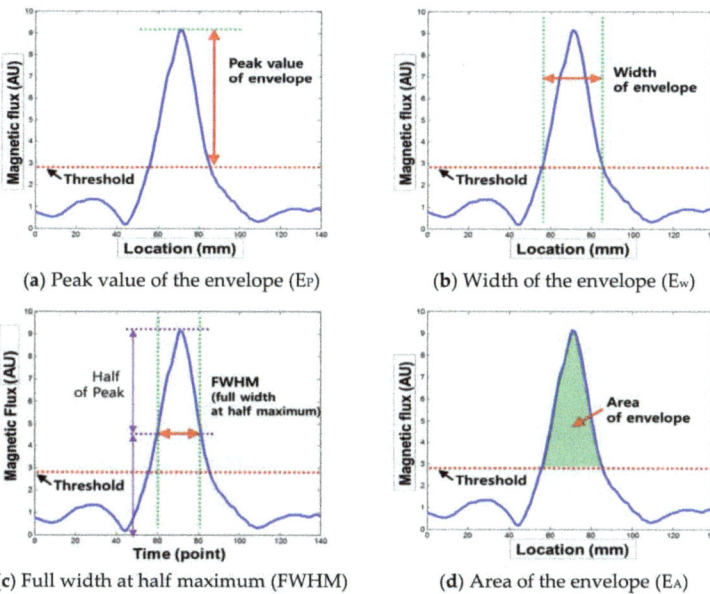

(a) Peak value of the envelope (E_P)

(b) Width of the envelope (E_w)

(c) Full width at half maximum (FWHM)

(d) Area of the envelope (E_A)

Figure 5. New damage indexes using the relationship between the envelope signal and the threshold [9].

Also, the width of the envelope (E_W) was extracted to supplement the P-P width by calculating the range where the envelope exceeds the threshold, which was then used to represent the x-component of the enveloped MFL signal, as shown in Figure 5b. The E_W extracted from under the peak is generally larger than the P-P width, in accordance with the triangular shape [9].

However, when the peak is too large or too small, E_W cannot reflect the magnetic properties of the peak due to the fixed, relatively low threshold. To deal with this limitation, the FWHM (full width at half maximum) was applied to stably represent the width of the peak. To extract the FWHM, the width value at half height of the peak was indexed when the peak was obtained from a signal, as shown in Figure 5c [19].

In addition, the area of the envelope (E_A) was extracted by integrating the amplitude of the signal in the excess range considering, the shape of the envelope signal, as shown in Figure 5d. Even if the height and width are the same, the shape of the peak may change. Thus, the area of the envelope can effectively represent the total energy of the magnetic flux leakage, as it can reflect the shape of the enveloped MFL signal.

These extracted damage indexes can be independently used to quantify the damage. However, since they have mutually complementary relationships, the accuracy of damage evaluation can be improved by using different combination of these indexes, such as developing complex multi-dimensional damage indexes or utilizing various parameters for pattern recognition.

2.5. ANN Based Pattern Recognition for Damage Quantification

An artificial neural network (ANN) can be used to make an approximation function by learning from collected data, and it can objectively classify unknown data based on this approximation function [20]. The ANN consists of a number of processing elements that are connected to form layers of neurons, although the networks may be complex. The missing links between sets of inputs and outputs were found by determining the optimal synaptic weights, based on the available training data of the inputs and outputs [21]. In this research, various damage indexes were extracted from the MFL signals and were used as training data to train the ANN classifier for the purposes of estimating the damage in a wire rope.

A supervised multi-layer feed-forward ANN with backpropagation is typically employed. The Levenberg-Marquardt (LM) algorithm, which is similar to the Newton method [22], is used for back propagation in ANN learning. The root mean squared error (RMSE) was used as the performance index in this study. In this research, various damage indexes were extracted from the MFL signal and were used as training data to teach the ANN how to estimate the damage in a wire rope.

3. Experimental Study

3.1. Experimental Setup & Procedure

A series of experimental studies was carried out to examine the capabilities of the proposed damage detection and quantification technique. To perform the experiments, five steel wire-rope specimens with a 10 mm diameter and 800 mm length were prepared. An intact wire rope was prepared and magnetic flux signals were measured from this wire rope, which were then used to establish the threshold for the outlier analysis. Four kinds of artificial damaged steel wire specimens were prepared for quantitative analysis according to the size of the damage, as shown in Figure 6 and Table 1.

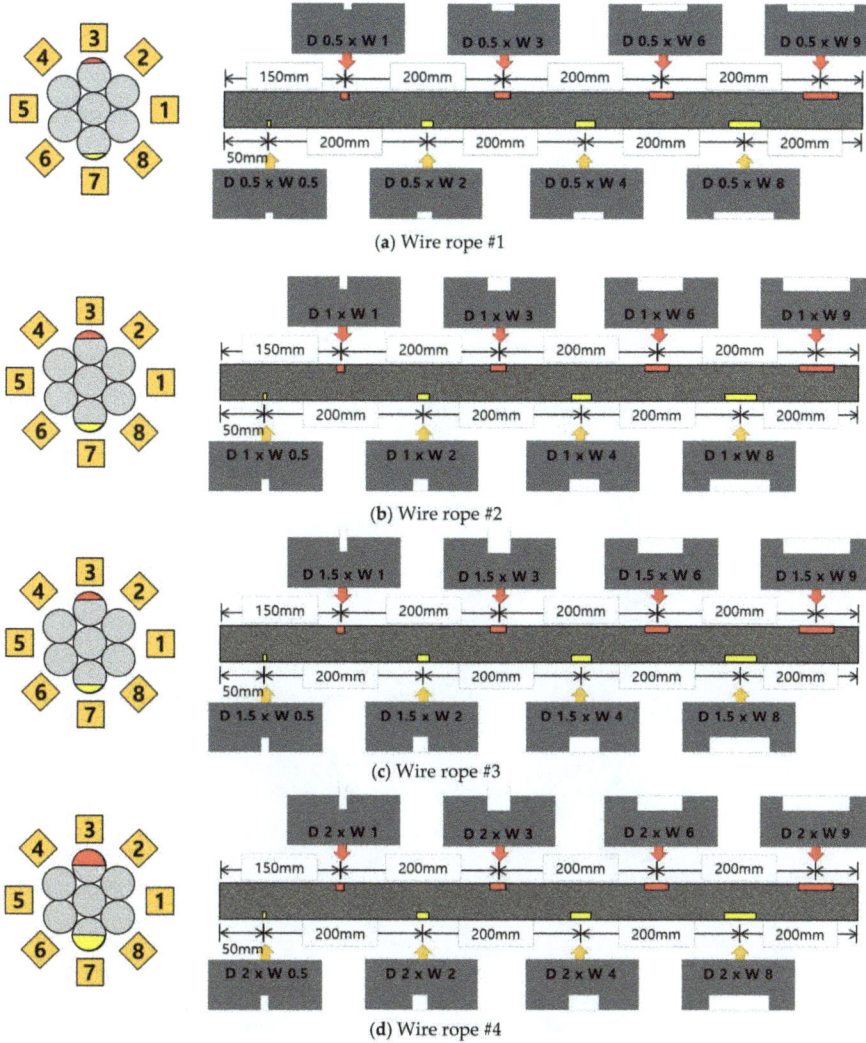

Figure 6. Specification of the wire rope specimens.

As shown in Tables 1 and 2 and Figure 6, stepwise formations of various damage sizes were considered for all wire ropes (#1 to #4), for a total of 32 different types of local damage. In each wire rope, eight types of damage with a width ranging from 0.5 mm to 9 mm were machined, with four in the upper parts of the rope and four in the lower parts with a 10 cm spacing. The depths of the damage were 0.5 mm for wire rope #1, 1 mm for wire rope #2, 1.5 mm for wire rope #3, and 2 mm for wire rope #4.

Table 1. Specification of the damage for wire ropes #1 and #2.

No.	Direction	Specification of the Damage			
Wire #1	Upper	Damage #1-2 Location 150 mm Depth 0.5 mm Width 1 mm 	Damage #1-4 Location 350 mm Depth 0.5 mm Width 3 mm 	Damage #1-6 Location 550 mm Depth 0.5 mm Width 6 mm 	Damage #1-8 Location 750 mm Depth 0.5 mm Width 9 mm
	Under	Damage #1-1 Location 50 mm Depth 0.5 mm Width 0.5 mm 	Damage #1-3 Location 250 mm Depth 0.5 mm Width 2 mm 	Damage #1-5 Location 450 mm Depth 0.5 mm Width 4 mm 	Damage #1-7 Location 650 mm Depth 0.5 mm Width 8 mm
Wire #2	Upper	Damage #2-2 Location 150 mm Depth 1 mm Width 1 mm 	Damage #2-4 Location 350 mm Depth 1 mm Width 3 mm 	Damage #2-6 Location 550 mm Depth 1 mm Width 6 mm 	Damage #2-8 Location 750 mm Depth 1 mm Width 9 mm
	Under	Damage #2-1 Location 50 mm Depth 1 mm Width 0.5 mm 	Damage #2-3 Location 250 mm Depth 1 mm Width 2 mm 	Damage #2-5 Location 450 mm Depth 1 mm Width 4 mm 	Damage #2-7 Location 650 mm Depth 1 mm Width 8 mm

Table 2. Specification of the damages for the wire rope #3 and #4.

No.	Direction	Specification of the Damage			
Wire #3	Upper	Damage #3-2 Location 150 mm Depth 1.5 mm Width 1 mm 	Damage #3-4 Location 350 mm Depth 1.5 mm Width 3 mm 	Damage #3-6 Location 550 mm Depth 1.5 mm Width 6 mm 	Damage #3-8 Location 750 mm Depth 1.5 mm Width 9 mm
	Under	Damage #3-1 Location 50 mm Depth 1.5 mm Width 0.5 mm 	Damage #3-3 Location 250 mm Depth 1.5 mm Width 2 mm 	Damage #3-5 Location 450 mm Depth 1.5 mm Width 4 mm 	Damage #3-7 Location 650 mm Depth 1.5 mm Width 8 mm

Table 2. *Cont.*

No.	Direction	Specification of the Damage							
		Damage #4-2		Damage #4-4		Damage #4-6		Damage #4-8	
		Location	150 mm	Location	350 mm	Location	550 mm	Location	750 mm
		Depth	2 mm	Depth	2 mm	Depth	2 mm	Depth	2 mm
		Width	1 mm	Width	3 mm	Width	6 mm	Width	9 mm
	Upper								
Wire #4		Damage #4-1		Damage #4-3		Damage #4-5		Damage #4-7	
		Location	50 mm	Location	250 mm	Location	450 mm	Location	650 mm
		Depth	2 mm	Depth	2 mm	Depth	2 mm	Depth	2 mm
		Width	0.5 mm	Width	2 mm	Width	4 mm	Width	8 mm
	Under								

In other words, the change in MFL signal was determined according to the change in damage width. The changes in MFL signal were also analyzed for various damage depths, which was accomplished by comparing the wire ropes (#1–#4) with different depths. The test setup for measuring MFL signals was composed of an MFL sensor head, a compact DAQ, and a terminal board, as shown in Figure 7.

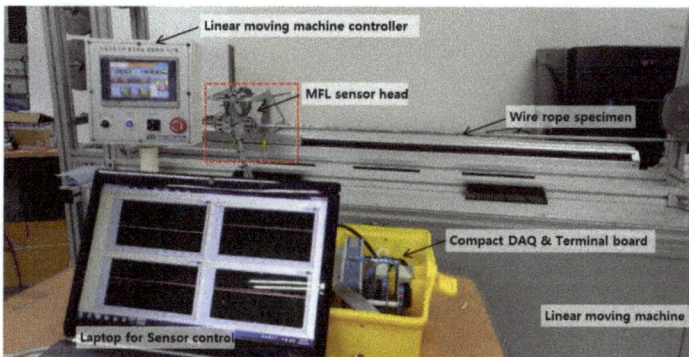

Figure 7. Experimental setup.

The MFL sensor head moves linearly along the fixed wire rope specimen with a constant velocity of 2 m/s to measure MFL signals using a linear moving machine. The data acquisition equipment, which consists of a terminal board and a compact DAQ, measures 8 magnetic flux signals simultaneously at the MFL sensor head.

The MFL sensor head is composed of 8 channels of the sensor module, and each sensing module contains a Hall sensor and a permanent magnet yoke that can independently measure the magnetic flux signal. Eight sensing modules were circumferentially arranged in a circular configuration to measure signals from the entire cross section of the wire rope, as shown in Figure 8.

Signals were repeatedly measured 40 times at each specimen, and the sampling rate was 10 kHz. The measured signals were then processed to facilitate effective damage detection throughout signal processing and the enveloping process based on the Hilbert transform (HT).

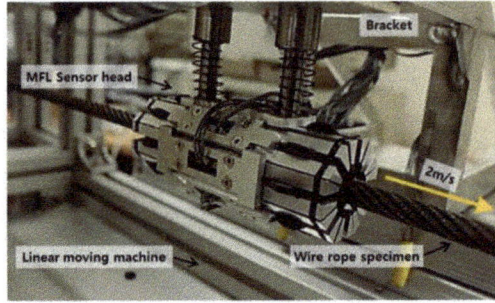

Figure 8. 8-channel MFL sensor head.

3.2. MFL Based Damage Detection Results

After signal processing, the enveloped MFL signals measured from wire ropes #1–#4 are displayed by overlapping in Figure 9.

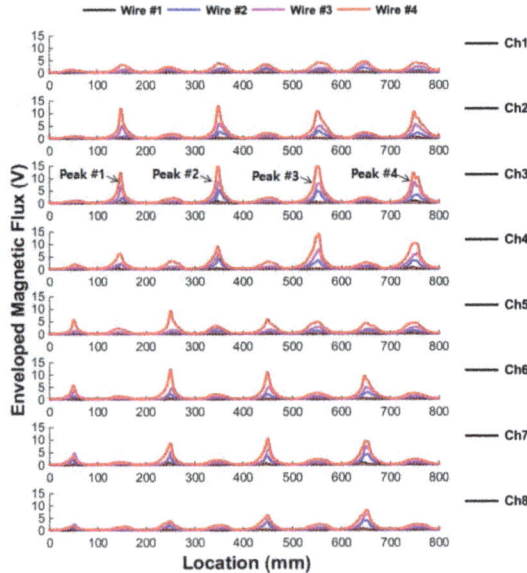

Figure 9. Overlapped graphs of the enveloped MFL signals for wire ropes #1–#4.

Figure 9 shows that leakage signals exceeding the threshold were generated at 150, 350, 550 and 750 mm, where the actual damage is located. It was also confirmed that leakage signals were generated at 50, 250, 450 and 650 mm, where damage was located. This shows that the detection of damage in the wire rope is possible through detection of the MFL signal.

The most sensitive sensing channel for upper damages is channel 3, which is nearest to the upper damage, and channel 7 is most sensitive for lower damages. A relatively weaker MFL signal was detected in the neighboring sensing channels. This shows that the closer the damage is to the sensing channel, the better the detection capability of the signal, indicating that the circumferential direction of the damage can be deduced using this method.

3.3. Quantitative MFL Signal Analysis Using Damage Indexes

3.3.1. Analysis and Quantification of the Leakage Flux Signal with Increasing Damage Depth

MFL signals collected from wires #1–#4 with different depths of damage were compared to analyze the change in patterns of the MFL signals with increasing damage depth. In this research, a quantitative analysis was conducted using the MFL signals measured from the top four damaged regions by utilizing the most sensitively measured signals from channel 3.

It can be seen from Figure 10 that the magnitude of the envelope signal increases with the depth of the defects in all peaks sections. In general, the height of the peak, which is known to be affected by the depth, increased stepwise with increasing depth, while the width of the envelope, which is known to be closely correlated with the damage width, increased with increasing depth. Comparing each peak section, however, it is seen that the size of the peak varies with the width, even though the depth damage is the same; therefore, it is confirmed that the height of the peak is not determined by the depth of the damage.

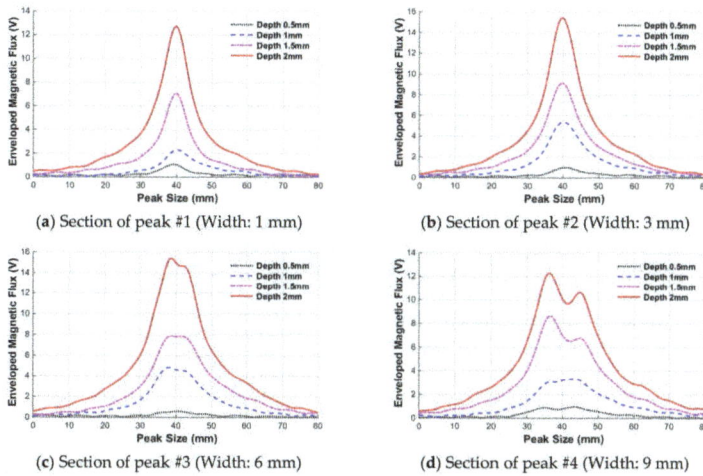

(a) Section of peak #1 (Width: 1 mm)

(b) Section of peak #2 (Width: 3 mm)

(c) Section of peak #3 (Width: 6 mm)

(d) Section of peak #4 (Width: 9 mm)

Figure 10. Variation of the enveloped MFL signal according to damage depth.

Next, the changes in envelope signal were quantitatively extracted according to the damage depth by using the damage indexes introduced in Section 2.4. First, the most representative index of the MFL signal, the P-P value (P-P$_V$), is presented in Figure 11. As previously seen in the peak signal in Figure 10, the P-P value increased as the damage depth increased in accordance with the width of the damage.

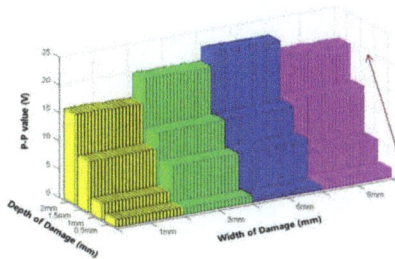

Figure 11. Variation of the P-P value (P-P$_V$) according to the damage depth.

Figure 12 shows the extraction results of the peak value of the envelope using the relationship between the envelope signal and the threshold value. The peak value of envelope (E_P), which is similar to the P-P value, steadily increased with increasing damage depth.

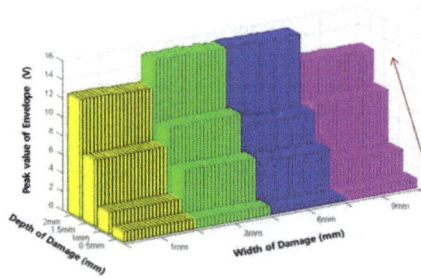

Figure 12. Variation of the peak value of envelope (E_P) according to the damage depth.

Figure 13 below shows the results of extracting the width of envelope (E_W), which is a damage index developed to estimate the original damage width. The width of the envelope exhibits a positive relationship with increasing damage depth, and was therefore determined to be appropriate both for estimating the damage width, and the damage depth.

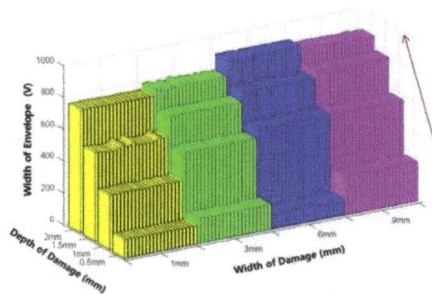

Figure 13. Variation of the width of envelope (E_W) according to the damage depth.

The area of the envelope (E_A) is shown in Figure 14. The peak value of envelope and the width of the envelope were determined based on how the envelope increased with increasing damage depth. Therefore, the area of the envelope also increased with increasing damage depth, exhibiting a large increase in magnitude close to the product of the two exponents.

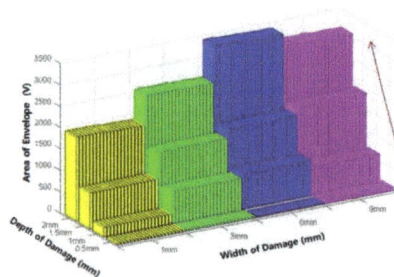

Figure 14. Variation of the area of envelope (E_A) according to the damage depth.

An analysis of the changes in the damage indexes according to damage depth showed that all four indexes, i.e., the P-P value, peak value of the envelope, width of the envelope, and area of the envelope, increased with increasing damage depth. Therefore, these indexes were used with the ANN to estimate the damage depth.

3.3.2. Analysis and Quantification of the Leakage Flux Signal with Increasing Damage Width

Figure 15 shows data for wire rope specimens with four different widths overlaid on top of each specimen. These data were used to examine the characteristics of the MFL signal according to damage width.

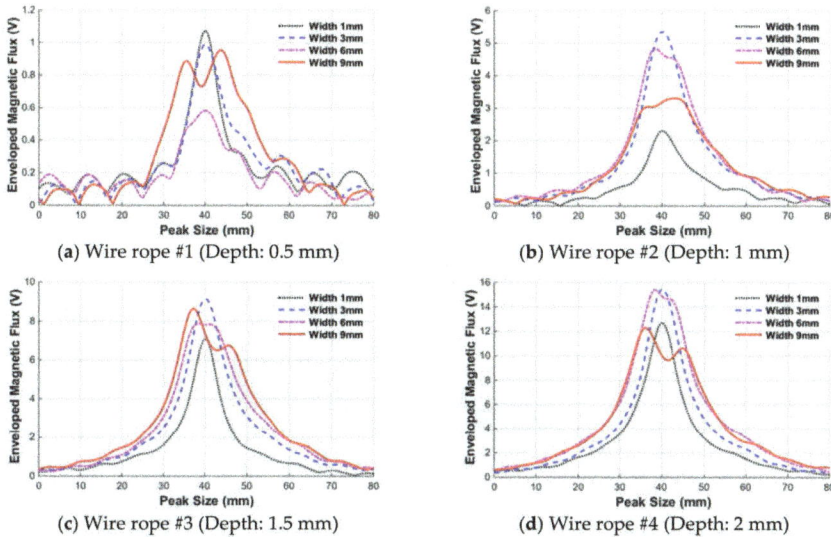

Figure 15. Variation of the enveloped MFL signals according to the damage width.

In the case of wire rope #1, the peak was not clear because the damage depth was very small (0.5 mm), and therefore no intuitive pattern for the height and width of the peak due to the damage width was identified. However, for a damage depth of 1 mm, the width of the peak increased gradually from 1 mm to 6 mm with increasing damage width. However, when the damage width was extended from 6 mm to 9 mm, the peak was distorted and the height of the peak decreased. Additionally, the width of the peak did not significantly increase either, and unlike damage depth, there was no consistent pattern, thus confirming that the characteristics vary slightly depending on the size of the peak.

The change in leakage flux signal with increases in the damage width was investigated by quantifying the damage index determined using various characteristics of the peak.

First, the P-P width (P-P$_W$), commonly known as a representative index that reflects the width of the damage, was extracted and is shown in Figure 16.

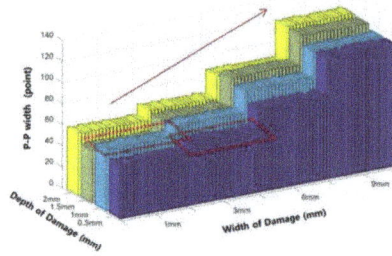

Figure 16. P-P width (P-P$_W$) according to the damage width.

Figure 16 shows that the P-P width at each depth experienced a gradual increase with increasing damage width. However, complete separation does not occur because the P-P width has a value similar to the P-P widths at damage widths and depths of (i) 1 mm and 1 mm, and (ii) 3 mm and 0.5 mm, respectively. Figure 17 shows the width of the envelope (E$_W$) as a function of the damage width variation.

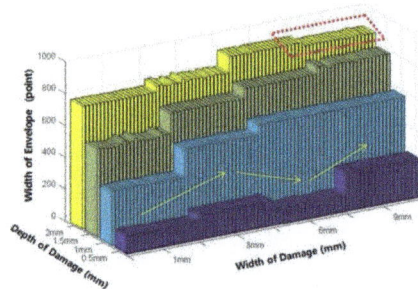

Figure 17. Width of the envelope (E$_W$) according to damage width.

The index decreased at a depth of 0.5 mm without increasing stepwise at a width of 6 mm. This is because the peak size itself is too small, as can be seen by comparing with the peaks in Figure 15. In addition, the index decreased in the section where the depth increased from 6 mm to 9 mm at a damage depth of 2 mm; this was concluded to be an error caused by peak distortion. In addition, the overall height of the peak decreased and the threshold value was smaller than the height of the peak.

To compensate for the effects of size reduction, the FWHM extracted at the half height of the peak was used instead, as shown in Figure 18.

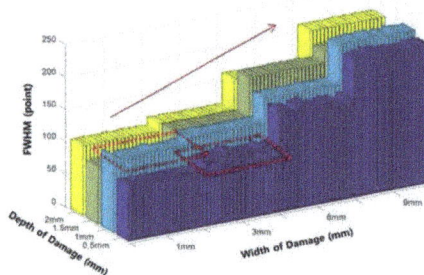

Figure 18. FWHM according to the damage width.

Extracted FWHM values exhibited a positive correlation with increasing width, similar to the P-P value. However, the values for the 1 mm width index and 3 mm width index were still not completely separated. However, since it was confirmed that changes in the index follow a certain pattern, this index can be effectively used when constructing the pattern recognition algorithm.

3.4. ANN Based Wire Rope Damage Size Estimation

3.4.1. Procedure of ANN Based Damage Size Estimation

To estimate the depth and width of damage in the wire ropes, an ANN based pattern recognition technique was applied in this study. Estimation of the damage size was performed using a two-step ANN pattern recognition process as shown in Figure 19. After estimating the depth of the damage using the 1st ANN classifier, the width was estimated by re-reflecting the estimated depth value. Among the various ANN algorithms, the Levenberg–Marquardt algorithm was used to estimate the damage size in this study.

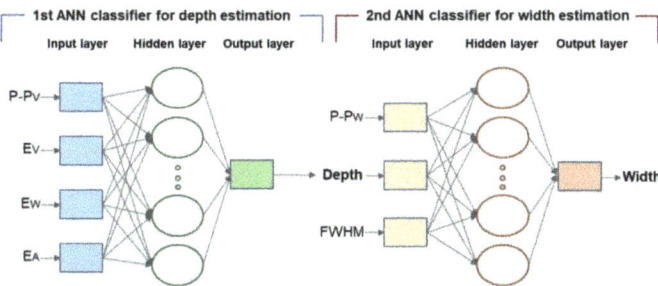

Figure 19. Two-step ANN pattern recognition process.

3.4.2. Depth Estimation of Wire Rope Damage Using the ANN

Various damage indexes were used for training the ANN classifier for damage depth estimation. Based on the previous section, it was confirmed that the P-P value, peak value of the envelope, area of the envelope, and width of the envelope have positive relationships with increasing damage depth. Therefore, the ANN classifier was trained using all four kinds of damage indexes.

The distribution of the learning data obtained by mapping it into three-dimensional space is displayed in Figure 20. In reality, four damage indexes were used, but only three indexes were used in the following graphs due to dimensional restrictions of visualization.

As shown in Figure 20, each of the damage indexes reflected the change in depth, and even when the damage indexes were mapped into three-dimensional space, they were clearly classified according to the damage depth.

As mentioned above, the ANN classifier was trained using four kinds of damage indexes, each extracted from 40 data points according to the damage depth.

To verify the performance of the ANN classifier, the damage indexes were extracted in the same way using the 40 test data collected under the same conditions, which were then substituted into the learned ANN classifier to estimate the depth of the damage. Figure 21 shows the results of the estimated damage depth according to the ANN classifier.

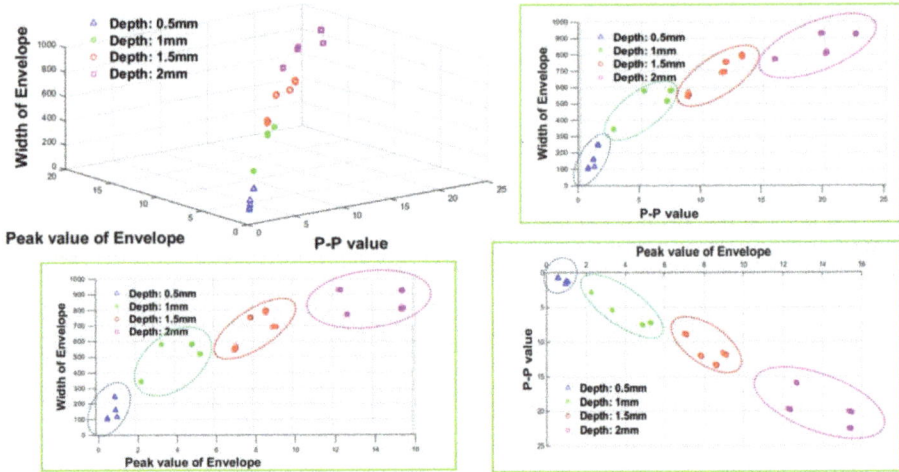

Figure 20. Three-dimensional distribution of the damage indexes for training the ANN for depth estimation.

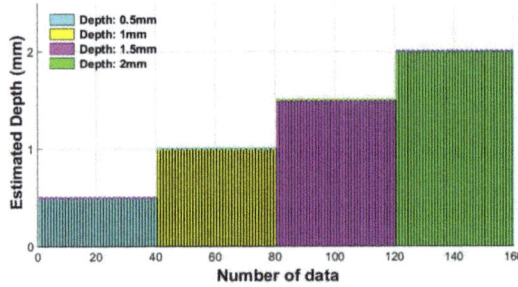

Figure 21. Estimated depth value using the ANN classifier.

Estimation of the damage using the ANN is shown in Figure 21, which accurately estimates the depth of damage for each of the four types, each with a difference of 0.5 mm at every step. The error between the 40th estimated times of each step was also close to zero. Therefore, this ANN based depth estimation algorithm can estimate the damage depth with high accuracy for evaluating the condition of the wire rope.

3.4.3. Width Estimation of Wire Rope Damage Using the ANN

Next, an algorithm for estimating the width of the damage via pattern recognition techniques using the ANN as well as damage depth estimation was investigated. The learning method for the ANN is similar to the method of damage depth estimation. Previously, the depth and width of the damage already confirmed that the damage indexes are sensitive to each other. Thus, the damage indexes used for depth estimation as well as other combinations of damage indexes were used for ANN learning for width estimation.

The FWHM and P-P width showed clear increasing patterns as the damage width increased. Therefore, this combination is a useful index for ANN classifier learning for the estimation of damage width.

In addition, for the MFL signal, it was already confirmed that the depth of damage can also affect the damage index in terms of the damage width. Therefore, the 'estimated depth value', which was previously constructed during the damage depth estimation step using the ANN, was used as an auxiliary index for considering the damage depth. Therefore, P-P width, FWHM, and estimated depth value were used as the ANN learning data for width estimation. These are presented in three-dimensional space in Figure 22.

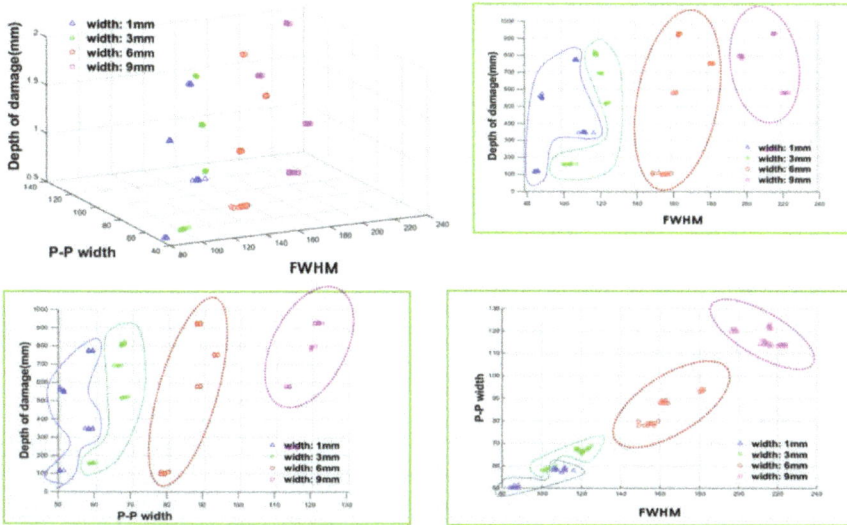

Figure 22. Three-dimensional distribution of the damage indexes for training the ANN for width estimation.

Figure 22 shows the distribution of the damage indexes for training the ANN for width estimation. There is an ambiguous boundary between the widths of 1 mm and 3 mm, but it is generally clustered according to increasing damage width.

As shown above, 40 sets of learning data were used for each damage width, and the ANN classifier was trained for width estimation.

Subsequently, in order to verify the performance of the ANN classifier with regard to width estimation, 40 data per damage width were collected under the same experimental conditions as the training data used to the classifier. The results of the damage width estimation by the ANN classifier are shown in Figure 23.

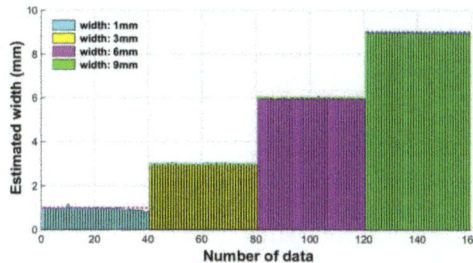

Figure 23. Estimated width value using the ANN classifier.

All 40 of the estimates for damage width closely matched the damage width of the actual wire rope. There was an error of less than 0.1 between the measurement intervals, as can be confirmed by Figure 23, and such a small estimation error is negligible. Therefore, we confirmed that highly accurate damage width estimation was possible through an ANN based pattern recognition algorithm.

Therefore, it is expected that effective quantitative wire rope inspection can be provided through application of the MFL based NDE technique and the ANN pattern recognition technique presented in this research.

4. Conclusions

The MFL-based NDT technique was used to detect damage in steel wire ropes. An MFL sensor head was fabricated and a series of experimental studies were performed to verify the feasibility of the proposed technique. In addition, damage indexes were extracted to quantify the size of the damage. An ANN-based pattern recognition method using the extracted damage indexes was used to automatically estimate the amount of damage. This approach to wire rope NDE was confirmed through the following observations:

(1) Magnetic flux leakage was detected at locations with actual damage by using a Hall sensor located near the damage.

(2) The MFL signals at the damaged areas became more apparent via the enveloping process based on the Hilbert transform.

(3) Envelopes of the MFL signal exceeded the thresholds based on the GEV distribution around areas with actual damage.

(4) Damage indexes were extracted to quantify the MFL signals; these damage indexes can classify the damage size according to increases in damage size.

(5) Four types of damage indexes based on the relationship between the envelope signal and the threshold were proposed. These damage indexes can improve the accuracy of quantification of the damage size.

(6) Two-step ANN based pattern recognition was applied to estimate the depth and width of the damage. The ANN classifier was trained using multi-dimensional damage indexes extracted from the MFL signals; the trained ANN classifier can successfully estimate the size of damage with little error.

Overall, these results demonstrated that the proposed damage detection and quantification method using MFL sensors and an ANN classifier is capable of diagnosing defects in steel wire ropes. This approach will be complemented and validated by further research performed on various types of damage and environments.

In addition, it is expected that the proposed wire rope NDE method can be utilized as an advanced inspection tool for real-time in situ wire rope monitoring in combination with the Internet of Things and robot technologies.

Acknowledgments: This research was supported by the Basic Science Research Program through the National Research Foundation of Korea (NRF) funded by the Ministry of Education [NRF-2017-R1A6A3A04011933] and [NRF-2017R1A2B3007607], and the Disaster Safety Technology Development & Infrastructure Construction Program funded by the Ministry of Public Safety and Security [MPSS-IS-2014-0115], and the Korea Ministry of Land, Infrastructure and Transport (MOLIT) as the 'u-City Master and Doctor Course Grant Program'.

Author Contributions: Ju-Won Kim and Seunghee Park conceived and designed the experiments; Ju-Won Kim performed the experiments; Ju-Won Kim and Seunghee Park analyzed the data; Seunghee Park contributed reagents/materials/analysis tools; Ju-Won Kim wrote the paper.

Conflicts of Interest: The authors declare no conflict of interest.

Sensors **2018**, *18*, 109

References

1. Weischedel, H.R. The inspection of wire ropes in service: A critical review. *Mater. Eval.* **1985**, *43*, 1592–1605.
2. Egen, R.A. Nondestructive testing of wire rope. In Proceedings of the 9th Annual Offshore Technology, Houston, TX, USA, 2–5 May 1977; pp. 375–382.
3. Kim, J.; Kim, J.-W.; Lee, C.; Park, S. Development of embedded EM sensors for estimating tensile forces of PSC girder bridges. *Sensors* **2017**, *17*, 1989. [CrossRef] [PubMed]
4. Wang, M.L.; Wang, G.; Zhao, Y. *Sensing Issues in Civil Structural Health Monitoring*; Springer: Dordrecht, The Netherlands, 2005.
5. Lenz, J.; Edelstein, A.S. Magnetic sensors and their applications. *IEEE Sens. J.* **2006**, *6*, 631–649. [CrossRef]
6. Mukhopadhyay, S.; Srivastava, G.P. Characterisation of metal loss defects from magnetic flux leakage signals with discrete wavelet transform. *NDT&E Int.* **2000**, *33*, 57–65. [CrossRef]
7. Zhang, J.; Tan, X. Quantitative inspection of remanence of broken wire rope based on compressed sensing. *Sensors* **2016**, *16*, 1366. [CrossRef] [PubMed]
8. Mandal, K.; Atherton, D.L. A study of magnetic flux-leakage signals. *J. Appl. Phys.* **1998**, *31*, 3211–3217. [CrossRef]
9. Kim, J.-W.; Park, S. MFL based local damage detection and quantification for steel wire rope NDE. *J. Intell. Mater. Syst. Struct.* **2017**. [CrossRef]
10. Lenz, J.E. A review of magnetic sensors. *Proc. IEEE* **1990**, *78*, 973–989. [CrossRef]
11. Park, S.; Kim, J.-W.; Lee, C.; Lee, J.J. Magnetic flux leakage sensing-based steel cable NDE technique. *Shock Vib.* **2014**, *2014*, 929341. [CrossRef]
12. Shi, Y.; Zhang, C.; Li, R.; Cai, M.; Jia, G. Theory and application of magnetic flux leakage pipeline detection. *Sensors* **2015**, *15*, 31036–31055. [CrossRef] [PubMed]
13. Kang, D.H.; Kim, J.-W.; Park, S.-Y.; Park, S. Non-contact Local Fault Detection of Railroad Track using MFL Technology. *J. KOSHAM* **2014**, *14*, 275–282. [CrossRef]
14. Ramsden, E. *Hall-Effect Sensors: Theory and Applications*; NEWNES: Oxford, UK, 2006.
15. Feldman, M. Time-varying decomposition and analysis based on the Hilbert transform. *J. Sound Vib.* **2006**, *295*, 518–530. [CrossRef]
16. Coles, S. *An Introduction to Statistical Modeling of Extreme Values*; Springer: Berlin, Germany, 2001.
17. Li, L.M.; Zhang, J.J. Characterizing the Surface Crack Size by Magnetic Flux Leakage Testing. In *Nondestructive Characterization of Materials VIII*; Springer: Boston, MA, USA, 1998.
18. Wilson, J.W.; Kaba, M.; Tian, G.Y. New techniques for the quantification of defects through pulsed magnetic flux leakage. In Proceedings of the 17th World Conference on Non-destructive Testing, Shanghai, China, 25–28 October 2008; pp. 25–28.
19. Wikipedia, the Free Encyclopedia. Full Width at Half Maximum. Available online: https://en.wikipedia.org/wiki/Full_width_at_half_maximum (accessed on 28 July 2017).
20. Schalkoff, R. *Pattern Recognition: Statistical, Structural and Neural Approaches*; John Wiley & Sons: New York, NY, USA, 1992.
21. Worden, K.; Tomlinson, G.R. Classifying Linear and Non-linear Systems using Neural Networks. In Proceedings of the 17th International Seminar on Modal Analysis, Leuven, Belgium, 23–25 September 1992; pp. 903–922.
22. Hagan, M.T. *Neural Network Design*, 1st ed.; PWS Pub.: Boston, MA, USA, 1996.

sensors

MDPI

Article

An Interdigital Electrode Probe for Detection, Localization and Evaluation of Surface Notch-Type Damage in Metals

Lanshuo Li [1] ⓘ, Xiaoqing Yang [1,*], Yang Yin [1], Jianping Yuan [2], Xu Li [3], Lixin Li [3] and Kama Huang [1]

[1] School of Electronics and Information Engineering, Sichuan University, Chengdu 610065, China; lls@stu.scu.edu.cn (L.L.); yangyin@stu.scu.edu.cn (Y.Y.); kmhuang@scu.edu.cn (K.H.)
[2] National Key Laboratory of Aerospace Flight Dynamics, Northwestern Polytechnical University, Xi'an 710129, China; jyuan@nwpu.edu.cn
[3] School of Electronics and Information, Northwestern Polytechnical University, Xi'an 710129, China; nwpu_lixu@126.com (X.L.); lilixin@nwpu.edu.cn (L.L.)
* Correspondence: xqyang@scu.edu.cn; Tel.: +86-181-8079-9278

Received: 6 November 2017; Accepted: 21 January 2018; Published: 27 January 2018

Abstract: Available microwave notch-type damage detection sensors are typically based on monitoring frequency shift or magnitude changes. However, frequency shift testing needs sweep-frequency data that make scanning detection becomes difficult and time-consuming. This work presents a microwave near-field nondestructive testing sensor for detecting sub-millimeter notch-type damage detection in metallic surfaces. The sensor is loaded with an interdigital electrode element in an open-ended coaxial. It is simple to fabricate and inexpensive, as it is etched on the RC4003 patch by using printed circuit board technology. The detection is achieved by monitoring changes in reflection amplitude, which is caused by perturbing the electromagnetic field around the interdigital structure. The proposed sensor was tested on a metallic plate with different defects, and the experimental results indicated that the interdigital electrode probe can determine the orientation, localization and dimension of surface notch-type damage.

Keywords: interdigital electrode (IDE); metallic materials; microwave near-field detection; surface notch-type damage

1. Introduction

Nondestructive testing and evaluation (NDT&E) for defects in metal surfaces are especially crucial to those working in conditions that require high safety, such as ships and aircraft fuselages. Upon prolonged exposure to air and external impacts, there is a higher probability of causing destruction of these structures. Therefore, regular NDT&E for their safety is necessary. Several techniques have been developed to detect defects, such as acoustics [1,2], eddy-currents [3,4], and the magnetic method [5]. However, there are restrictions in practice. For instance, it is ineffective to detect defects hidden under coatings or under paint by these methods. While a class of technique based on microwave and millimeter wave developed rapidly, they have some irreplaceable superiority such as noncontact, non-pollution and the ability to penetrate non-metallic media [6–8], compared with other methods.

However, some limitations still exist in microwave NDT&E systems. For instance, high cost [9], complication [10] and long term consumption. In order to reduce detection cost, a sensor based on the frequency resonance of a complementary split-ring resonator (CSRR) for crack detection has been presented in [11], but the contact measurement process reduced its efficiency. To make the probe light and maneuverable, an open-ended substrate-integrated waveguide probe has been proposed in [12]. The probe was based on amplitude testing, and its depth-detection capability was 3–6 mm. However, it cannot be widely used because of its narrow detection range. An array waveguide probe has been put forward for the orientation and sizing of surface cracks [13]. It can avoid large inspection times and unwanted noise, but it is costly compared to printed circuit board (PCB) technology. In addition, a sensor based on a ground plane defect has been presented in [14], which detected defects by monitoring resonance frequency shifts. The sensor was simple and highly sensitive, however, sweep-frequency detection made it more time-consuming. Otherwise, some defect detection systems have been improved by introducing artificial intelligence [15,16]. Undoubtedly, these methods were effective in solving some specific problems, but there is still some space to improve in cost, miniaturization and operational aspects.

Based on the available literature about microwave near-field testing methods in metallic surfaces, one can conclude the development trend. First, the microwave near-field sensors based on rectangular waveguides [7,17], substrate-integrated waveguides [12] and coaxial lines [18] have been implemented as amplitude detection probes that are simple, light and applicable to various conditions. Second, research has focused on loading artificially engineered electromagnetic materials (metamaterials) to improve the sensitivity and resolution of microwave near-field sensors [9,11]. Furthermore, research that utilized metamaterials for microwave near-field testing was mainly directed towards frequency shift detection but not for amplitude detection. In the meantime, the microstrip line sensor-loaded special resonance structure enormously improved the sensitivity by monitoring shifts in resonance frequencies [14]. However, research progress on reflection amplitude detection with high sensitivity developed slowly. In consideration of the advantages of handiness, easy integration and low cost, it is essential to design a highly sensitivity probe based on reflection amplitude detection [19].

Interdigital electrodes (IDEs) are mainly used for filter [20] and antenna [21] designs that are compact and light. Recently, IDE sensors have been used to monitor changes in dielectric materials [22] and gas sensors [23] and can also be used for the surface defect detection of metallic materials [24], due to the extraordinary feature of IDE sensors. The IDE sensors are sensitive enough to distinguish the variation of closed regions. It is particularly advantageous for defect detection, as it significantly reduces undesirable factors influencing sensor response.

Previously, there has been extensive use of the IDE structure for band-stop filters [20,25], and the microstrip-line-excited IDE exhibits band-stop characteristics. The minimum transmission coefficient frequency depends on the resonance frequency of the IDE. In order to prove that the IDE filter has the ability to detect surface defects in metals, an IDE microstrip band-stop filter, which was perpendicular to an aluminum plate with notch-type damage, was designed by the finite element method. The resulting shift in the minimum transmission frequency demonstrated that the IDE filter can be used for detecting defects. Therefore, when a plate surface is placed close to an IDE, various dimensions of damage disturb the electromagnetic field, resulting in various shifts in resonance frequency. Based on above facts, the IDE structure gives us discernment into designing sensitive surface probes for defect detection in metals.

This paper presents a novel sensor for notch-type damage detection based on the reflection amplitude in optimum frequency of the IDE structure. Essentially, the IDE probe works as a near-field sensor. It operates from 7.0 GHz to 14.5 GHz for notch-type damage detection in metals, including orientation and dimension detection. In addition, the depth and angle of the notch-type damage was detected through significant changes of the reflection coefficient, respectively. The IDE probe has a simple structure, is easy to operate, is multifunctional, and has a low fabrication cost.

2. Probe Design

In order to prove the IDE filter can detect surface damage in metals, we designed an IDE microstrip band-stop filter that is perpendicular to an aluminum plate with notch-type damage in ANSYS HFSS software, as shown in Figure 1. A thin Teflon film with a thickness of 0.05 mm is laid on the aluminum surface before the defect under film being scanned, where w and d are the width and depth of the defect, respectively. The filter adopts a rectangular IDE structure in the middle of the microstrip line, which makes it operate around 6 GHz. To achieve this goal, one can use the approximate model described in [26].

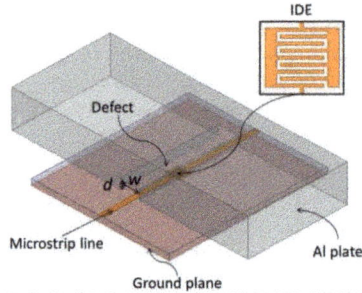

Figure 1. Schematic drawing of the microstrip IDE sensor monitoring an aluminum plate with surface notch-type damage.

When the sensor detects a plate with or without defects, the frequency of minimum transmission changes as shown in Figure 2. According to Figure 2, 6.33 GHz, 7.18 GHz, 7.95 GHz and 8.05 GHz are the resonance frequencies when the width of the notch-type damage is 1.2 mm, 0.6 mm, 0.2 mm, and 0.1 mm, respectively, and 8.31 GHz is the resonance frequency of the sensor when the surface is perfect. When the sensor passed over the notch-type damage with a width and depth of 0.1 mm and 2 mm, a shift of more than 260 MHz was observed with respect to the case without the defect, while a band-stop filter based on a complementary split-ring (CSRR) resonator with a shift of 210 MHz was observed in the same case. The notch-type damage was 0.2 mm in width and 2 mm in depth, and the sensor in this paper gave a resonance frequency shift of 360 MHz, which is more obvious than the resonance frequency shift realized by the CSRR sensor of 275 MHz [14]. The obvious shift in the minimum transmission frequency makes it a strong candidate for the detection of submillimeter-size defects in metallic surfaces. Based on the above analysis, the IDE structure is ideal for the design of sensitive surface probes. However, the IDE filter can only detect regions near the IDE structure, and the results will be disturbed when the defect is close to the micro-strip line. Thus, the micro-strip IDE filter is not adaptable as a scanning probe.

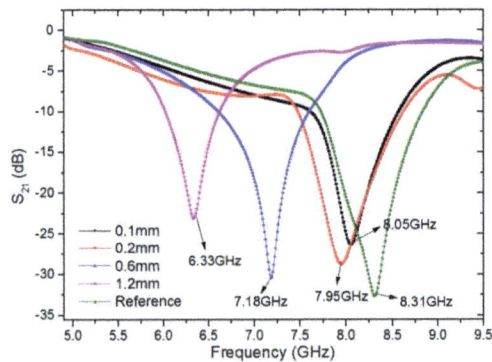

Figure 2. Scattering parameter of the IDE filter from full-wave simulation when encountering five different situations: aluminum blocks without a defect and with defects of 0.1 mm, 0.2 mm, 0.6 mm, and 1.2 mm in width, and 2 mm in depth for all cases.

In this study, a single IDE structure was etched on an RC4003 substrate to avoid an over-complicated structure and undesirable interference. In order to detect the angle of the defect, the classic rectangular IDE is transformed into a round structure that makes the probe more compact. The IDE probe shown in Figure 3 has a cross section of 12 mm × 12.8 mm. A Sub-Miniature-A (SMA) connector is soldered on the back, and the seven interdigital strip lines of rounded contour are etched on the other side. The IDE probe is fed with 50 Ω coaxial line using the via-technology.

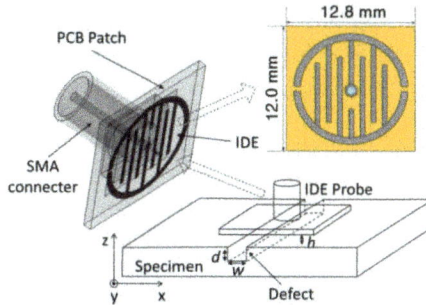

Figure 3. Schematic drawing of the IDE probe interrogating a metallic plate with long notch-type damage.

In order to illustrate the behavior of the IDE structure and predict its resonance frequency, the simulation model setup in HFSS is shown in Figure 3. The SMA connector is included in electromagnetic simulation for more accurate results. Figure 4 presents the simulation results of reflection coefficients with and without defect in the plate surface within 6.5–15.5 GHz. A significant difference at around 13.78 GHz can be found. Figure 5 shows the electric field distribution on the IDE surface where the working frequency is 13.78 GHz. There is a noticeable difference between the aluminum board with a defect and that without a defect in the intermediate area of the electric field distribution. Keeping in view the above results, we may safely draw the conclusion that the defect can perturb the electromagnetic field around the IDE probe.

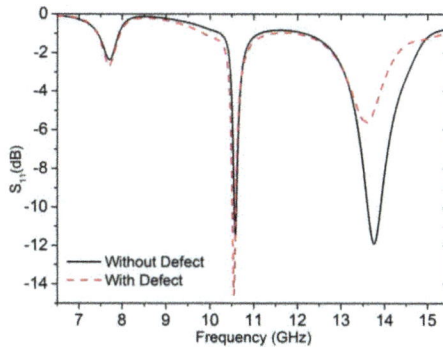

Figure 4. Reflection coefficients of the IDE probe with and without defect. The width and depth of the notch-type damage are both 4 mm. The standoff distance between the probe and the aluminum surface is 1mm for both cases.

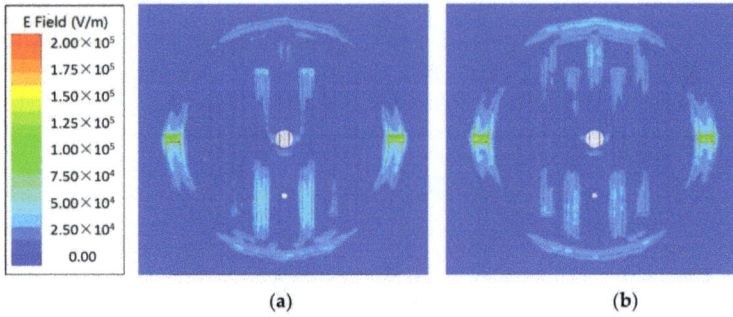

Figure 5. Electric field distributions on the surface of the PCB patch for different metallic surfaces, where the working frequency is 13.78 GHz. (**a**) Aluminum surface with defect; (**b**) Aluminum surface without defect.

3. Detection Theory

Researchers have detected surface defects in metals using the microwave near-field method by the detection principle through the standing-wave shift [11] or the reflection coefficient change [13]. The IDE probe detects defects by monitoring the reflection coefficient change to reduce the detection time. Therefore, in order to explore the effect of defects in metallic surfaces on the behavior of the IDE probe, we considered this detection system as a capacitance model to explain the optimal frequency. When the specimen is regarded as a capacitor, it has complex relative permittivity. In this case, the reflection coefficient at the end of the sensor is expressed as

$$\Gamma = \frac{1 - j\omega Z_0 \times C_{(\varepsilon_s)}}{1 + j\omega Z_0 \times C_{(\varepsilon_s)}} \tag{1}$$

where Z_0 is the characteristic impedance of the sensor and $C(\varepsilon_s)$ is the capacitance of complex relative permittivity of the specimen, and ω is the angular frequency.

From Equation (1), the reflection coefficient is determined according to combinations of the sensor, the specimen, and the measuring frequency. If the metallic specimen surface has a defect under the sensor, this defect becomes another capacitor (as shown in Figure 6). Then, Equation (1) becomes more complicated:

$$\Gamma = \frac{1 - j\omega Z_0 [C_{(\varepsilon_s)} + C_c]}{1 + j\omega Z_0 [C_{(\varepsilon_s)} + C_c]}$$

where C_c is the capacitance of the defect.

In this model, it has a standoff distance (h), and this air gap could change the reflection coefficient that is attributed to the system error. Therefore, when we use the microwave near field method to detect surface defects in metals, the standoff distance should be stationary [9,16].

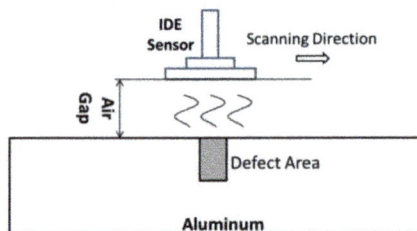

Figure 6. Configuration of specimen with surface defect, air gap and IDE sensor.

4. Measurements Procedure

The experimental setup is shown in Figure 7. The IDE probe is fixed on the fixator which can be used to adjust the standoff distance; the air gap is finally fixed at 1 mm. The specimen is put on the *x–y* stage. The *x*-axis knob is used to move the specimen along the probe, while the turntable is employed to rotate to a particular angle. In this work, the N5230A vector network analyzer (VNA) is used to record the magnitude of reflection coefficient. The IDE probe attached to an SMA connector is connected to the VNA. Figure 8 shows the artificial defects of the investigated specimen. The depth of defects increases with the same width on one side of the specimen, and the width of defects increases with the same depth (except the last one) on the other side. Before notch-type damage scanning (as shown in Figure 7), the IDE probe is located at a distance of 3 mm or further away from the defect, and the relative location of the probe is recorded. Then, the knob is rotated to move the plate along the probe, and the results of VNA for each step are recorded. A level instrument was used to ensure that their horizontal relative positions remain constant. When the plate moved away from the probe by about 3 mm, we finished the detection of this defect. Other defects are measured in the same way. When notch-type damage is encountered, a significant change in the reflection coefficient curve can be registered.

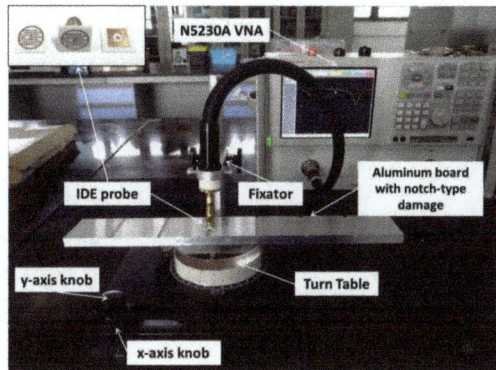

Figure 7. Photograph of the experimental configuration. The IDE probe is fixed on the fixator and the aluminum plate can be moved along the x-axis by turning the knob. The precision is 0.1 mm/step.

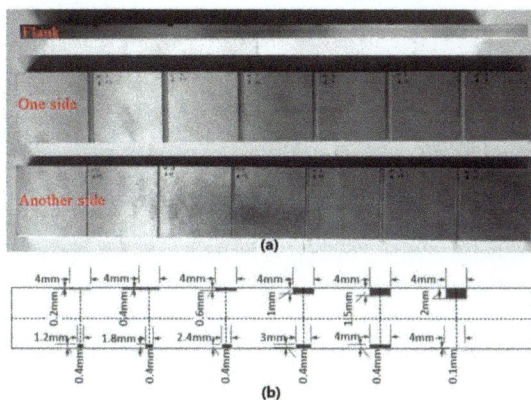

Figure 8. Schematic showing the aluminum sample. (**a**) Two sides of the aluminum plate; (**b**) Structure diagram of the aluminum plate.

5. Result and Discussion

5.1. Scanning Detection

Figure 9 shows the experimental results of scanning the aluminum plate with long notch-type damage. The width and depth are 4 mm and 0.4 mm respectively, and the defect locates at 0 mm. According to Figure 9, there is a symmetry change in magnitude relative to the position of the notch-type damage at 13.86 GHz when the defect is encountered. The distance between the two troughs in the middle is close to the real value of the width.

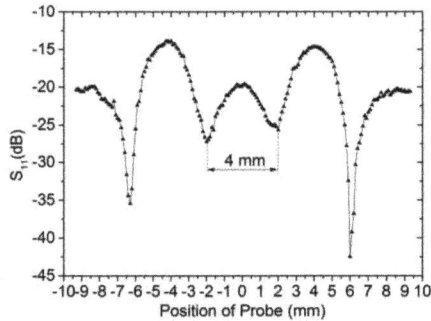

Figure 9. Experimental results of scanning the notch-type damage (w = 4 mm and d = 0.4 mm). The operation frequency is 13.86 GHz.

In order to verify the capability of the IDE probe in width detection, other four defects in different width are scanned. The depth of the defects is 0.4 mm for all cases. The measured values are got by computing the distance between the two troughs in the middle. The actual width and measured value are shown in Table 1. The measured results are very close to the true value when the width is larger than 1.8 mm. However, the probe predicted width is 0.8 mm, while the actual width is 1.2 mm.

Table 1. Comparison of actual width and measured value.

Actual Width (mm)	Measured Value (mm)	Relative Error
3.0	3.2	6.7%
2.4	2.6	8.3%
1.8	2.0	11.1%
1.2	0.8	33.3%

5.2. Orientation and Depth Detection

To confirm the results associated with the orientation of notch-type damage, various angles (0°–90°) are detected. Referring to Figure 3, the defect was placed at the center of the probe's interrogation aperture, and the w and d were set to 4 mm and 2 mm, respectively. The experimental results of various angles corresponding to various S-parameters are shown in Figure 10a, which shows that the probe output signals are indeed affected by orientation.

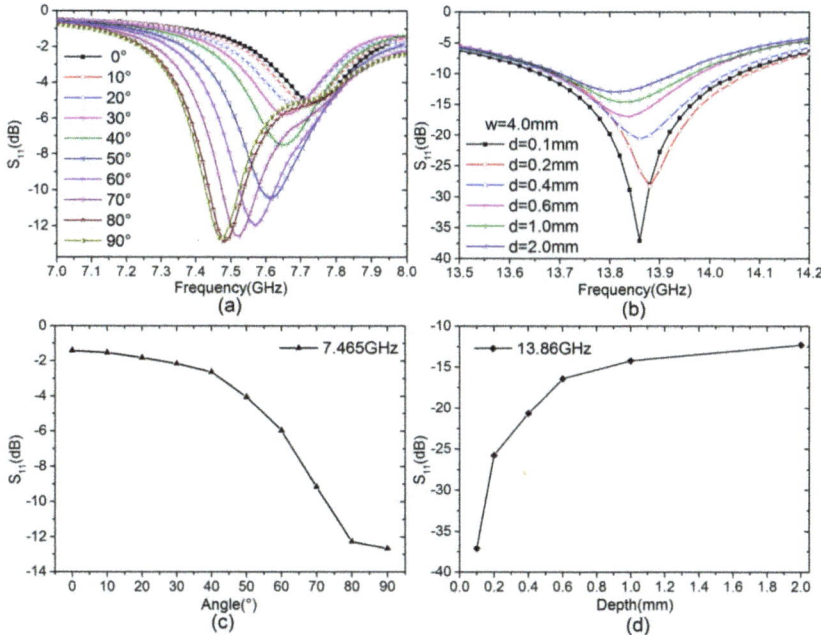

Figure 10. Experimental results for different condition. (**a**) Different angles (w = 4 mm and d = 2 mm); (**b**) Various depths (w = 4 mm); (**c**) Different angles and its amplitude at 7.465 GHz; (**d**) Various depths and its amplitude at 13.86 GHz.

Prior to depth detects, its orientation can be done by rotating the plate until the amplitude of S_{11} attain the values corresponding to 0°. Various depths (0.1–2 mm) have been detected as shown in Figure 10b, while the width is 4 mm for all cases. Probe output signals are affected by depth, and various amplitudes of S-parameters corresponded to various depths.

From the results in Figure 10a, each angle (0–90°) corresponded to the unique amplitudes of the reflection coefficient. The magnitude decreases while angle increases within 7.2–7.47 GHz. As shown in Figure 10b, different depths correspond to a unique amplitude of the reflection coefficient; the magnitude decreases while depth decreases within 13.7–13.86 GHz. By adopting such simple monotonic relations, the orientation and depth value can be predicted by monitoring the magnitude of S-parameters at one frequency point. In order to select the frequency in which the amplitude change is monotonous and significant, we designed an optimization program in MATLAB based on the largest variances algorithm. In this program, we first pick out the frequency in which amplitude change is monotonous. Then, the variance coefficient for each frequency point is calculated and recorded. Finally, the biggest variance coefficient frequency was determined for detection. This is demonstrated by the experimental results shown in Figure 10c,d, and their estimated values show relative errors are less than 8% and 10% for angle and depth detection, respectively.

6. Conclusions

In this paper, a novel microwave IDE probe for detection, orientation and sizing of surface notch-type damage in metallic surfaces has been outlined. The probe utilizes IDE technology to make the probe compact and sensitive. The detection results indicated that scanning detection based on an IDE probe can be used to predict the width when the width is larger than 1.8 mm. In addition, the angle and depth of the defect can be detected by recording the change in magnitude at a single frequency point. Compared to the other microwave methods presented in literature, the IDE probe is multifunctional, relatively inexpensive and sensitive in amplitude detection. To summarize, the microwave IDE probe is promising for the nondestructive evaluation of metallic structure surfaces, and for other applications such as material detection and characterization.

Acknowledgments: This work was supported by the National Natural Science Foundation of China under Grant 61372043.

Author Contributions: Lanshuo Li and Xiaoqing Yang conceived and designed the experiments; Lanshuo Li and Yang Yin performed the experiments; Jianping Yuan and Kama Huang analyzed the data and developed the mathematic model; Xu Li and Lixin Li contributed materials tools; Lanshuo Li wrote the paper.

Conflicts of Interest: The authors declare no conflict of interest.

References

1. Ludwig, R.; Roberti, D. A nondestructive ultrasonic imaging system for detection of flaws in metal blocks. *IEEE Trans. Instrum. Meas.* **1989**, *38*, 113–118. [CrossRef]
2. Bechou, L.; Dallet, D.; Danto, Y.; Danto, P.; Daponte, P.; Ousten, Y.; Rapuano, S. An improved method for automatic detection and location of defects in electronic components using scanning ultrasonic microscopy. *IEEE Trans. Instrum. Meas.* **2003**, *52*, 135–142. [CrossRef]
3. Yun, T.G. Design and Implementation of Distributed measurement systems using fieldbus-based intelligent sensors. *IEEE Trans. Instrum. Meas.* **2001**, *50*, 1197–1202.
4. Bernieri, A.; Ferrigno, L.; Laracca, M.; Molinara, M. Crack shape reconstruction in eddy current testing using machine learning systems for regression. *IEEE Trans. Instrum. Meas.* **2008**, *57*, 1958–1968. [CrossRef]
5. Betta, G.; Ferrigno, L.; Laracca, M. GMR-based ECT instrument for detection and characterization of crack on a planar specimen. *IEEE Trans. Instrum. Meas.* **2012**, *61*, 505–512. [CrossRef]
6. Kharkovsky, S.; Ryley, A.C.; Stephen, V.; Zoughi, R. Dual-polarized near-field microwave reflectometer for noninvasive inspection of carbon fiber reinforced polymer-strengthened structures. *IEEE Trans. Instrum. Meas.* **2008**, *57*, 168–175. [CrossRef]
7. Zoughi, R.; Kharkovsky, S. Microwave and millimetre wave sensors for crack detection. *Fatigue Fract. Eng. Mater. Struct.* **2008**, *31*, 695–713. [CrossRef]
8. Zarifi, M.H.; Deif, S.; Abdolrazzaghi, M.; Chen, B.; Ramsawak, D.; Amyotte, M.; Vahabisani, N.; Hashisho, Z.; Chen, W.X.; Daneshmand, M. A Microwave Ring Resonator Sensor for Early Detection of Breaches in Pipeline Coatings. *IEEE Trans. Ind. Electron.* **2018**, *65*, 1626–1635. [CrossRef]
9. Albishi, A.; Ramahi, O.M. Detection of surface and subsurface cracks in metallic and non-metallic materials using a complementary split-ring resonator. *Sensors* **2014**, *14*, 19354–19370. [CrossRef] [PubMed]
10. McClanahan, A.; Kharkovsky, S.; Maxon, A.R.; Zoughi, R.; Palmer, D.D. Depth Evaluation of Shallow Surface Cracks in Metals Using Rectangular Waveguides at Millimeter-Wave Frequencies. *IEEE Trans. Instrum. Meas.* **2010**, *59*, 1693–1704. [CrossRef]
11. Albishi, A.M.; Boybay, M.S.; Ramahi, O.M. Complementary Split-Ring Resonator for Crack Detection in Metallic Surfaces. *IEEE Microw. Wirel. Compon. Lett.* **2012**, *22*, 330–332. [CrossRef]
12. Mazlumi, F.; Gharanfeli, N.; Sadeghi, S.H.H.; Moini, R. An open-ended substrate integrated waveguide probe for detection and sizing of surface cracks in metals. *NDT E Int.* **2012**, *53*, 36–38. [CrossRef]
13. Ahanian, I.; Sadeghi, S.H.H.; Moini, R. An array waveguide probe for detection, location and sizing of surface cracks in metals. *NDT E Int.* **2015**, *70*, 38–40. [CrossRef]
14. Alibishi, A.M.; Ramashi, O.M. Microwaves-Based High Sensitivity Sensors for Crack Detection in Metallic Materials. *IEEE Trans. Microw. Theory Tech.* **2017**, *65*, 1864–1872. [CrossRef]
15. Gao, B.; Woo, W.L.; Tian, G.Y.; Zhang, H. Unsupervised Diagnostic and Monitoring of Defects Using Waveguide Imaging with Adaptive Sparse Representation. *IEEE Trans Ind. Inform.* **2016**, *12*, 406–416. [CrossRef]
16. Ali, A.; Hu, B.; Ramahi, O.M. Intelligent detection of cracks in metallic surfaces using a waveguide sensor loaded with metamaterial elements. *Sensors* **2015**, *15*, 11402–11416. [CrossRef] [PubMed]

17. Gao, B.; Zhang, H.; Woo, W.L.; Tian, G.Y.; Bai, L.; Yin, A. Smooth Nonnegative Matrix Factorization for Defect Detection Using Microwave Nondestructive Testing and Evaluation. *IEEE Trans. Instrum. Meas.* **2014**, *63*, 923–934. [CrossRef]

18. Yang, S.H.; Kim, K.B.; Kang, J.S. Detection of surface crack in film-coated metals using an open-ended coaxial line sensor and dual microwave frequencies. *NDT E Int.* **2013**, *54*, 91–95.

19. Mirala, A.; Shirazi, R.S. Detection of surface cracks in metals using time-domain microwave non-destructive testing technique. *IET Microw. Antennas Propag.* **2017**, *11*, 564–569. [CrossRef]

20. Yang, R.Y.; Weng, M.H.; Hung, C.Y.; Chen, H.J.; Houng, M.P. Novel compact microstrip interdigital bandstop filters. *IEEE Trans. Ultrason. Ferroelectr. Freq. Control* **2004**, *51*, 1022–1025. [CrossRef] [PubMed]

21. Sam, S.; Kang, H.; Lim, S. Frequency Reconfigurable and Miniaturized Substrate Integrated Waveguide Interdigital Capacitor (SIW-IDC) Antenna. *IEEE Trans. Antennas Propag.* **2014**, *62*, 1039–1045. [CrossRef]

22. Bao, X.; Ocket, I.; Kil, D.; Bao, J.; Puers, R.; Nauwelaers, B. Liquid measurements at microliter volumes using 1-port coplanar. In Proceedings of the 2017 First IEEE MTT-S International Microwave Bio Conference (IMBIOC), Gothenburg, Sweden, 15–17 May 2017; pp. 1–4.

23. Alcantara, G.P.; Andrade, C.G.M. A short review of gas sensors based on interdigital electrode. In Proceedings of the 2015 IEEE 12th International Conference on Electronic Measurement & Instruments, Qingdao, China, 16–18 July 2015; pp. 1616–1621.

24. Abdullah, B.M.; Cullen, J.; Mason, A.; Al-Shamma'a, A.I. A Novel Method for Monitoring Structural Metallic Materials Using Microwave NDT. In *Sensing Technology Current Status and Future Trends I*; Springer International Publishing: Berlin, Germany, 2014; Volume 7, pp. 161–180.

25. Lancaster, M.J.; Huang, F.; Porch, A.; Avenhaus, B.; Hong, J.-S.; Hung, D. Miniature Superconducting Filters. *IEEE Trans. Microw. Theory Tech.* **1996**, *44*, 1339–1346. [CrossRef]

26. Aboush, Z.; Porch, A. Compact, Narrow Bandwidth, Lumped Element Bandstop Resonators. *IEEE Microw. Wirel. Compon. Lett.* **2005**, *15*, 524–526. [CrossRef]

Communication

Application of Negative Curvature Hollow-Core Fiber in an Optical Fiber Sensor Setup for Multiphoton Spectroscopy

Maciej Andrzej Popenda [1] (iD), Hanna Izabela Stawska [1], Leszek Mateusz Mazur [2],
Konrad Jakubowski [2] (iD), Alexey Kosolapov [3], Anton Kolyadin [3] and Elżbieta Bereś-Pawlik [1,*]

[1] Department of Telecommunications and Teleinformatics, Wroclaw University of Science and Technology, 50-370 Wroclaw, Poland; maciej.popenda@pwr.edu.pl (M.A.P.); hanna.stawska@pwr.edu.pl (H.I.S.)
[2] Advanced Materials Engineering and Modelling Group, Wroclaw University of Science and Technology, 50-370 Wroclaw, Poland; leszek.mazur@pwr.edu.pl (L.M.M.); kudzu.jakubowski@gmail.com (K.J.)
[3] Fiber Optics Research Center of Russian Academy of Sciences, Moscow 119333, Russia; kaf@fo.gpi.ru (A.K.); kolyadin@fo.gpi.ru (A.K.)
* Correspondence: elzbieta.pawlik@pwr.edu.pl; Tel.: +48-71-320-21-19

Received: 4 September 2017; Accepted: 1 October 2017; Published: 6 October 2017

Abstract: In this paper, an application of negative curvature hollow core fiber (NCHCF) in an all-fiber, multiphoton fluorescence sensor setup is presented. The dispersion parameter (D) of this fiber does not exceed the value of 5 ps/nm × km across the optical spectrum of (680–750) nm, making it well suited for the purpose of multiphoton excitation of biological fluorophores. Employing 1.5 m of this fiber in a simple, all-fiber sensor setup allows us to perform multiphoton experiments without any dispersion compensation methods. Multiphoton excitation of nicotinamide adenine dinucleotide (NADH) and flavin adenine dinucleotide (FAD) with this fiber shows a 6- and 9-fold increase, respectively, in the total fluorescence signal collected when compared with the commercial solution in the form of a hollow-core photonic band gap fiber (HCPBF). To the author's best knowledge, this is the first time an NCHCF was used in an optical-fiber sensor setup for multiphoton fluorescence experiments.

Keywords: negative curvature fiber; hollow core fiber; multiphoton fluorescence; photonic crystal fiber sensor

1. Introduction

Multiphoton microscopy, first presented by Webb et al. [1], has become a standard imaging procedure for many laboratories, as it allows for greater imaging depths and reduced biological phototoxicity when compared to single-photon methods. Other non-linear optical phenomena (NLOP), such as Second Harmonic Generation (SHG) or Coherent Anti-Stokes Raman Scattering (CARS), are talso being implemented in the imaging systems, forming powerful diagnostic tools [2–6]. Ultrashort, high-energy laser pulses are essential for efficient induction of the NLOP, and the transmission of such pulses through optical fibers is problematic because of the pulse's temporal broadening due to the dispersion. Currently, this problem can be addressed in two ways—either by using dispersion compensation systems [7–9] or by using hollow-core, photonic bandgap fibers (HCPBF's) [10], or both [11]. However, dispersion compensating systems are bulky setups, while HCPBF's keep their dispersion parameters only within relatively narrow optical bandwidths, which becomes a problem when considering the optical spectrum of two photon absorption cross sections of biological fluorophores [12].

Negative curvature hollow-core fibers (NCHCF) are a relatively new type of microstructured hollow-core fibers, guiding the light in their air-filled core. They are significantly different from

HCPBFs—their optical structure is much simpler, and the ARROW (Antiresonant Reflecting Optical Waveguides) guiding mechanism [13] allows their use at different optical bandwidths, usually very distant from each other [14,15]. Additionally, because of a very low coupling of core and cladding modes, as well as their hollow-core architecture, they are a promising possibility for the transmission of high-power, ultra-short optical pulses [16,17]. Up to the present day, Sherlock et al. [18] were the first and only ones to present the potential of this type of fiber in multiphoton microscopy. In this paper, we present an application of NCHCF fibers in an optical fiber sensor setup for the purpose of multiphoton fluorescence experiments. By creating a simple fiber probe setup we were able to perform a multiphoton excitation (MPE) of two endogenous fluorophores—NADH and FAD—and collect their fluorescence spectra. Unique dispersion parameters of the NCHCF occur in the range of 680–750 nm [19], which is optimal for efficient excitation of the aforementioned fluorophores. When comparing the suitability of two different fibers — NCHCF and HCPBF — a large increase in the total fluorescence intensity was observed when using the NCHCF as the excitation fiber.

2. Hollow-Core Fibers

Two different types of optical fibers were used during this research for the purpose of femtosecond pulse delivery—the custom-made NCHCF, which was previously characterized in [19], and a commercial HCPBF—the HC-800-02 (NKT Photonics, Birkerød, Denmark). The cross sections of both fibers are presented in Figure 1.

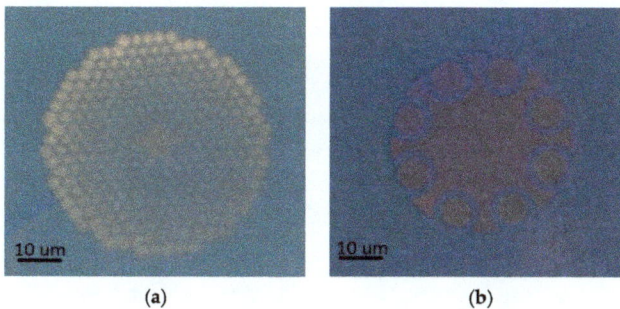

Figure 1. Cross sections of microstructured, photonic bandgap fiber—HC-800-02 from NKT Photonics (**a**) and NCHCF with eight separated capillaries (**b**). An approximate scale bar of 10 μm is presented in the bottom left corner of each picture. Central part of each fiber is its core, surrounded with either a large number of small air holes (HC-800-02, (**a**)) or a single row of larger capillaries (NCHCF, (**b**)).

HC-800-02 has a microstructured cladding consisting of a large number of small air holes, which creates a photonic band gap across its operating spectrum, effectively trapping the light inside its air-filled core. According to this fiber specification [20], its core diameter is 7.5 μm, while its D (dispersion parameter) values range from -100 ps/nm $\times$ km at 760 nm to ~200 ps/nm $\times$ km at 870 nm, crossing zero at ~775 nm. The other fiber used —NCHCF—contains a microstructured cladding as well. However, this cladding construction is much simpler as it consists of only one row of large (when compared to the used HCPBF) capillaries. Light in this type of fibers propagates along the core walls due to the mechanisms described by the ARROW model. The core of this fiber is surrounded by eight separate capillaries, creating the negative curvature condition. The diameter of the core is 21 μm, while the capillaries have their walls 828 nm thick, according to [19]. The HC-800-02 has already proven useful for the multiphoton experiments [11], while the presented NCHCF has not been applied for this purpose so far.

3. Multiphoton Fluorescence Sensor Design and Construction

A schematic cross section and a photograph of the sensor head are presented in Figure 2. Although the idea of this type of sensor has already been presented by many other researchers [21–26], never before has it been used with a combination of two types of fibers so distinct from each other—hollow core fibers and classic, solid core ones.

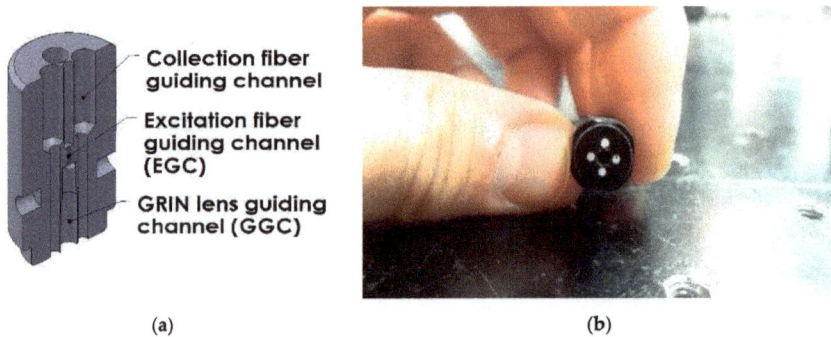

(a)　　　　　　　　　　　　　　　　　(b)

Figure 2. Fiber sensor head cross section (**a**) and a photograph of the sensor detection face (**b**). In the picture of the sensor face, the central part is a hole for the GRIN lens, while the four smaller circles surrounding it are collection fiber faces. The sensor's outer diameter is 10 ± 0,05 mm. Sensors head was fabricated in the university CNC (Computerized Numerical Control) center.

The central part of the sensor consists of two concentric channels—a GRIN lens guiding channel (GGC) and an excitation fiber guiding channel (EGC), with diameters of 1.8 mm and 0.25 mm, respectively. Channels were drilled at the two opposite planes, parallel to the sensor's main axis. This type of construction allows for easy exchange of excitation fibers, which makes it a flexible solution, easily adaptable for different experiments requiring different excitation wavelengths. The concentricity of GGC and EGC was at a level of 20 μm, meaning the excitation fiber (either NCHCF or HC-800-02) and the GRIN lens (GRIN 2908, Thorlabs, Newton, NJ, USA) could have had their main axes shifted by 20 μm. As the GRIN lens outer diameter was 1.8 mm, we were expecting a minimal focal point shift due to this misalignment, and in the case of our experiments it did not pose a problem. The GGC was surrounded by four smaller, symmetrically placed (1 mm diameter) adjacent holes, where the collection fibers (GH-4001 Eska^{TM}, Mitsubishi Rayon, Tokyo, Japan) were inserted. These fibers present many advantages in terms of remote fluorescence collection—the large numerical aperture (NA_{COL} = 0.5) and core diameter (Φ_{COL} = 980 μm), combined with good transmission in the visible region [27] make them a suitable choice for remote fluorescence sensing. Although the collection fiber ring was not fully filled because of the collection fiber's large stiffness, their large NA and core diameter compensated for this loss.

The initial design of this simple probe was imperfect because of large GRIN lens diameter, which resulted in ~1.5 mm offset between the main axes of the collection fiber and GRIN lens. This made the collection fiber's base aperture too small to collect the fluorescence occurring at the focal point of the GRIN lens. This problem was overcome by polishing the collection fiber's end-face at a proper angle, which increased the original collection aperture angle. According to Snell's law, the fiber's collection cone angle is related to its end-face polishing angle by the following formula:

$$\beta' = \arcsin(n_{col} \times \sin(\alpha + \gamma)) - \gamma, \tag{1}$$

where β' is the numerical aperture angle of the collection fiber after polishing, n_{col} is the collection fiber's core refraction index, α is the maximum angle of refraction at the air–fiber core boundary, with respect to the axis normal to the fibers collection face (in our case—$\alpha \approx 19.61°$, derived from the Snell's law), and γ is the polishing angle.

One can easily conclude from the equation above that the fiber acceptance cone angle, β', increases with the angle of tip-polishing, γ, which also causes the area from which the signal can be collected to rise. According to the above, we have polished the collection faces of each of the collection fibers at the angle of 15°, which was the largest we could afford. Due to such modification, the NA_{COL} changed from 0.5 to 0.68 (the collection fiber's acceptance angle increased from 30° to 42.8°). The effect of aperture shift before and after the polishing at the chosen angle is presented in Figure 3. The sensor head was made from PMMA (polymethylmethacrylate) to match the collection fibers material, allowing for convenient polishing of the collection fibers tips. The outer diameter of the sensor was 10 mm, which is quite compact and allows it to be used in limited spaces.

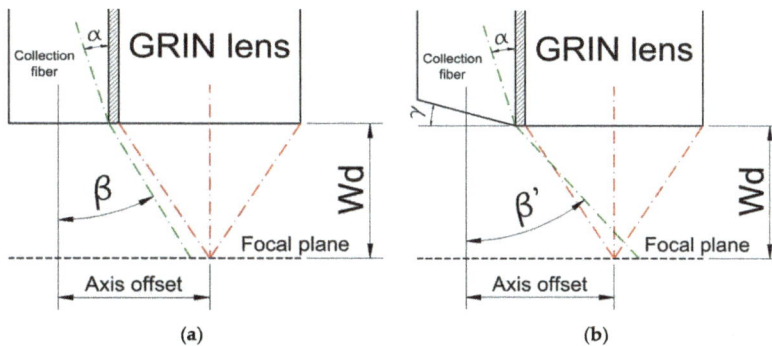

(a)   (b)

Figure 3. A scheme for the fiber sensor collection tip. In cases of omitting the 15° polish (**a**), the collection fiber's original aperture angle, β (represented as the green dot-dash line), does not cover the GRIN lens focal point, thus virtually no fluorescence signal can be collected. Due to the polishing at the $\gamma = 15°$ angle (**b**), the collection fiber's acceptance angle changes to β', according to Equation (1). Visible aperture shift occurs, allowing to collect fluorescence signal due to the covering of the GRIN lens focal point. In both pictures, α represents the maximum refraction angle at the fiber core-air boundary (approximately 19.61°), while W_d stands for the working distance of the GRIN lens. The dimension of the axis offset is approximately 1.5 mm, while $W_d = 1.449$ mm.

4. Optical Setup for the Measurements of Multiphoton Fluorescence and Autocorrelation of Ultrashort Laser Pulses

A scheme of the measurement setup is presented in Figure 4. Ultrafast laser pulses, provided by a Ti:Sapphire oscillator (Chameleon Ultra II, Coherent, Santa Clara, CA, USA), were focused by a 10×/0.24 NA microscope objective and coupled into the excitation fiber (HC-800-02 or NCHCF) placed on a 3D translation stage (MBT616D, Thorlabs, Newton, NJ, USA). Coupling efficiencies of ~70% were achieved for both fibers, and an output power of 70 mW (for both excitation fibers) was used during multiphoton fluorescence experiments. Although that may seem relatively high, we did not have to reduce it as no real-life biological samples were investigated. The pulse repetition rate, f_{rep}, and output pulse width, τ_{pulse}, were 80 MHz and 161.3 fs, respectively. Each of the excitation fibers was 1.5 m long, and each was used for the transmission of different excitation wavelengths—NCHCF for 730 nm and HCPBF for 780 nm. The other end of the excitation fiber was placed in the center of the fiber sensor body. The remaining ends of the collection fibers (fluorescence signal delivery tips) were bundled in a single ferrule, and the fluorescence signal was focused by an 8×/0.2 NA microscope objective on the entrance slit of the spectrometer (S2000, OceanOptics, Dunedin, FL, USA). This 8× objective was additionally mounted on a custom-made, 3D translation stage to ensure proper alignment of the focused signal. Although this spectrometer model is designed for use with optical fibers, this type of setup was necessary for two reasons: firstly, it allowed us to overcome the problem of a large mismatch between the numerical apertures of the spectrometer ($NA_{spectrometer}$ = 0.22) and the collection fibers (NA_{COL} = 0.5) and, secondly, it provided space for the convenient use of optical filters (FESH700, Thorlabs, Newton, NJ, USA), necessary in this epi-fluorescence sensor setup. Initial alignment of the focused signal was performed by coupling a broadband white light source into the collection fibers and finding the position at which the maximum intensity was observed at the spectrometer. After that, a sample cuvette (111-QS, Hellma Analytics, Müllheim, Germany) with the investigated solution was placed in front of the sensor's face, and its emission spectra were collected. Autocorrelation measurements were performed via an autocorrelator (pulseCheck, A.P.E., Berlin, Germany). Both the spectrometer and the autocorrelator were controlled via a PC. All the data analysis was performed via the MATLAB® custom-written software.

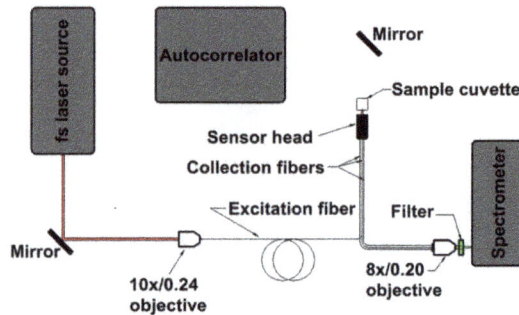

Figure 4. Optical setup schematic. Two mirrors (PF10-03-P01, Thorlabs) were used for the convenience of guiding the laser beam – first one allowed to couple the fundamental beam into the fiber-coupling objective, and the second one was used during the autocorrelation function (ACF) measurements. When measuring the ACF of the setup without the fiber, the GRIN lens was placed directly after the 10× objective, and the beam collimated in this way was directed via the second mirror at the autocorrelator's aperture. For the ACF measurements of the excitation fibers output beam, the sample cuvette was removed, and the mirror used in the same way as above.

5. Fluorescence Measurements

Because the fluorescence was detected in the epi direction, it was necessary to determine the influence that a large amount of reflected light may have on the collected spectra (unwanted emission peaks, spectral shifts, etc.). Two control samples were prepared for this reason—a colloidal silica solution (LUDOX® HS-40, Sigma-Aldrich, St. Louis, MO, USA) and a fluorescein solution ($C_{fluorescein}$ = 10^{-5} M, diluted in 5×10^{-2} M NaOH). NADH and FAD (Sigma-Aldrich) were diluted in 10^{-2} M NaOH and H_2O, respectively. The concentration of each was at the level of 10^{-3} M to compensate for their very low two-photon absorption cross sections [12], which will be further explained in the next section.

6. Results and Discussion

6.1. Dispersion Effects of Coupling Fundamental Laser Beam into Hollow-Core Fibers

For both excitation fibers, their best wavelength (dispersion- and attenuation-wise) was chosen, resulting in very good performance in terms of temporal pulse broadening. The decision to pick only these two wavelengths was based on the fact that wavelengths of 780 nm and above are widely used for the purpose of multiphoton spectroscopy and microscopy [7–9,11], which allows for good imaging results; however, as one can conclude from the work of Denk et al. [1], both the excitation wavelength and pulse duration have a direct influence on the number of photons absorbed during multiphoton absorption process. Thus, it was our goal to show that combining good dispersion parameters and proper excitation wavelength can lead to significantly better fluorescence yields when compared with solutions employing only one of the latter, i.e., fibers transmitting ultrashort pulses with no temporal broadening, but away from the optimal excitation wavelength, or the other way—proper excitation wavelength is at the cost of temporal pulse broadening.

Based on the autocorrelation function (ACF) fitting procedure, the Gaussian signal fit was chosen to estimate the temporal shape of the excitation pulses as presenting the smallest fit root mean square error (RMSE) at the level of 5×10^{-3}. Thus, the temporal pulse width, τ_{pulse}, was determined to be $(\sqrt{2})^{-1} \times \tau_{FWHM}$ of the fitted pulses ACF. The values of τ_{pulse}, obtained for the incident laser and at the outputs of both excitation fibers (730 nm pulses transmitted through the NCHCF and 780 nm pulses transmitted through the HCPBF), were measured to be $\tau_{pulse} = 161.3$ fs, $\tau_{NCHCF} = 162.0$ fs, $\tau_{HCPBF} = 162.2$ fs. The total temporal pulse broadening for the NCHCF and HCPBF was estimated at 0.7 fs and 0.9 fs, respectively, which can be considered negligible, suggesting that both fibers should perform similarly in terms of multiphoton experiments. However, D values of the HC-800-02 increase rapidly for wavelengths below 775 nm, reaching -100 ps/nm $\times$ km at 760 nm. The D slope of this fiber at this spectral region can be roughly estimated at the level of 7 ps/nm^2 $\times$ km, which is a few orders of magnitude higher than in the case of the NCHCF (0.01 ps/nm^2 $\times$ km) [18]. As a result of this, the spectral region of 760 nm and below is unavailable for the HCPBF under testing, making it an inferior choice for the purpose of multiphoton excitation of endogenous fluorophores. Additionally, in the case of NCHCF, such a low value of the D slope allows for avoiding higher-order dispersion compensation systems, which simplifies the optical setup.

6.2. Two-Photon Induced Fluorescence of Biological Fluorophores Excited with NCHCF and HCPBCF Fibers

Due to the fact that the sensor detected epi-fluorescence from the solution of various fluorophores placed in a quartz cuvette, the light of the fundamental laser beam was partially reflected from the front surface of the cuvette and had high intensity. It was difficult to eliminate it completely, even when using a proper cut-off filter. However, no detector saturation or unexpected emission peaks were observed. The TPEF (two-photon excited fluorescence) spectra of fluorescein, excited at 730 nm and 780 nm, peaked at 520.1 and 519.7 nm, respectively. These results, as well as the observed spectra shapes, are in agreement with literature [28], proving that this setup could be used for fluorescence measurements. The fluorescence emission spectra of the fluorophores of interest, NADH and FAD, are presented in Figure 5.

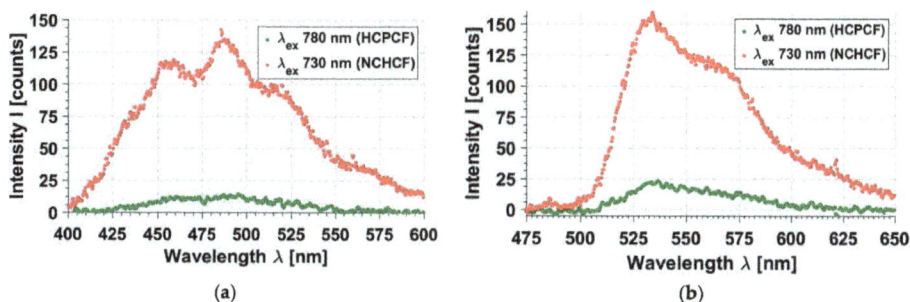

Figure 5. Emission spectra of (**a**) NADH solution in 0.01 M NaOH; and (**b**) FAD solution in water. A visible increase in the total emission signal for both the compounds can be observed for the excitation wavelength of 730 nm. HCPCF stands for Hollow-Core, Photonic Crystal Fiber—HC-800-02 in our case.

As was mentioned earlier, both the biological fluorophores had high concentrations of 10^{-3} M. Because of that, a large spectral shift, as well as the overall distortions of both emission spectra, can be observed. Concentrations had to be so high because of very weak two-photon absorption cross sections of both fluorophores at the 780 nm wavelength, making the signal hard to detect in the presented sensor setup, proving the need for its further optimization in terms of the collection efficiency. However, comparing the fluorescence emission intensities for both excitation fibers, HCPBF and NCHCF, one can notice an approximate 6- (FAD) and 9- (NADH) fold increase in the total signal intensity in the case of NCHCF excitation, keeping the power of fundamental beam at the same level. This can be explained by two major factors: firstly, these fluorophores have much larger two-photon absorption cross sections for the 730 nm wavelength, and secondly, the NCHCF fiber allows for temporally and spectrally undistorted transmission of ultrashort pulses of 730 nm. These results prove our initial assumption that fibers combining both good dispersion parameters and transmission bandwidth suitable for the specific fluorophore excitation may become a good choice for non-linear endoscopy devices.

7. Conclusions

The application of a new type of fiber—a negative curvature, hollow-core fiber in the multiphoton fluorescence setup—has been presented. A comparison between this fiber and a commercial solution for fiber-based, femtosecond pulse delivery was performed. The total temporal broadening of femtosecond pulses coupled into both fibers was estimated to be at the level of 0.7 fs for the NCHCF fiber at the 730 nm wavelength and 0.9 fs for the HC-800-02 fiber at the 780 nm wavelength. However, the NCHCF has two main advantages over the HCPBF—first, its dispersion-free spectral bandwidth occurs at 680–750 nm, making it a superior solution for the multiphoton fluorescence excitation of endogenous fluorophores. Indeed, the total fluorescence emission signal of NADH and FAD solutions, induced by the 730 nm pulses transmitted through NCHCF, is nearly 6 and 9 times stronger, respectively, when compared to the fluorescence induced with 780 nm pulses transmitted through the commercial HCPBF. Another advantage over the HCPBF is the extremely low D slope of the NCHCF—0.01 ps/nm^2 × km compared to the 7 ps/nm^2 × km of HCPBF, which makes it a perfect solution for a fuller utilization of the tunability of current ultrafast laser sources, very important for multiphoton spectroscopy and microscopy. The presented optical fiber sensor still requires improvements in terms of collection efficiency. Its largest drawbacks are the absence of additional fibers in the fiber ring and very large dead volume between the GRIN lens and collection fibers. However, the easily modifiable construction and flexibility provided by the possibility of excitation fiber exchange make the authors believe that this sensor may, eventually, become an interesting alternative to the currently used sophisticated solutions.

Acknowledgments: The authors acknowledge support from the Polish National Science Centre grants DEC-2013/10/A/ST4/00114, DEC-2013/09/B/ST5/03417 and Wroclaw University of Science and Technology, Grant No. 0401/0105/2016. Calculations have been carried out in Wroclaw Centre for Networking and Supercomputing (http://www.wcss.wroc.pl), Grant No. 184. We would also like to thank Marek Samoć and Evgeny M. Dianov for supporting this research and helpful remarks concerning the manuscript.

Author Contributions: Maciej Andrzej Popenda, Hanna Izabela Stawska and Elżbieta Bereś-Pawlik developed and designed the sensor setup; Maciej Andrzej Popenda, Hanna Izabela Stawska and Leszek Mateusz Mazur designed and prepared the measurement setup; Maciej Andrzej Popenda, Leszek Mateusz Mazur and Konrad Jakubowski conducted the optical and spectroscopic measurements; Anton Kolyadin and Alexey Kosolapov designed and provided the NCHCF fiber; Maciej Andrzej Popenda wrote the manuscript; all the authors contributed to the manuscripts edition and revision; Elżbieta Bereś-Pawlik supervised the research.

Conflicts of Interest: The authors declare no conflict of interest.

References

1. Denk, W.; Strickler, J.H.; Webb, W.W. Two-photon laser scanning fluorescence microscopy. *Science* **1990**, *248*, 73–76. [CrossRef] [PubMed]
2. Campagnola, P.J.; Loew, L.M. Second-harmonic imaging microscopy for visualizing biomolecular arrays in cells, tissues and organisms. *Nat. Biotechnol.* **2003**, *21*, 1356–1360. [CrossRef] [PubMed]

3. Lee, Y.J.; Kim, S.-H.; Moon, D.W.; Lee, E.S. Three-color multiplex CARS for fast imaging and microspectroscopy in the entire CHn stretching vibrational region. *Opt. Express* **2009**, *17*, 22281–22295. [CrossRef] [PubMed]

4. Lukic, A.; Dochow, S.; Bae, H.; Matz, G.; Latka, I.; Messerschmidt, B.; Schmitt, M.; Popp, J. Endoscopic fiber probe for nonlinear spectroscopic imaging. *Optica* **2017**, *4*, 496–501. [CrossRef]

5. Chen, H.; Wang, H.; Slipchenko, M.N.; Jung, Y.; Shi, Y.; Zhu, J.; Buhman, K.K.; Cheng, J. A multimodal platform for nonlinear optical microscopy and microspectroscopy. *Opt. Lett.* **2009**, *17*, 1282–1290. [CrossRef]

6. Hoover, E.E.; Squier, J.A. Advances in multiphoton microscopy technology. *Nat. Photonics* **2013**, *7*, 93–101. [CrossRef] [PubMed]

7. Field, J.J.; Carriles, R.; Sheetz, K.E.; Chandler, E.V.; Hoover, E.E.; Tillo, S.E.; Hughes, T.E.; Sylvester, A.W.; Kleinfeld, D.; Squier, J.A. Optimizing the fluorescent yield in two-photon laser scanning microscopy with dispersion compensation. *Opt. Express* **2010**, *18*, 13661–13672. [CrossRef] [PubMed]

8. Katona, G.; Szalay, G.; Maák, P.; Kaszás, A.; Veress, M.; Hillier, D.; Chiovini, B.; Vizi, E.S.; Roska, B.; Rózsa, B. Fast two-photon in vivo imaging with three-dimensional random-access scanning in large tissue volumes. *Nat. Methods* **2012**, *9*, 201–208. [CrossRef] [PubMed]

9. Tang, S.; Jung, W.; McCormick, D.; Xie, T.; Su, J.; Ahn, Y.; Tromberg, B.J.; Chen, Z. Design and implementation of fiber-based multiphoton endoscopy with microelectromechanical systems scanning. *J. Biomed. Opt.* **2009**, *14*. [CrossRef] [PubMed]

10. Tai, S.P.; Chan, M.C.; Tsai, T.H.; Guol, S.H.; Chen, L.J.; Sun, C.K. Two photon fluorescence microscope with a hollow core photonic crystal fiber. *Opt. Express* **2004**, *12*, 6122–6128. [CrossRef] [PubMed]

11. Choi, H.; So, P.T.C. Improving femtosecond laser pulse delivery through a hollow core photonic crystal fiber for temporally focused two-photon endomicroscopy. *Sci. Rep.* **2014**, *4*. [CrossRef] [PubMed]

12. Huang, S.; Heikal, A.A.; Webb, W.W. Two-Photon Fluorescence Spectroscopy and Microscopy of NAD(P)H and Flavoprotein. *Biophys. J.* **2002**, *82*, 2811–2825. [CrossRef]

13. Duguay, M.A.; Kokubun, Y.; Koch, L.T.; Pfeiffer, L. Antiresonant reflecting optical waveguides in SiO$_2$-Si multilayer structure. *Appl. Phys. Lett.* **1986**, *49*, 13–15. [CrossRef]

14. Yu, F.; Knight, J. Negative curvature hollow core optical fiber. *IEEE J. Sel. Top. Quantum Electron.* **2016**, *22*, 1–11. [CrossRef]

15. Pryamikov, A.D.; Kosolapov, A.F.; Alagashev, G.K.; Kolyadin, A.N.; Vel'miskin, V.V.; Biriukov, A.S.; Bufetov, I.A. Hollow-core microstructured 'revolver' fibre for the UV spectral range. *Quantum Electron.* **2016**, *46*, 1129–1133. [CrossRef]

16. Kolyadin, A.N.; Kosolapov, A.F.; Pryamikov, A.D.; Biriukov, A.S.; Plotnichenko, V.G.; Dianov, E.M. Light transmission in negative curvature hollow core fiber in extremely high material loss region. *Opt. Express* **2013**, *21*, 9514–9519. [CrossRef] [PubMed]

17. Meng, F.; Liu, B.; Li, Y.; Wang, C.; Hu, M. Low Loss Hollow-Core Antiresonant Fiber with Nested Elliptical Cladding Elements. *IEEE Photonics J.* **2017**, *9*. [CrossRef]

18. Sherlock, B.; Yu, F.; Stone, J.; Warren, S.; Paterson, C.; Neil, M.A.; French, P.M.; Knight, J.; Dunsby, C. Tunable fibre-coupled multiphoton microscopy with a negative curvature fibre. *J. Biophotonics* **2016**, *9*, 715–720. [CrossRef] [PubMed]

19. Kolyadin, A.N.; Alagasheva, G.K.; Pryamikova, A.D.; Mouradianb, L.; Zeytunyanb, A.; Toneyanb, H.; Kosolapova, A.F.; Bufetova, I.A. Negative curvature hollow-core fibers: Dispersion properties and femtosecond pulse delivery. *Phys. Procedia* **2015**, *73*, 59–66. [CrossRef]

20. NKT® Photonics HC-800-02 Photonic Crystal Fiber Data Sheet. Available online: http://www.nktphotonics.com/wp-content/uploads/sites/3/2015/01/HC-800.pdf (accessed on 2 October 2017).

21. Utzinger, U.; Richards-Kortum, R.R. Fiber optic probes for biomedical optical spectroscopy. *J. Biomed. Opt.* **2003**, *8*, 121–147. [CrossRef] [PubMed]

22. Flusberg, B.A.; Cocker, E.D.; Piyawattanametha, W.; Jung, J.C.; Cheung, E.L.M.; Schnitzer, M.J. Fiber-optic fluorescence imaging. *Nat. Methods* **2005**, *2*, 941–950. [CrossRef] [PubMed]

23. Munzke, D.; Saunders, J.; Omrani, H.; Reich, O.; Loock, H.P. Modeling of fiber-optic fluorescence probes for strongly absorbing samples. *Appl. Opt.* **2012**, *51*, 6343–6351. [CrossRef] [PubMed]

24. Bhowmick, G.K.; Gautam, N.; Gantayet, L.M. Design optimization of fiber optic probes for remote fluorescence spectroscopy. *Opt. Commun.* **2009**, *282*, 2676–2684. [CrossRef]

25. Ma, J.; Chiniforooshan, Y.; Hao, W.; Bock, W.J.; Wang, Z.Y. Easily fabricated, robust fiber-optic probe for weak fluorescence detection: Modeling and initial experimental evaluation. *Opt. Express* **2012**, *20*, 4805–4811. [CrossRef] [PubMed]

26. Coony, T.F.; Skinner, H.T.; Angel, S.M. Comparative study of some fiber-optic remote Raman probe designs. Part I: model for liquids and transparent solids. *Appl. Spectrosc.* **1996**, *50*, 836–848. [CrossRef]

27. Mitsubishi Rayon ESKA^{TM} Plastic Optical Fibre GH-4001 Attenuation and Transmission Datasheet. Available online: http://fiberopticpof.com/pdfs/Plastic_Fiber_Optics_&_Cable/ESKA_Testing_Data/AttenuationLossChart.pdf (accessed on 2 October 2017).

28. Sjöback, R.; Nygren, J.; Kubista, M. Absorption and fluorescence properties of fluorescein. *Spectrochim. Acta Part A* **1995**, *51*, 7–21. [CrossRef]

MDPI AG
St. Alban-Anlage 66
4052 Basel
Switzerland
Tel. +41 61 683 77 34
Fax +41 61 302 89 18
www.mdpi.com

Sensors Editorial Office
E-mail: sensors@mdpi.com
www.mdpi.com/journal/sensors

www.ingramcontent.com/pod-product-compliance
Lightning Source LLC
Chambersburg PA
CBHW051722210326
41597CB00032B/5566